Midwest Studies in Philosophy
Volume XVIII

MIDWEST STUDIES IN PHILOSOPHY

EDITED BY PETER A. FRENCH, THEODORE E. UEHLING, JR.,
HOWARD K. WETTSTEIN

Many papers in MIDWEST STUDIES IN PHILOSOPHY are invited and all are
previously unpublished. The editors will consider unsolicited manuscripts that
are received by January of the year preceding the appearance of a volume. All
manuscripts must be pertinent to the topic area of the volume for which they
are submitted. Address manuscripts to MIDWEST STUDIES IN PHILOSOPHY,
Department of Philosophy, University of California, Riverside, CA 92373.

The articles in MIDWEST STUDIES IN PHILOSOPHY are indexed in THE
PHILOSOPHER'S INDEX.

Forthcoming Volumes

Volume XIX   1994 Naturalism
Volume XX   1995 Moral Concepts

Available Previously Published Volumes

Volume XIII   1988 Ethical Theory: Character and Virtue
Volume XIV   1989 Contemporary Perspectives in the Philosophy of Language II
Volume XV   1990 The Philosophy of the Human Sciences
Volume XVI   1991 Philosophy and the Arts
Volume XVII   1992 The Wittgenstein Legacy

Midwest Studies in Philosophy Volume XVIII

Philosophy of Science

Editors

Peter A. French
Trinity University

Theodore E. Uehling, Jr.
University of Minnesota, Morris

Howard K. Wettstein
University of California, Riverside

University of Notre Dame Press ● Notre Dame, Indiana

Published by the University of Notre Dame Press
Notre Dame, IN 46556
Printed in the United States of America

Library of Congress Cataloging-in-Publication Data

Philosophy of science / editors, Peter A. French, Theodore E.
Uehling, Jr., Howard K. Wettstein.
        p.   cm. — (Midwest studies in philosophy : v. 18)
    ISBN 0-268-01406-X
    1. Science—Philosophy.   I. French, Peter A.
II. Uehling, Theodore Edward   III. Wettstein, Howard K.
IV. Series.
Q175.P5125
501—dc20   93-8496
                                                   CIP

Midwest Studies in Philosophy
Volume XVIII
Philosophy of Science

Fictionalism . . . . Arthur Fine   1

Underdetermination, Realism, and Reason . . . . John Earman   19

Epistemology for Empiricists . . . . Elliott Sober   39

Wittgensteinian Bayesianism . . . . Paul Horwich   62

Carnapian Inductive Logic for a Value Continuum . . . Brian Skyrms   78

How to Defend a Theory Without Testing It:

   Niels Bohr and the "Logic of Pursuit" . . . . Peter Achinstein   90

Empiricism, Objectivity, and Explanation . . . . Elisabeth A. Lloyd and

                                 Carl G. Anderson   121

Theoretical Explanation . . . . R. I. G. Hughes   132

Selective Scientific Realism, Constructive

   Empiricism, and the Unification of Theories . . . . Steven Savitt   154

Essentially General Predicates . . . . Peter Railton   166

On Philosophy and Natural Philosophy

   in the Seventeenth Century . . . . Howard Stein   177

There's a Hole and a Bucket, Dear Leibniz . . . . Mark Wilson   202

The Transcendental Character of Determinism . . . . Patrick Suppes   242

Idealization and Explanation: A Case Study

   from Statistical Mechanics . . . . Lawrence Sklar   258

The Fabric of Space: Intrinsic vs

   Extrinsic Distance Relations . . . . Philip Bricker   271

Against Experimental Metaphysics . . . . Martin R. Jones and

                                  Robert K. Clifton   295

Scientific Realism and Quantum

   Mechanical Realism . . . . Linda Wessels   317

Vacuum Concepts, Potentia, and the Quantum

   Field Theoretic Vacuum Explained for All . . . . Paul Teller   332

Genic Selection, Molecular Biology, and

   Biological Instrumentalism . . . . Alex Rosenberg   343

Could There Be a Science of Economics? . . . . John Dupré   363

Function and Design . . . . Philip Kitcher   379

Midwest Studies in Philosophy
Volume XVIII

Fictionalism

ARTHUR FINE

> One can [say] that the atomic system behaves "in a certain relation, 'As
> If'. . . " and "in a certain relation, 'As If'. . .," but that is, so to speak,
> only a legalistic contrivance which cannot be turned into clear thinking.
> Letter from E. Schrödinger to N. Bohr, October 23, 1926

O ver the last few years the realism-antirealism debate has produced a few
arguments (see Fine 1986) and a somewhat larger number of epithets.
The residual realist groups, while divided among themselves into competing
factions, are still set against views generically labeled antirealist. These antirealist views are (sometimes) differentiated as being empiricist, constructivist,
instrumentalist, verificationist, or one of the three 'P's—phenomenalist, positivist, or pragmatist; and so on. When an especially derisive antirealist label is
wanted, one can fall back on the term "fictionalist," coupled with a dismissive
reference to Vaihinger and his ridiculous philosophy of "As If."[1] But what is
"fictionalism" or the philosophy of "As If," and who was this Vaihinger?

1. SCHEFFLER'S FICTIONALISM

Within the standard literature in the philosophy of science (excluding the
philosophy of mathematics[2]) Israel Scheffler (1963) contains perhaps the last
extended discussion of a position called "fictionalism." In the context of exploring the problems and resources of a syntactical approach to scientific theories,
especially the problems of the significance and justification of theoretical
terms, Scheffler distinguishes between a Pragmatic Attitude (or pragmatism)
and a Fictionalist Attitude (or fictionalism). Pragmatism takes the existence of
effective and systematic functional relations as sufficient for rendering discourse
significant, that is, capable of truth or falsity. Thus pragmatism, in Scheffler's
terms, holds that there may be a language which is significant throughout and

1

also capable of expressing all of science. Fictionalism, by contrast, employs some criterion of intuitive clarity as necessary for significant discourse. Where this criterion is not satisfied we have a "fiction."

One kind of fictionalism is *eliminative;* it seeks to eliminate and/or to replace the fictional part of scientific discourse. The eliminative project is to construct a language that is thoroughly significant and also capable of expressing all of science. The unreconstructed language of science will not do. Another kind of fictionalism is *instrumentalist.* Instrumentalism withholds the language of truth, evidence, and belief from the fictional part of science, but holds that considerations relating to interest and utility are nevertheless sufficient for retaining that fictional component. Thus instrumentalism would suppose that any language capable of expressing all of science has a fictional (i.e., nonsignificant) component. On this view there is not much difference between pragmatism and instrumental fictionalism except over the question of whether significant discourse has to be intuitively clear. What is more important is that they both would tolerate discourse that fails to meet the criterion of clarity, provided the discourse were suitably functional.

Scheffler's discussion of fictionalism is bound to the 1960s project of telling a philosophical story about the regimentation of science as a whole. Specifically, it is tied on to that project's preoccupation with the syntax of the regimentation and the problem of meaning for so-called "theoretical terms." Thus Scheffler redrafts the concepts of pragmatism, instrumentalism, and fictionalism to suit his global, syntactic approach and his meaning-related needs. Those familiar with the American pragmatism of Peirce or James, or with the instrumentalism of Dewey, will find that Scheffler's terminology points at best to shadows of fragments of the originals. The same is true with respect to the fictionalism that was Hans Vaihinger's.

2. VAIHINGER AND LOGICAL POSITIVISM

We can begin to appreciate the importance of Vaihinger and his fictionalism for the philosophy of science by looking at the connection with the Vienna Circle that formed around Schlick, Carnap, and Neurath in the 1920s and the similar "Vienna Circle" of a decade earlier, also involving Neurath, along with the mathematician Hans Hahn and the physicist Philipp Frank. The term "logical positivism" became the catch word of the movement represented by these circles (and also the Berlin group around Reichenbach) following the article with that title by Blumberg and Feigl (1931). But that term did not originate in the manifestos of Neurath, nor in the writings of other dominant figures associated with these circles. Rather, as Philipp Frank notes (1949, 43) the first use of "logical positivism" comes from Hans Vaihinger's (1911), *The Philosophy of "As If."*[3]

Vaihinger described his philosophy using a number of different terms, including *critical positivism, idealistic positivism* and also (163) *logical positivism.* Here the modifier "logical" (as also "idealistic") refers to a logical or mental construct and the "positivism" insists on there being a suitable observational or experimental demonstration before one associates any reality with the construct. In describing his view as a kind of positivism, Vaihinger was trying to associate himself with an empiricist approach to positive, scientific knowledge and to disassociate his view from rationalism or Platonism; indeed, from any view that would presume some reality to correspond to whatever the mind logically constructs. Thus the logical positivism that Vaihinger introduced was an anti-metaphysical position of the same general kind as the logical positivism (or empiricism) usually associated with Vienna. Despite this similarity of philosophical orientation, however, Vaihinger is seldom cited by the Vienna positivists as a precursor, and never as an ally. To the contrary, as with today's realists, the writings of the logical positivists generally contain only the most curt and disparaging references to Vaihinger's central ideas.

Thus in Schlick's (1932) well-known reply to Max Planck's attack on Machian positivism, Schlick writes "if . . . Hans Vaihinger gave to his 'Philosophy of As If' the subtitle an 'idealistic positivism' it is but one of the contradictions from which this work suffers" (Schlick, 85). At the end of that essay Schlick says explicitly that the logical positivism to which he subscribes "is not a 'Theory of As If' (Schlick, 107).[4] With these faultfinding remarks Schlick attempts to distance the Vienna school from Vaihinger, and misleadingly identifies Vaihinger's "idealistic positivism" and his "As If" as a species of idealism, rather than as marking a positivist or empiricist attitude toward *ideas*. Similarly, Philipp Frank, who acknowledges Vaihinger's view as "the school of traditional philosophy that was nearest to [Vienna positivism] in spirit and in time" (Frank, 1949, 42), is still at pains to distinguish the two, which he does by denigrating Vaihinger as having a "complete lack of understanding" of what Frank regards as the important distinction between a coherent conceptual system and the operational definitions that connect it with the world of facts (Frank, 42–43). One can hardly read these protestations of the significant differences between Vaihinger and the logical positivists without getting a strong feeling that, indeed, they protest too much.

3. VAIHINGER AND THE PHILOSOPHY OF "AS IF"

Hans Vaihinger was born in Nehren, Württemberg in 1852. He received a religious and philosophical education at Tübingen and then did postgraduate work at Leipzig, where he also studied mathematics and natural science. He was Professor of Philosophy at Halle from 1884 until 1906, when he had to resign his position due to failing eyesight. He died at Halle in 1933. His *Philosophy of "As If"* first took shape as his dissertation of 1877. It was not published, however, until 1911. Vaihinger says that earlier he considered the time for it not ripe. The book is subtitled "A System of the Theoretical, Practical and Religious Fictions of Mankind" and the central topic of the whole work is an account of what Vaihinger calls "fictions" and "fictive judgments."

The book underwent extensive revisions and extensions in a very large number of editions. The sixth edition was translated into English by C. K. Ogden in 1924, with a second English edition in 1935, and several reprintings thereafter. In 1919 Vaihinger (and Raymund Schmidt) initiated the *Annalen der Philosophie*, a journal originally devoted entirely to the philosophy of "As If," and also committed to fostering interdisciplinary contributions, especially from the special sciences. The positivist Joseph Petzoldt joined as a third editor in 1927. By 1929, with Vaihinger blind and Schmidt effectively in control, the *Annalen* had fallen on hard times. After prolonged and delicate negotiations, the journal was taken over by Carnap and Reichenbach, to be reborn as *Erkenntnis*, the official organ of the logical positivist movement.[5] Earlier Vaihinger had founded *Kantstudien*, still today a leading journal for Kant scholarship. In addition to his "As If," Vaihinger wrote a well-known, two volume commentary on Kant's first critique, where he developed the so-called "patchwork thesis" that distinguishes four layers relating to the subjective sources of knowledge in the Transcendental Deduction. He also wrote a work on Nietzsche and completed several other philosophical studies. Vaihinger is usually regarded as a neo-Kantian, although his reading of Kant was very idiosyncratic. For example, where Kant generally considers scientific principles as providing the possibility of objective knowledge (i.e., as constitutive) for Vaihinger in large measure (although not totally) scientific principles are fictions, functioning as regulative ideals. Overall, Vaihinger's work, in fact, shows a strong British influence—especially due to Berkeley on the philosophy of mathematics and Hume on impressions and the imagination. We shall see that, in many respects, Vaihinger is closer to American pragmatism than to the transcendental idealism of Kant. Indeed, Ralph B. Perry, and other keepers of the pragmatic tradition, identify Vaihinger, Williams James, and Poincaré as leading pragmatist thinkers of their era.[6]

Thus, despite its current eclipse, in his own time Vaihinger's fictionalism was widely known and, like the related work of Mach and Poincaré, it had its own strong following. Moreover, judging by the overall reaction, I am inclined to think that the impact of Vaihinger's work then was not unlike the impact of Thomas Kuhn's work in our time. In the 1960s and 1970s most philosophers of science reacted to Kuhn with strident criticism. The response of their logical positivist forbearers to Vaihinger, as we have seen, displayed a similarly hostile tone. Today, however, notwithstanding that critical rejection, most English-speaking scientists, and commentators on science, are familiar with Kuhn's basic ideas about "paradigms" in science and freely employ his language.The same was true of German-speaking scientists and commentators with respect to Vaihinger's fictions and his language of "As If" up until the Second World War.[7] (Note, in the epigraph, Schrödinger's use of the "as if" idiom and his only slightly veiled reference to what Vaihinger called "legal fictions.") Except in discussions of legal philosophy,[8] however, Vaihinger did not survive the intellectual sea change that followed the war and restructured the philosophical canon. What I should like to do here is to review the main features of Vaihinger's work on fictions and then focus on a few central issues, showing their relevance to contemporary science and discussions in the philosophy of science.

4. FICTIONS

Vaihinger's general concern is with the role of fictional elements (or "fictions") in human thought and action. He begins with an elaborate classification of these elements and proceeds to illustrate their variety and use in virtually every field of human endeavor that involves any degree of reflective thought. Thus he illustrates the use of fictions in economics, political theory, biology, psychology, the natural sciences, mathematics, philosophy, and religion. Vaihinger acknowledges earlier treatment of fictions by Jeremy Bentham and he completes the survey by examining related conceptions that he finds in Kant, Forberg, F. A. Lange, and finally in Nietzsche. This survey is designed to achieve an effect. Vaihinger is trying to show that fictions are everywhere, that fictive thinking (like deductive and inductive thought) is a fundamental human faculty, and that—however dimly or partially—the importance of fictions has been recognized by the great thinkers. The extent to which Vaihinger's work was consumed by the philosophical public is evidence that, in good measure, he achieved the desired effect. Even his critics, like Morris R. Cohen in this country (see section 6, below), had to admit that Vaihinger had succeeded in getting people to realize the importance of fictions—reluctantly, even in science.

Vaihinger uses the term "fiction" loosely, sometimes for concepts and sometimes for propositions. He begins his account by distinguishing what he calls "real" or "genuine" fictions from what he calls "semi-fictions." Real fictions are characterized by three features: (1) they are in contradiction with reality, (2) they are in contradiction with themselves, or self-contradictory, and (3) they are generally understood to have these features when they are introduced. The semi-fictions satisfy (1) and (3) but not (2); that is, they are generally understood to be in contradiction with reality, but are not also self-contradictory. Following what one might call the Puritan principle,[9] Vaihinger also distinguishes between virtuous and vicious fictions, those that are scientific and those that are unscientific. The scientific fictions are an effective means to certain ends; they are useful and expedient. Where this utility is lacking the fictions are unscientific. An example of the unscientific kind would be the introduction of a so-called "dormative power."

The two primary examples of virtuous and real fictions with which Vaihinger begins are atoms and the Kantian "thing-in-itself." With respect to atoms his problem seems to be the difficulty of reconciling the conception that he takes from Cauchy, Ampère, and others, that atoms are centers without extension, with the idea that they are also the substantial bearer of forces. Vaihinger regards this as "a combination . . . with which no definite meaning can be connected" (219) and so, presumably, as contradictory. Moreover, since he thinks that the idea of a vacuum is itself contradictory (following Leibniz here) the fact that there is supposed to be a vacuum between atoms further implicates the atom in contradictions as well. Still, Vaihinger recognizes the usefulness of the chemical atom, for example, in organizing chemical combinations by definite proportions; hence the expedient and scientific nature of the atomic fiction. I will pass over the "thing-in-itself" to give some examples of semi-fictions.[10] Vaihinger suggests that in eighteenth-century France the limitations and inaccuracies of Cartesian vortex theory were already known, at least to some, yet it remained a useful way to organize the motions of bodies. This may not be a very good example, since it is not clear that anyone ever understood how vortex theory was supposed to work. (So maybe this fiction is not scientific at all!) But then Vaihinger also calls attention to the status of Ptolemaic astronomy, the limitations of which he says were known by Arab scholars in the Middle Ages. Still, the Ptolemaic system was a useful (and so scientific) way to deal with the heavens—and so a scientific (semi-) fiction. Both of these examples are of what Vaihinger calls "heuristic" fictions, and they may help us understand what Vaihinger means when he says that a fiction is in contradiction with reality; namely, that it is in some measure not true to what it purports to refer. It is not so easy to understand what he has in mind, however, when he regards the genuine fictions as involving a self-contradiction. More on this shortly. First let me run through some of the fundamental distinctions that Vaihinger develops.

To begin with he classifies fictions into ten primary kinds, with a few subkinds.

1. Abstractive
2. Of the Mean
3. Schematic
    Paradigmatic, Rhetorical, Utopian, Type
4. Analogical
5. Legal (Juristic)
6. Personificatory
    Nominal
7. Summational
8. Heuristic
9. Practical (Ethical)
10. Mathematical

Generally the name that he uses indicates the salient fictional feature. For example, abstractive fictions neglect important elements of reality. Thus we have Adam Smith's assumption that all human action is dictated by egoism, which looks at human action as if the sole driving force were egoism; or the treatment of extended bodies as if all their mass or gravity were concentrated at a point—and so the fictions of point masses and centers of gravity. Another abstractive fiction is what we would call the fiction of Robinson Crusoe worlds. This treats language as if it developed in worlds containing only a single individual. Vaihinger, however, holds to a social conception of language which makes this idea of a private language quite impossible of realization. Other kinds of fictions yield similar examples. These ten categories are not disjoint. Thus a personificatory fiction involves an analogy to a person, as in God the father, but this is also an analogical fiction. There are countless other overlaps as well. Nor is it clear whether this list is intended to be complete; presumably not, since for Vaihinger even the Kantian categories are not to be regarded as fixed.

Vaihinger does suggest, however, that as we move down the list we are likely to move away from semi-fictions and toward the genuine thing. By the time we get to mathematics, with its various number systems, and the limiting geometrical concepts of points, lines, surfaces—not to mention the fluxions and differentials of the calculus—we have arrived at a really fictional realm. Echoing Berkeley (and Dewey) Vaihinger writes, "Mathematics, as a whole, constitutes the classical instance of an ingenious *instrument,* of a *mental expedient* for facilitating the operation of thought" (57).

I have remarked that Vaihinger has a dynamic view of categories and classifications (as did later neo-Kantians like Cassirer, Reichenbach, and even Einstein). For example, at one point he says that the distinction between semi-fictions and genuine fictions is not stable in time. Often we learn that what we took for a contradiction with reality also involves a self-contradiction, so the semi-fictions may become real ones. Of greater importance for him is the distinction between hypotheses and semi-fictions,[11] for where we talk of Descartes's vortices or Ptolemaic astronomy (or Newton's laws, for that matter), it may seem that we are concerned with hypotheses and not with fictions at all. Vaihinger recognizes that when a scientific idea is first introduced we may not know whether it is a fiction or a hypothesis. We may begin by believing it to be the one and learn later that it is the other. What is the difference?

For Vaihinger, hypotheses are in principle *verifiable* by observation. We choose among hypotheses by selecting the *most probable.* In this way we *discover* which are true. By contrast, fictions are *justifiable* to the extent to which they prove themselves useful in life's activities; they are not verifiable. We select among fictions by choosing the most expedient with respect to certain ends. (He does not say that we maximize utility nor use Reichenbach's notion of *vindication,* but he might have.) Finally, fictions are the product of human invention; they are not discovered. In this connection he would have us contrast Darwin's hypothesis of descent with Goethe's schematic semi-fiction of an original animal archetype.

Showing his scientism, Vaihinger suggests that Goethe's fiction "prepares the way" for Darwin (86), whose conception of evolution and the survival of the fittest is in many ways the linchpin for Vaihinger's whole system. That system treats human thought functionally, as an evolving biological phenomenon driven by the struggle for existence. In the introduction, prepared specially for the Ogden translation, Vaihinger writes, "[A]ll thought processes and thought-constructs appear a priori to be not essentially rationalistic, but biological phenomena. . . . Thought is originally only a means in the struggle for existence and to this extent a biological function" (xlvi). Here Vaihinger shows his engagement with Nietzsche, and the will to power, just as similar evolutionary commitments expressed by John Dewey show his engagement with Hegel.

5. WOULD IT BE A MIRACLE?

Throughout the whole discussion concerning a shift in status between hypotheses and fictions Vaihinger proceeds, as a constructivist might, without attending to the difference between something *being* an hypothesis or *being* a fiction, and our *believing* it to be so. He also does not face up to the question of what becomes of a false hypothesis; does that make it a fiction?

This much is clear. If we knowingly retain a false but useful hypothesis, we have a fiction. The informed use, for example, of Galileo's law of free fall, which postulates a constant gravitational attraction, to calculate free-fall time (or distance) would amount to a fiction, since gravity is not really constant. Similarly with the law of the simple pendulum, the perfect gas law, and so on. In all these cases where local approximations are used, in Vaihinger's terms we have a fiction. In the case of the perfect gas law, we may even have a genuine fiction.

In highlighting the idealizations and approximations commonly used in modeling physical phenomena, Vaihinger's central concern is to undo the opinion that if constructs are devoid of reality they are also devoid of utility. Put the other way around, Vaihinger regards the inference from utility to reality as fundamentally incorrect. Thus, despite his pragmatic emphasis on thought as a tool for action, he wants to distinguish his position from the Jamesian form of pragmatism that regards truth to be whatever turns out to be "good" by way of belief, for all the scientific fictions satisfy this formula. On the other hand, what concerns him more is to demonstrate, by the sheer number and range of his examples, that the inference from scientific success at the instrumental level to the literal truth of the governing scientific principles is thoroughly fallacious. I would conjecture that part of the revolutionary impact of Vaihinger's work, and a source of some antagonism to it, was to wean his generation away from what we now call the explanatory argument for realism, or (what Putnam, I believe, dubbed) the "wouldn't it be a miracle argument." As Vaihinger puts it, "Man's most fallacious conclusion has always been that because a thing is *important* it is also *right*" (69). Vaihinger was concerned that we see through this way of thinking not only in the scientific domain, but more importantly when it comes to religion. For Vaihinger took the idea of God, and the related conceptions of salvation, judgment, immortality, and the like, as fictional (indeed as genuine fictions) and yet of supreme importance in our lives, arguing that this was Kant's way too. In this area, as in the scientific, what Vaihinger preached was critical tolerance, and not skepticism.

Vaihinger, then, offers no general method, no magic criterion, that will enable one to tell whether a construct corresponds to reality or whether a principle is a hypothesis. To the contrary, his examples show that favorites of the realism game (in his time and ours), like fertility or unifying power, are not a reliable guide. Each case has to be looked at on its own, and only time and sensible judgment will tell, if anything will. Vaihinger does, however, suggest a rule of procedure. Namely, that we begin by supposing that we are dealing with a fiction and then go from there. Thus Vaihinger would put the burden of proof on the non-fictional foot. Part of what dictates this strategy is Vaihinger's view that we generally find it hard to tolerate the ambiguity and resultant tension that comes from acting on what we acknowledge to be a fiction. There is, he thinks, a natural psychological tendency to discharge this tension (and achieve "equilibrium") by coming to believe that some reality may actually correspond to our useful fictive constructs—"the *as if* becomes *if*"(26). Intellectual integrity, however, requires that we recognize the tendency to believe too readily and suggests a countervailing strategy to help keep us honest. Intellectual growth of the species requires that we come to terms with the tension and learn to tolerate the ambiguity. As with the historic changes that involve a public understanding of the functions and abuses of social and religious dogma, to acknowledge fictive thinking and adopt the strategy of placing the burden on the realist foot would also be a move toward liberation. Although Vaihinger never makes reference to James's will to believe, it is clear that he would have opposed it.

6. CONTRADICTIONS AND GENUINE FICTIONS

There is a sliding line between hypotheses and the semi-fictions. But the genuine fictions are something else, since they are supposed to involve a contradiction and so could not possibly be exemplified in nature. It is this conception of Vaihinger's that has generated the most controversy. M. R. Cohen simply asserts that "in every case the claim of self-contradiction rests on positive misinformation." Indeed, Cohen believes this must be so for otherwise, "no fruitful consequences could be drawn from them [i.e., the so-called genuine fictions] and they would not have the explanatory power which makes them so useful in science" (Cohen 1923, 485).

There are, I think, two related problems here. The first is how the use of a contradiction could be fruitful. The second problem concerns what it means to treat something "as if" it had contradictory properties. Thus the first issue, say for the fiction of atoms, is how they could be useful in treating chemical compounds. The second concerns what it means to treat matter as if it were composed of atoms, when the very concept of an atom is supposed to be contradictory.

These problems are not new, and certainly not peculiar to Vaihinger. For instance, in the realm of mathematical entities, Vaihinger acknowledges the lead of Berkeley who held that it was impossible for lines and figures to be infinitely divisible. Yet Berkeley also held that the applicability of these geometrical notions required that we speak of lines on paper, for example, as though they contained parts which they really do not; i.e., that we treat them as though they were divisible. This is precisely Vaihinger's "as if," marking in this case a genuine fiction.

Cohen's way of dealing with these problems is by denial. That is, he just takes it as given that nothing fruitful could be obtained from a contradiction, and that no sense attaches to treating something "as though" it had contradictory properties. (Beineberg: "It's as though one were to say: someone always used to sit here, so let's put a chair ready for him today too, even if he has died in the meantime, we shall go on behaving as if he were coming." Törless: "But how can you when you know with certainty, with complete mathematical certainty, that it's impossible."[12]) Cohen concludes that Vaihinger's examples of genuine fictions, insofar as they are useful, *must* all be mistaken. At best, they are all semi-fictions. This conclusion is drawn too quickly, and apart from the brief examination of one case (that of $\sqrt{-1}$, which is the subject of the parenthetical dialogue) Cohen offers no argument whatsoever for it.

Vaihinger is well aware of the tendency to see difficulties here, which he describes as a "pardonable weakness" (67). Again he refers to the "psychical tension" that may be created by the "as if" (83). (Törless: "[I]t's queer enough. But what is actually so odd is that you *can really* go through quite ordinary operations with imaginary or other impossible quantities, all the same, and come out at the end with a tangible result! . . . That sort of operation makes me feel a bit giddy.") In mathematics, especially, the use of genuine fictions, according to Vaihinger, requires that we compensate for the contradiction by what he calls (109 f.) the method of "antithetic errors." Vaihinger repeats Berkeley's slogan that "thought proceeds to correct the error which it makes" (61). After drawing what Vaihinger calls "the necessary consequences" the impossible fictional premise just drops out, like the middle term in a syllogism. (Beineberg: "Well, yes, the imaginary factors must cancel each other out in the course of the operation. . . .") We use the "as if," Vaihinger also says, to construct a scaffolding around reality which we then cast off when its purpose has been fulfilled (68–69). (Compare, Törless: "Isn't that like a bridge where the piles are there only at the beginning and at the end, with none in the middle, and yet one crosses just as surely and safely as if the whole of it were there?")

With these various caveats, Vaihinger is not trying to explain how contradictions can be fruitful, so much as to reject the idea that some one generic explanation of their fertility is required. (Beineberg: "I'm not going to rack my brains about it: these things never get one anywhere.") He only wants to help free us from the idea that fictional thought is diseased, and that "impossible" concepts are somehow beyond the rational pale. This practice is not much different from what Wittgenstein came to later. Like the later Wittgenstein, Vaihinger thinks if we keep in mind our human purposes we will see that by and large our ordinary ways of thinking, which involve a large amount of fictive activity, are all right. (He does remark that the wish to understand the world *as a whole* is childish, because when we apply the usual categories outside of their customary home in human experience they engender illusory problems—like that of the purpose of it all (172–73). Language on holiday?) Thus Vaihinger tries to slip out of the knot posed by the use of contradictions without suggesting a general answer to the sense of the "as if," when a contradiction is involved, or proposing a theory to answer the Kantian question of how genuine fictions could be useful.

What Vaihinger seems to be suggesting is that one ought to reject these questions about contradictions. Instead of pursuing these, "how is it possible that" questions, simply pay detailed attention to reflective practice and notice how fictions are actually used. This optimistic naturalism is tutored by the fruitful history of a long (and, arguably, inconsistent) mathematical practice in the use of infinitesimals. The lesson that Vaihinger takes from this history is that we human animals learn by practice what can be done in a particular domain, and what must be avoided, in order to obtain results that are useful for our purposes. These days one could point to the (inconsistent) use of delta functions in quantum calculations (despite their *ex post facto* rationalization in terms of Schwartz distributions), or the hierarchy of classifications that allow for the cancellation of infinities (antithetical errors!) in quantum field theory. So what Vaihinger suggests, by example, is that we ought not demand a general theory concerning the fruitful use of contradictions as such. It is sufficient to attend to the variety of successful practices, each of which, literally, shows how it is done.[13]

Although no logic of the use of contradictions comes out of Vaihinger's procedures, there is a conception of rationality. The picture that Vaihinger suggests is that human thinking is circuitous and will try many "roundabout ways and by-paths" (xlvii) to find something that works. Rationality, then, simply constrains this activity by imposing the general criterion of fertility. As John Dewey put it, "To suggest that man has any natural propensity for a reasonable inference or that rationality of an inference is a measure of its hold on him [is] grotesquely wrong" (Dewey 1916, 425). For Vaihinger and Dewey, there is no pre-established harmony between rationality and reality. Rather, rationality is a canon which we assemble bit by bit from experience in order to check our propensity for varied and erroneous inference.

7. ALLIES AND DIFFICULTIES

Once the new positivism took a sharply logical turn (in the hands of Schlick and, more especially, Reichenbach and Carnap) it is not difficult to understand logical positivism's impatience with Vaihinger. His distinctions are many (I count thirty types of fictions, but that is probably not all) and scarcely crisp. Indeed, one can readily picture his relaxed tolerance toward the use of contradictions driving Carnap to distraction, all by itself—as Wittgenstein's similar attitude seems to have done. One wonders, however, about Neurath who always had more historicist leanings. Neurath's work shares several aspects with Vaihinger's, including: (1) a community or social conception of language, which Neurath began to champion in the period when Vaihinger's book was first published, and which Neurath later urged on Carnap, (2) Neurath's opposition to "pseudo-rationalism," what we today would describe as a penchant for global meta-narratives about science—his charge against Popper's falsificationism and also against the generalizing thrust of his positivist colleagues, (3) Neurath's naturalism, which made taxonomy part of his philosophical practice, i.e., recording, classifying, and (provisionally) accepting the beliefs and behavior of scientists, repairing the boat while at sea, and (4) Neurath's pragmatic tolerance for the use of expedient means and his positive inclination to find some good in each among competing scientific ideas.[14]

Like Neurath, in this positive inclination, I have been trying to emphasize the good things in Vaihinger, in part by way of trying to understand what made his work of such interest in its time. In so doing, however, I need to acknowledge that there are difficulties in Vaihinger's work beyond his fuzziness. Two are conspicuous. One is that Vaihinger frequently relies on secondary sources in place of giving his own, original analysis. This problem stands out, as Cohen suggests, in the rationale he provides for the alleged contradictions in his examples of genuine fictions. For example, a case can be made to sustain Vaihinger's judgment that the number systems constitute genuine fictions, but not by citing Berkeley, or others, without supplementing that with more analysis than Vaihinger himself provides (for he does provide some). A second prominent difficulty is the way in which Vaihinger sets his discussion of fictive judgments in the context of a universal psychologistic logic. Although he is careful to avoid the conception of the mind as a substance, opting for a more behaviorist picture, he does advance the idea that there is a general set of individual, associationist mechanisms involved in the production of fictions. Indeed he sketches some, including a "law of ideational shifts" (124 ff.). Clearly the conception of a special psychological faculty for fictive judgment suited his purpose of promoting the value (indeed the necessity) of fictions, and it outweighed his otherwise sensible resistance to proposing perfectly general theories. No doubt, quite apart from Vaihinger's special interest, the idea of universal mechanisms of cognition is also independently appealing. In Vaihinger's time (as in our own), few managed to break free of this sort of psychologism. Vaihinger was not one of them.

There is a third charge against Vaihinger, and the one most frequently encountered. It is that, besotted by the topic of fictions, he simply saw them everywhere (forces, atoms, the classical laws of motion, virtually all of mathematics—not to mention substance, free will, God, and so on) and that, taken to its logical conclusion, his system actually demands this. Concerning this latter charge, consider his distinction between hypotheses and fictions. That depends on contrasting discovery (hypothesis) with invention (fiction), verification (hypothesis) with utilitarian justification (fiction). But why, one might ask, should we take these contrasts as anything more than fictional themselves, as useful expedients in the labor of the mind?[15] What gives the question its bite is that Vaihinger provides no firm grounds for sorting and grading into fictional versus nonfictional. It is part of his own scheme that there are no general answers to such questions; that is, answers that can be derived from general subject-neutral principles. His answer about how to grade and sort, if there is one, has to be narrowly tailored, topic specific, and historicist. That is what his system demands. To charge, on that account, that Vaihinger has no answer and that therefore we are free to assimilate all thought to fiction is to impose on Vaihinger the nihilistic standard according to which if a question cannot be answered on the basis of perfectly general principles, then we are free to answer it just how we please. Since Vaihinger's whole enterprise stands in opposition to this nihilism, however, this is not an appropriate standard for the interpretation of his work. Thus, instead of showing up a deep flaw in Vaihinger, this objection exposes the tenacity with which one can cling to misplaced presuppositions about generality. This nicely illustrates what Vaihinger calls "the preponderance of the means over the ends" (which we know as Parkinson's Law). In this case, that preponderance accounts for how generality, a means to certain ends, gets displaced in the course of time into an end-in-itself. The remaining charge, that Vaihinger simply finds altogether too many fictions, has, of course, to be carefully defended and responded to on a case by case basis. Vaihinger would ask no more, nor should we damn him on the basis of anything less.

8. VAIHINGER'S REPUTATION

Despite its moderately historical veneer, my treatment of Vaihinger has been rather unhistorical. I have not asked who his readers were, with what themes of theirs his work resonated, into what larger cultural phenomena, trends, and institutions it fit, and so on. Moreover, although I believe that Vaihinger's ideas on fictions were widely influential and I can readily point to similar ideas similarly expressed over a broad intellectual terrain, I have not really traced the vectors by virtue of which those ideas were spread there. Similar social-historical work needs to be pursued to understand why Vaihinger fell out of favor just when he did, and I have not done that work either. Nevertheless, I want to address this last issue, as it relates to logical positivism, where some things can be said for sure and some conjectures can at least be raised.

For sure, Vaihinger's flaws alone do not explain the disrepute into which he has fallen. After all, of what is he guilty? He makes a lot of distinctions, difficult to keep in mind, which are not even exhaustive and complete. He is somewhat fuzzy, sometimes. He sometimes comes up short on analysis. Naively, he accepts a lot of what scholars have said, and scientists done, as correct. He subscribes to an individualistic picture of cognition that is more universal that he ought to allow. He may have made some factual errors in what he took to be fictions, and what not. I might add that he has a somewhat old-fashioned penchant for coining "laws" and for diagnosing historical trends. He also likes to moralize. Sometimes he repeats himself. In a large work, no doubt there are some inconsistencies. (I have not found a good example of this, unlike Schlick, apparently; but let us grant it.)

Examining this list, it looks like what Vaihinger mostly needed was a competent editor! Yet what he received from the logical positivists, and his legacy from them to us, has been something rather different. Put most simply, the positivists set about making Vaihinger a marginal figure. They succeeded. In today's literature, when he is mentioned at all, it is almost always in the margins—in a footnote one-liner or a parenthetical thrust. One will protest, perhaps, that in compiling my list of flaws I have not discussed the originality, quality, or viability of his ideas. Surely, in the end, that is what counts. Indeed it does, and that is my point. My omission tracks Vaihinger's positivist critics, for they do not discuss the originality, quality, or viability of his ideas either. Mainly they mock him and promptly place his "As If" in the trash. This is especially striking behavior in figures like Carnap, Reichenbach, Schlick, and allies, who were formidable critics. Precisely by not bringing Vaihinger into their critical discussions, the logical positivists made it appear that his ideas were simply not worth discussing. Since, as we have seen, many of his ideas were also theirs, that explanation of why Vaihinger was not treated seriously cannot be correct. So why did they do it?

One clue to the dynamic may be found by reflecting on Neurath. Neurath was the organizer of the logical positivist movement, the "big engine" whose energy and political skills kept the group growing and alive. Until a very recent revival of interest in Neurath, however, few have promoted him as a serious thinker on par with the others. Neurath's influence on the intellectual development of logical positivism has lain in the shadows. His ideas, especially those in opposition to the mainstream, were mostly lost from general view. I would suggest two things that might help explain this neglect. Firstly, although Neurath wrote some slogans about verifiability and meaning, he never really made the commitment to an analysis of scientific language that became a characteristic feature of logical positivism.[16] Secondly, Neurath mostly stood in opposition to those tendencies in the movement that set the projects for understanding science in general terms; the projects of a general account of explanation, a general theory of confirmation, a general account of laws, or theories, and so on. Indeed, what has come down as the modern agenda of philosophy of science includes, for the most part, the very things that Neurath thought not sensible to pursue. As indicated above, the approach that Neurath advocated was more historicist, taxonomic, and naturalistic than the items on that agenda. His methodology, then, did not give pride of place to formal methods, meaning, and language. His orientation was piecemeal and particularist, self-consciously not global (the approach he mocked as "pseudo-rational"). Obviously, Neurath also had a great deal in common with his positivist colleagues. But, I suggest, by rejecting a linguistic and global orientation Neurath was easily cast as an outsider, and his reservations about these features of the program were by and large just not discussed. This was certainly not due to spite or bad feelings, but because the others had their own projects to pursue and could not be always engaged in justifying their whole approach. Like Einstein with respect to the quantum theory, Neurath suffered the fate of one who keeps questioning fundamentals. After a while, his reservations were set aside in order to get on with things.

My point about Vaihinger should be plain; it is the same point. In contrast with Scheffler's post-war setting for fictionalism (see section 1), Vaihinger was more like Neurath in not having a linguistic and meaning-related orientation, despite some occasional philological excursions. Unlike Scheffler, he was also by and large a confirmed taxonomist and naturalist of science, not subscribing to the global projects so dear to the neo-positivist hearts. Like Neurath, these aspects of Vaihinger could not be confronted without bringing into question the whole logical positivist project. So, Vaihinger was not discussed. He was written off. In Vaihinger's case, however, there was a further problem. For, as we have seen, Vaihinger had actually anticipated and set up institutional structures to pursue a program of philosophical reform and reevaluation uncomfortably close to the project of logical positivism. Logical positivism, however, proclaimed itself a new program in the history of thought, the vanguard of a new en-lightenment, the cutting edge of a new modernity. To be sure, the movement claimed the heritage of Mach and Poincaré, of Helmholtz, Hilbert, Einstein, Bohr, and (sometimes) Freud. These, however, were leading scientists of the era. With regard to existing philosophical schools, however, logical positivism acknowledged no peers, presenting itself as a fresh starting point. (Indeed, even their debt to Kant and their neo-Kantian contemporaries has to be excavated.) To cultivate the image of new-philosophical-man, they had to downplay the continuities between them and the popular Vaihinger, distancing themselves from him. So they did.

9. VAIHINGER'S LEGACY

Perhaps the dismissive attitude that logical positivism adopted to Vaihinger was overdetermined by the social and political circumstances suggested above, and probably by others as well. Whatever a better historical treatment would show, our attitude need not be bound by the judgment of those other times. With respect to general philosophical orientation, Vaihinger (and Neurath) point us toward a more naturalistic and particularistic approach to understanding science, an approach (like Vaihinger's to religion) that is at once critical and tolerant (with strong emphasis on both). This approach calls into question the viability of the universal projects of philosophy of science, demanding a hearing for a non–theory-dominated way. It also moves us away from a preoccupation with language and meaning. The approach is, broadly speaking, pragmatic in its emphasis on the importance of scientific practice in relation to scientific theory, although it is not reductive in the Jamesian way with respect to truth. This general sort of orientation is already being explored in various contemporary naturalisms, in my NOA, and also in some constructivist and broadly deflationist programs. I take these to be part of Vaihinger's legacy. What, then of fictions and the "As If"?

Vaihinger's emphasis on fictions exalts the role of play and imagination in human affairs. He finds no realm of human activities, even the most serious of them, into which play and imagination fail to enter. Surely he is right. These faculties are part of the way we think ("constructively"), approach social and intellectual problems ("imaginatively"), employ metaphor and analogy in our language, and relate to others every single day.

Within science, idealizations and approximations are an integral part of ordinary everyday procedure. The representation of three dimensions on two (that is, graphing), the conceptualization of four (or twenty-seven!) in terms of three, all call on the imagination to create a useful fiction—as does any pictorial presentation of data. The images by virtue of which whole fields are characterized ("black hole," "strings," "plates," "bonds," "genetic code," "software," "systems," "chaos," "computable," "biological clock," and so on) have the same character. Indeed, new techniques are constantly being developed for the creation of scientific fictions. Game and decision theory come readily to mind. Computer simulation, in particle physics or weather forecasting, is also a significant postwar example. Preeminently, the industry devoted to modeling natural phenomena, in every area of science, involves fictions in Vaihinger's sense. If you want to see what treating something "as if" it were something else amounts to, just look at most of what any scientist does in any hour of any working day.

In these terms, Vaihinger's fictionalism and his "As If" are an effort to make us aware of the central role of model building, simulation, and related constructive techniques, in our various scientific practices and activities. Vaihinger's particularist attitude over the question of whether and to what extent any model captures an element of the truth, warns us to be wary of overriding arguments about how to interpret (useful) scientific constructs in general. History shows us that there are no magical criteria that fix the interpretation, and no simple answers here. (History also shows that in the puzzle cases that exercise philosophers, where ordinary scientific procedures do not seem to settle the issue, science gets along perfectly well when these realist questions are not pursued. But emphasizing this fact would be to push Vaihinger toward NOA.)

By distancing itself from Vaihinger, logical positivism missed an opportunity that would have kept it in the mainstream of scientific thought throughout this century. Those who would dismiss a view by associating it with Vaihinger and his "As If" make the same error. For the dominant self-conception of postwar science has been that of science as the builder of useful models. In our century Vaihinger was surely the earliest and most enthusiastic proponent of this conception, the preeminent twentieth-century philosopher of modeling.

NOTES

An earlier version of this paper was presented to a conference at Princeton University organized by Bas van Fraassen in the spring of 1992, where there were many useful comments. Joseph Pearson and Thomas Uebel helped with the history of Vaihinger and his fictions. Mara Beller provided me with a copy of the Schrödinger letter from which the epigraph is drawn and with encouragement to pursue my interest in Vaihinger. Thomas Ryckman offered good criticism and advice. Thanks all!

1. Thus Horwich (1991) conjures up Vaihinger's fictionalism in order to set the tone for a criticism of van Fraassen's constructive empiricism.

2. For fictionalism in mathematics see Papineau (1988), whose position resembles that of Berkeley and Vaihinger, although Papineau is not concerned to trace these antecedents.

3. Page number references to Vaihinger are to the Ogden translation (Vaihinger, 1924).

4. Ryckman (1991) notes some reversals in Schlick's attitude, which was receptive to Vaihinger early on but then turned hostile.

5. For details about the history of *Erkenntnis*, see Hempel (1975), and for its relation with Vaihinger and the *Annalen*, see Hegselmann and Siegwart (1991).

6. See Gould (1970).

7. Spariosu (1989) discusses Vaihinger and draws out parallels between his fictionalism and the ideas expressed by leading thinkers in German-speaking physics prior to the Second World War. Unfortunately, many of Spariosu's philosophical or scientific conclusions and generalizations seem unsound. Still, the textual record of Vaihingerisms that he compiles is impressive evidence of the resonance of fictionalism among prominent scientists of the time. One could add the complementarity of Niels Bohr to Spariosu's list. See the epigraph.

8. The legal realism of Jerome Frank (1970), for example, makes extensive use of Vaihinger's ideas.

9. "That which is not useful is vicious." Attributed to Cotton Mather.

10. For discussions of Vaihinger and the Kantian "thing-in-itself," especially with respect to the third *Critique*, see Schaper (1965, 1966).

11. Also important are what he calls dogmas, especially given his religious and ethical concerns. I will omit his treatment of dogmas, given our concerns.

12. This citation and the others running parenthetically through this section are from Musil (1906, 106–107) whose dialogue between Beineberg and Törless struggles with exactly the issues that divide Cohen from Vaihinger, and even over Cohen's one case of imaginary numbers. Musil provides illustrations for arguments that Cohen did not find. (My thanks to Joseph Pearson for calling my attention to these passages in Musil.)

13. David Lewis (1982; 1983, 276–78) pursues a contrary tack, suggesting a quite general approach via "disambiguation" for how contradictions may fruitfully be used without harm.

14. See Uebel (1992) for a discussion of these and other features of Neurath's work.

15. This is a line pursued by Spariosu (1989).

16. Appearances to the contrary, his proposals about protocol sentences were not in aid of regimenting the language of science. Rather, Neurath was trying to display the complex range of objective, subjective, and social factors that enter into any scientific report, in part to undercut the idea that they *could* usefully be regimented. See Uebel (1991).

REFERENCES

Blumberg, A. E., and H. Feigl. 1931. "Logical Positivism, A New Movement in European Philosophy." *Journal of Philosophy* 28: 281–96.

Cohen, M. R. 1923. "On the Logic of Fictions." *Journal of Philosophy* 20: 477–88.

Dewey, J. 1916. *Essays in Experimental Logic*. Chicago.

Fine, A. 1986. "Unnatural Attitudes: Realist and Instrumentalist Attachments to Science." *Mind* 95: 149–79.

Frank, J. 1970. *Law and the Modern Mind*. Gloucester, Mass.

Frank, P. 1949. *Modern Science and Its Philosophy*. Cambridge, Mass.

Gould, J. A. 1970. "R. B. Perry on the Origin of American and European Pragmatism." *History of Philosophy* 8: 431–50.

Hegselmann, R., and G. Siegwart. 1991. "Zur Geschichte der 'Erkenntnis'." *Erkenntnis* 35: 461–71.

Hempel, C. G. 1975. "The Old and the New 'Erkenntnis'." *Erkenntnis* 9: 1–4.

Horwich, P. 1991. "On the Nature and Norms of Theoretical Commitment." *Philosophy of Science* 58: 1–14.

Lewis, D. 1982. "Logic for Equivocators." *Noûs* 16: 431–41.

Lewis, D. 1983. *Philosophical Papers*. Oxford.

Musil, R. 1906. *Young Törless*. Translated by E. Wilkins and E. Kaiser. New York.

Papineau, D. 1988. "Mathematical Fictionalism." *International Studies in the Philosophy of Science* 2: 151–75.

Ryckman, T. 1991. "Conditio sine qua non: 'Zuordnung' in the Early Epistemologies of Cassirer and Schlick." *Synthèse* 88: 57–95.

Schaper, E. 1965. "The Kantian 'As If' and its Relevance for Aesthetics." *Proceedings of the Aristotelian Society* 65: 219–34.

Schaper, E. 1966. "The Kantian Thing-in-Itself as a Philosophical Fiction." *Philosophical Quarterly* 16: 233–43.

Scheffler, I. 1963. *The Anatomy of Inquiry*. New York.

Schlick, M. 1932. "Positivism and Realism." Reprinted in *Logical Positivism*, edited by A. J. Ayer, 82–107. New York, 1959.

Spariosu, M. 1989. *Dionysus Reborn: Play and the Aesthetic Dimension in Modern Philosophical and Scientific Discourse*. Ithaca, N.Y.

Uebel, T. 1992. *Overcoming Logical Positivism from Within*. Amsterdam.

Vaihinger, H. 1911. *Die Philosophie des Als Ob*. Leipzig.

Vaihinger, H. 1924. *"The Philosophy of 'As If'."* Translated by C. K. Ogden. London.

Underdetermination, Realism, and Reason

JOHN EARMAN

The issue of underdetermination rears its head again and again in discussions of scientific realism. The reason behind these multiple appearances can be traced to the fact that most of the parties to the debate about realism share the following picture. There are two main components to scientific realism: the *semantic component*, which enjoins us to read the theories of science literally, and the *epistemic component*, which assumes that the evidence of observation and experiment can provide us with good reasons to believe what a theory, read literally, says about the world. Underdetermination (so the story goes) generates a tension between these two components. The anti-realist produces (or at least asserts the existence of) theories that are empirically equivalent in that they say the same things about the observable. But if we take semantic realism seriously, these theories say different and, indeed, incompatible things about the unobservable. The anti-realist then claims that it is not possible to be an epistemic realist with respect to such theories since any observation that provides reason to believe one of the empirically equivalent theories will give equally good reason to believe each of the other theories.

If this is a correct picture of the threat of underdetermination, then one can predict the shape of various reactions and, thus, how the philosophical community will divide in response to the threat. The prediction is in fact borne out by the literature, which displays two categories of reaction. The first category contains *retreat*, which can take two main forms. (i) Maintain epistemic realism by backing down on semantic realism. Admit that empirically equivalent theories are just different ways of describing the same factual states of affairs. Poincaré, Reichenbach, and others have opted for this form of retreat in the philosophy of science. But the pattern of reasoning involved is to be found in much broader contexts. Perhaps the most notorious example is Kripke's (1982) reading of the philosophy of the later Wittgenstein. Kripke finds in the *Philosophical Investigations* a radical skepticism based on the following underdetermination argument. Nothing about the past behavior of a person or her past mental states establishes whether by 'plus' she meant $+$ or $\oplus$ (where $x \oplus y = x + y$ if $x, y < 57$ and 5 otherwise). Therefore (!), there is no fact of the matter about whether she meant $+$ or $\oplus$. And more generally, there is no objective fact of the matter about the meaning of any word. (ii) Maintain semantic realism by backing down on epistemic realism. Claim that we can have good reason to believe that theories, read literally, are empirically adequate but not that what they say about the unobservable is true. Van Fraassen's (1980, 1985) constructive empiricism is the leading contemporary example of this form of retreat.

The second category of reaction contains *defense* rather than retreat. Like retreat, defense can also can take two main forms. (iii) Resolve or at least mitigate the tension between semantic and epistemic realism by showing that it is not so easy to construct interesting examples of empirically equivalent theories. (iv) Resolve the tension by arguing that even if there are such examples, empirical evidence can provide good reasons to believe one of the empirically equivalent theories over the others.

In a recent article Laudan and Leplin (1991) opt for both (iii) and (iv). Laudan and Leplin are to be commended for providing one of the few systematic attempts to come to grips with the underdetermination issue. However, I believe that their defense of (iv) fails to appreciate how empirical equivalence— or, as I would prefer to say, empirical indistinguishability—of theories does serve to undermine claims of justified belief. But more generally, there is a widespread failure to appreciate how the underlying issues transcend questions about the status of theoretical entities. These issues concern the reliability of inductive methods in general, whether they are supposed to get us from the observed to the unobservable or from the observed to unobserved observables. Since there is no agreement on the proper stance to take on these issues, or even how to best pose them, it is hardly surprising that debates about scientific realism tend to end in unproductive squabbling.

Without attempting to offer any final resolution of the underlying issues, I will try to pose them in what I hope to be a fruitful form and to relate them to the realism debate. My presupposition throughout is that of semantic realism; thus for present purposes I am setting aside (i) as an acceptable form of retreat. My primary concern here is whether in agreement with constructive empiricism, underdetermination arguments do serve to undermine epistemic realism with respect to scientific theories.

1. EMPIRICAL INDISTINGUISHABILTY

The set-up I will use to investigate the underdetermination problem is fairly simple. It contains three elements. (1) The hypotheses $\mathfrak{H} = \{H_1, H_2, \ldots\}$ under investigation. (2) The empirical evidence matrix $\mathfrak{E} = \{E_1, E_2, \ldots\}$. $\mathfrak{E}$ contains statements that are supposed to express whatever is regarded in the context of investigation as empirical evidence. If you believe in an observational–non-observational distinction, you may wish to confine $\mathfrak{E}$ to observational statements. If further you think that what observations give us directly are singular judgments, then you will want to confine $\mathfrak{E}$ to atomic observational statements. But I want to emphasize as strongly as possible that the issues to be discussed do not depend upon the observational-theoretical distinction. $\mathfrak{E}$ could consist of statements about the behavior of quarks, and still the problems to be posed here would not disappear. (3) A set of worlds W. The members of W encode the possibilities relevant to the semantics and epistemology. For $w \in W$ and $E \in \mathfrak{E}$, E^w stands for E or $\neg E$ according as $\models_w E$ or $\models_w \neg E$ (respectively, E is true in w or E is false in w). In most of the concrete examples to be studied below $\mathfrak{H}$ and $\mathfrak{E}$ consist of sentences of a first-order language, and the elements of W are models of the language. With respect to W the elements of $\mathfrak{H}$ are assumed to be self-consistent, pairwise exclusive, and mutually exhaustive.

With the help of this apparatus several senses of empirical distinguishability–indistinguishability can be codified. The first definition is supposed to capture the notion that empirically distinguishable hypotheses have incompatible empirical consequences. Towards this end, let $\overline{\mathfrak{E}}$ stand for the closure of $\mathfrak{E}$ under finite truth-functional combinations. (Of course, if $\mathfrak{E}$ is rich enough, then $\overline{\mathfrak{E}} = \mathfrak{E}$.)

Def 1. The $\mathfrak{H}$ are *empirically distinguishable*$_1$ (ED$_1$) with respect to $\mathfrak{E}$ and W iff $\forall H_i, H_j \in \mathfrak{H}(i \neq j)$ $\exists E \in \overline{\mathfrak{E}}$ such that $\forall w_i, w_j \in W$ such that $\models_{w_i} H_i$ and $\models_{w_j} H_j$, $E^{w_i} = E$ and $E^{w_j} = \neg E$.

If the elements of $\mathfrak{E}$ are finitely verifiable and finitely falsifiable, then ED$_1$ requires in effect that the H's be distinguishable by "crucial tests."

The next definition is supposed to capture the notion that empirically distinguishable theories have different empirical consequences.

Def 2. The $\mathfrak{H}$ are *empirically distinguishable*$_2$ (ED$_2$) with respect to $\mathfrak{E}$ and W iff $\forall H_i, H_j \in \mathfrak{H}(i \neq j)$ $\exists E \in \overline{\mathfrak{E}}$ such that (a) $\forall w_i \in W$ such that $\models_{w_i} H_i$, $E^{w_i} = E$, and (b) it is not the case that $\forall w_j \in W$ such that $\models_{w_j} H_j$, $E^{w_j} = E$ (or vice versa with and H_i and H_j interchanged).

The third definition is supposed to capture the idea that hypotheses are empirically distinguishable just in case any pair of worlds in which two distinct hypotheses are true are separated by some piece of empirical evidence.

Def 3. The $\mathfrak{H}$ are *empirically distinguishable*$_3$ (ED_3) with respect to $\mathfrak{E}$ and W iff $\forall H_i, H_j \in \mathfrak{H}(i \neq j)$ and $\forall w_i, w_j \in W$ such that $\models_{w_i} H_i$ and $\models_{w_j} H_j$, $\exists E \in \mathfrak{E}$ such that $E^{w_i} \neq E^{w_j}$.

Empirical indistinguishability can then be defined by negation: EI$_i \equiv$ $\neg$ED$_i$, $i = 1, 2, 3$. Whatever the choice of $\mathfrak{E}$ and W, we have ED$_1 \Rightarrow$ ED$_2$

and $ED_1 \Rightarrow ED_3$, so that $EI_3 \Rightarrow EI_1$ and $EI_2 \Rightarrow EI_1$. Under plausible restrictions on $\mathfrak{E}$ and W other entailment relations among the notions of empirical distinguishability-indistinguishability will hold; some of these relations will be studied below.

By the "empirical equivalence" of hypotheses, Laudan and Leplin (1991) seem to have in mind some species of EI_2. They concentrate on this notion because they want to demolish an imagined opponent who operates with an essentially hypothetico-deductive account of the relation between hypotheses and evidence. No such commitment to hypothetico-deductivism is made here. I argue below that when hypotheses are empirically indistinguishable in the sense of EI_3, *no* inductive methodology can deliver good reasons to believe one of the competing hypotheses over the others. Furthermore, if the inductive methodology can be cast in a probabilistic format, it can provide good reasons for believing hypotheses about unobserved observables without being capable of providing good reasons for believing hypotheses about unobservables. Before I can formulate these arguments, it is first necessary to fix some more terminology.

2. UNDERDETERMINATION OF THEORY BY EMPIRICAL EVIDENCE

Underdetermination (U) is a code word for the idea that knowledge of the truth values in a world $w \in W$ of any number of elements of $\mathfrak{E}$ does not suffice to provide good reason to believe that one rather than another element of $\mathfrak{H}$ is true in w. With this understanding, the main question before us is whether or not empirical indistinguishability in one of the above senses—or some other sense that needs to be articulated—entails underdetermination. The answer depends, of course, on what is required of "good reasons" for belief. If, for example, one adopts Bayesian personalism as an account of scientific reasoning, the answer is obviously no.[1] For even in the face of empirical indistinguishability in any of its senses, one can be a good Bayesian and assign a high degree of belief to one of the competing hypotheses and low degrees of belief to the others. But being a good Bayesian personalist strikes many as being insufficient to guarantee that the degrees of belief have any normative force. For example, a good Bayesian personalist can assign a very high degree of belief to, say, the latest version of quark theory, or to Velikovsky's *Worlds in Collision* scenario, on the basis of no relevant evidence whatsoever.

In what follows I will explore the idea that in order for an inductive methodology to supply good reasons for belief, the methodology must exhibit reliability. I will assume that the methodology follows Bayesian lines in that it is couched in terms of degrees of belief that at every moment of time satisfy the axioms of probability, and that change of belief over time takes place via conditionalization on pieces of empirical evidence that have been learned by observation or experiment. What reliability means for a methodology of this type, and for more general types as well, will be discussed in the next section.

3. RELIABILITY

Most of the versions of traditional epistemology that worry about reliability of methodology tacitly assume that reliability should obtain over all logically possible worlds, or at least over a rich subset of such worlds. I will also begin discussion with this presupposition but will consider modifications of it as the discussion progresses. The kind of reliability I will be concerned with here is long-run or limiting reliability. More stringent forms of reliability will have even more dire consequences for underdetermination.

An *evidence sequence* e^w for a world $w \in W$ is generated by checking the truth values in w of the elements in $\mathfrak{E}$ in any order you like. More formally, e^w is a sequence $E_{1'}^w, E_{2'}^w, \ldots$ where each $E_{i'}$ belongs to $\mathfrak{E}$ and where for each element of $\mathfrak{E}$ either the element or its negation occurs in the sequence. e_n^w stands for the conjunction of the first n elements of the evidence sequence e^w. $[H](w) = 1$ or 0 according as $\models_w H$ or $\models_w \neg H$.

> *Def. 4.* The probability function Pr is *completely reliable in the long run* for $\mathfrak{H}$ with respect to $\mathfrak{E}$ and W iff $\forall \ w \in W, \forall \ e^w, \forall \ H_i \in \mathfrak{H}, Pr(H_1/e^w) \rightarrow [H_{i'}](w)$ as $n \rightarrow \infty$.

> *Def. 5.* Pr is *completely .5 reliable in the long run* for $\mathfrak{H}$ with respect to $\mathfrak{E}$ and W iff $\forall \ w \in W, \forall \ e^w, \forall \ H_i \in \mathfrak{H}, \exists \ N$ such that $\forall \ n \geqslant N, Pr(H_i/e_n^w) > .5$ or $< .5$ according as $\models_w H_i$ or $\models_w \neg H_i$.[2]

Of course, the basic idea of reliability has nothing to do with probability or Bayesianism. A *truth identification method* for $\mathfrak{H}$ can be equated with a function F from $\{e_n^w\}$ to $\mathfrak{H}$. (Think of $F(e_n^w)$ as the conjecture, based on the evidence seen so far, as to which of the hypotheses is true.) F is said to be completely reliable for $\mathfrak{H}$ with respect to $\mathfrak{E}$ and W iff $\forall \ w \in W, \forall \ e^w, \exists \ N$ such that $\forall \ n \geqslant N, F(e_n^w) = H_i$ for the H_i true in w.[3] If there is a completely reliable Pr, even in the .5 sense, then obviously there must be a completely reliable F—just set $F(e_n^w) = H_i$ iff $Pr(H_i/e_n^w) > .5$. This provides a means of generating negative reliability results for Bayesian methodologies; for where it can be proved that no reliable F exists, *a fortiori* Bayesianism is not a reliable methodology.

4. A PROOF (?) OF THE UNDERDETERMINATION THESIS FOR THEORETICAL HYPOTHESES

If the hypotheses $\mathfrak{H}$ are EI_3 with respect to $\mathfrak{E}$ and W then obviously there is no reliable truth-identification method. Thus, no Pr can be completely reliable in the long run, even in the .5 sense, for $\mathfrak{H}$ with respect to $\mathfrak{E}$ and W. So if long-run reliability is a necessary condition for Bayesian methodology to supply good reasons to believe that one of the competing hypotheses is true, and if EI_3 holds for the kinds of cases typically discussed under the label of "underdetermination of theory by evidence," then indeed $EI_3 \Rightarrow U$.

One way to try to discharge the antecedent of the last mentioned conditional is to argue, first, that either EI_1 or EI_2 will hold in cases that deserve to be brought under the underdetermination label and, second, that in an important subclass of cases, EI_1 and EI_2 each entails EI_3. The subclass I have in mind is described by a number of assumptions. Suppose to begin that the descriptive vocabulary of the (first order) language of $\mathfrak{H}$ and $\mathfrak{E}$ is bifurcated into two parts, V_O and V_T. The subscripts 'O' and 'T' are supposed to recall the observational-theoretical distinction that is presupposed by the traditional realism–anti-realism debate. But for present purposes it is sufficient that the members of $\mathfrak{H}$ are stated in a language that outruns the language of the evidence $\mathfrak{E}$, whatever that happens to be. W consists of all logically possible worlds (i.e., all models of the language) . Suppose further that $\mathfrak{E}$ is rich enough that for any sentence E, singular or quantified, whose descriptive terms belong entirely to V_O, $\mathfrak{E}$ contains a sentence that is logically equivalent to either E or to $\neg E$. Now posit that EI_2 holds, i.e., that there are $H_i, H_j \in \mathfrak{H}(i \neq j)$ which have in common the same class C of $\mathfrak{E}$ consequences. Consider any model $\overset{\circ}{w}$ of the V_O sublanguage such that $\models_{\overset{\circ}{w}} C$. Then although there may be no extension of $\overset{\circ}{w}$ to a model of either H_i or H_j, there will be models $\overset{\circ}{w}_i$ and $\overset{\circ}{w}_j$ elementarily equivalent to $\overset{\circ}{w}$ such that $\overset{\circ}{w}_i$ is the V_O reduct of a model w_i of the full language and $\models_{w_i} H_i$ while $\overset{\circ}{w}_j$ is the V_O reduct of a model w_j of the full language and $\models_{w_j} H_j$. Since w_i and w_j make exactly the same $\mathfrak{E}$ sentences true, H_i and H_j are EI_3 with respect to the set W of logically possible worlds. And since this is so for any choice of $\overset{\circ}{w}$ that satisfies C, EI_3 is rampant among the logically possible worlds. A further argument which uses the compactness property of standard first-order logic shows that, in the setting under discussion, EI_1 entails EI_3 (see Earman 1992, chap. 6).

One reason why the above line of argumentation is not satisfying for present purposes is that it is sensitive to the details of what goes into $\mathfrak{E}$ and W. To illustrate, suppose as before that W consists of all models of the language whereas $\mathfrak{E}$ consists of all atomic observation sentences. Take $\mathfrak{H}$ to be $\{H, \neg H\}$ where $H = (\forall x)Px$ and 'P' belongs to V_O. Then the $\mathfrak{H}$ are EI_1, EI_2 and EI_3 with respect to $\mathfrak{E}$ and W. Next, keep $\mathfrak{H}$ and $\mathfrak{E}$ the same as before but pare down the possible worlds so that W now contains only models with countable domains, the elements of which are named by a fixed list $a_1, a_2, \ldots$. This would be the appropriate W to consider, for instance, if we were concerned with a process which successively produces an endless string of objects and we were concerned with formulating and testing hypotheses about the color of the objects (e.g., "All the objects are red"). Models containing non-standard objects would seem to be irrelevant to questions of reliability of inductive procedures for assessing the truth of such color hypotheses. In this case we have EI_1, ED_2 (because relative to the chosen W, H implies Pa_i for all i), and ED_3 (because of the fact that all the objects in the domains of the models W are named).

What needs to be argued, if the underdetermination thesis is to be established, is that when the $\mathfrak{H}$ are stated in a language that outruns the language of $\mathfrak{E}$, EI_3 is to be expected for any interesting choice of W. This is a matter I will return to at various points below. But more immediately I need to turn to the concern that arguments from reliability considerations prove too much because they not only lead to skepticism with respect to theoretical hypotheses but to a blanket skepticism as well.

5. RELIABILITY AND SKEPTICISM

Reliability considerations were used to generate one form of the underdetermination thesis for hypotheses couched in a language that outruns the language of the evidence. This result would not be very interesting if reliability considerations led to a general form of skepticism that applies to hypotheses formulated in the language of $\mathfrak{E}$. This worry is best investigated in terms of some concrete examples that instantiate the apparatus developed above.

> *Ex. 1.* This example recapitulates one given in section 4. $H = (\forall x)Px$, where 'P' can be considered to be an observational predicate. $\mathfrak{H} = \{H, \neg H\}$. W consists of models with a countable domains, the elements of which are named by a fixed list of names $a_1, a_2, \ldots$. $\mathfrak{E} = \{Pa_i\}, i = 1, 2, \ldots$. Here we have EI_1, ED_2, and ED_3. Now let Pr be a probability function that assigns non-zero priors to both H and $\neg H$ and that is countably additive in the sense that $\text{Pr}((\forall x)Px) = \lim_{n \to \infty} \text{Pr}(Pa_1 \& Pa_2 \& \ldots \& Pa_n)$.

Then Pr is completely reliable in the long run for $\mathfrak{H}$ with respect to $\mathfrak{E}$ and W.[4]

> *Ex. 2.* Let W be as in Ex. 1. Now $H = (\forall x)(\exists y)Rxy$, where we may think of 'R' as an observable relation. $\mathfrak{H} = \{H, \neg H\}$. And $\mathfrak{E} = \{Ra_i a_j\}, i, j = 1, 2, \ldots$. We now have $\text{EI}_1, \text{EI}_2, \text{ED}_3$. While ED_3 is a necessary condition for reliability, it is not sufficient. For it can be shown that in this case there is no truth-identification function that is completely reliable for W and *a fortiori*, no Pr can be completely reliable for all W in even the .5 sense.[5]

> *Ex. 3.* Let W consist of models with countable domains, the elements of which are named by $b_i, i = 0, \pm1, \pm2, \ldots$. (For sake of concreteness, think of a doubly infinite sequence of days stretching infinitely into the future and into the past.) $\mathfrak{E}_n = \{Qb_j : j \leqslant n\}, n = 0, \pm1, \pm2, \ldots$. $H_k = Qb_k$, and $\mathfrak{H}_k = \{H_k, \neg H_k\}$. (Again for sake of concreteness, think of Qb_j as asserting that the sun rises on day j.)

Reliability considerations now generate Humean skepticism about the future. With $k > n$, the $\mathfrak{H}_k$ are EI_3 with respect to $\mathfrak{E}_n$ and W so that there can be no completely reliable truth-identification function and, hence, Bayesian methodology cannot be completely reliable in even the .5 sense in fixing beliefs about whether the sun will rise tomorrow on the basis of when it has and has not risen in the past. This, I take it, is the essence of Hume's complaint about induction dressed up in modern jargon.

Two morals can be drawn from these toy examples. The first is that strict reliability considerations, even when applied to a rich set of possible worlds, do not generate blanket skepticism. Ex. 1 shows that complete reliability can be exhibited by Bayesian methodology for interesting hypotheses and rich sets of worlds. But the second and darker moral, drawn from Ex. 2 and Ex. 3, is that strict reliability considerations cannot be used to underwrite an underdetermination argument that generates skepticism about theoretical hypotheses without also generating skepticism about quantified and singular observational hypotheses. There are, for all I know, empiricists who would embrace all of these skeptical consequences of reliability demands. But this version of empiricism is certainly not van Fraassen's constructive empiricism since, for example, a warranted belief in the empirical adequacy of a theory will surely involve a belief in multiply quantified assertions about the observable, such as in Ex. 2.

In any case, what is in a label is not the important issue. What is clear is that strict reliability must be relaxed if the skepticism it generates about the theoretical realm is not to have a major spillover into the realm of the observable. In the next section I will explore one form the relaxation might take.

6. ALMOST SURE RELIABILITY

For those willing to employ probabilistic methodology, "almost sure" reliability or reliability over a subset of worlds of "measure one" may be good enough. This almost sure reliability may be either objective or subjective.

The former refers to cases where the measure is induced by a statistical mechanism that characterizes the worlds W. For instance, in Ex. 3 we can imagine that each day constitutes a statistically independent trial and that on each day the sun has the same objective chance $p, 0 < p < 1$, of rising. The "cylinder sets" consist of worlds which agree on some day n as to the rising–non-rising of the sun. The cylinder set corresponding to rising is given a measure p while the cylinder set corresponding to no rising is given a measure $1 - p$. The measurable sets $\mathfrak{M}$ are those in the σ-algebra generated by taking the cylinder sets as the basis sets. It can then be shown that there is a unique countably additive measure $\mathfrak{P}r$ on $\mathfrak{M}$ agreeing with the assignments on the basis sets.

Even if an objective statistical mechanism is at work, what we have to operate with in decision making and epistemology are our own personal estimates of the objective chances. Thus, for our purposes the operative notion of reliability is subjective. If S is a sentence of the language on which $\Pr$ is defined, $\mathrm{mod}(S) \equiv \{w \in W : \models_w S\}$. These subsets now form the basis for the σ-algebra of measurable sets $\mathfrak{M}$ of W. If $\Pr$ is countably additive (in the sense explained in Ex. 1), it can be shown that there is a unique countably additive measure $\mathfrak{P}r$ on $\mathfrak{M}$ such that $\Pr(S) = \mathfrak{P}r(\mathrm{mod})(S)$

for each sentence S.[6] The worry that this approach is overly subjective can be assuaged if questions of reasonableness of belief are relativized to a scientific community and if a community is identified with a set of mutually absolutely continuous probability functions (i.e., functions that assign O's to exactly the same sentences). There will then be community agreement on measure one reliability.

In Ex. 2 almost sure subjective reliability seems to come for free. That is, if Pr is any countably additive probability function, then Pr is reliable for $\mathfrak{H} = \{(\forall x)(\exists y)Rxy, (\exists x)(\forall y)\neg Rxy\}$ with respect to $\mathfrak{E} = \{Ra_i a_j\}, i, j = 1, 2, \ldots$, and almost any W (as measured in the Pr induced measure on W).[7] This almost sure reliability helps to remove the specter of skepticism with respect to multiply quantified observational hypotheses, at least if the evidence is allowed to range over all of the relevant individuals.

However, almost sure subjective reliability does not come for free in the Hume example Ex. 3. Here one form or another of dogmatism is the price to be paid for the sought-after reliability. The most blatant and intolerable form of dogmatism consists of assigning a prior probability 1 to one of the elements of $\mathfrak{H}_k = \{Qb_k, \neg Qb_k\}$ and 0 to the other. An agent using such a Pr will be almost surely reliable by her own lights. But since in her own mind she has settled *ab initio* the matter to be investigated and since no amount of empirical evidence will change her mind, the price she has paid for her narcissistic almost sure reliability is too high.

A somewhat less blatant form of dogmatism is exhibited by a Pr which assigns non-zero priors to both elements of $\mathfrak{H}_k$ and which assigns measure one to a subset $D \subset W$ on which the $\mathfrak{H}_k$ are ED3 with respect to the $\mathfrak{E}_n, n < k$. The subset is labeled D since it is in fact a subset of worlds that is (futuristically) deterministic with respect to the property 'Q': that is, $\forall w_1, w_2 \epsilon D$, if w_1 and w_2 agree as to whether Qb_i for $i \leqslant n$, then w_1 and w_2 agree as to whether Qb_k. So the price here for almost sure subjective reliability is the complete conviction that, prior to any empirical investigation, the world is deterministic as regards 'Q'. Though not as brazen as the first form of dogmatism, this second form is still hard to swallow. Prior to any investigation we are in no position to rule out the possibility that the rising of the sun is governed by a statistical mechanism of independent and identically distributed trials with probability $0 < p < 1$. The objective measure induced by such a mechanism gives D measure zero (see Earman 1986, 146–47). So to announce *ab initio* that one's subjective measure gives D measure one and that, consequently, no amount of empirical evidence will lead to a change of mind on this matter is to declare a kind of dogmatism that makes it impossible for one to join in common cause with other investigators that are not so *a priori* cocksure.

If one agrees that this second form of dogmatism is too high a price to pay for almost sure reliability, there are still two options open. One is to claim that Humean skepticism is intolerable and that, therefore, even almost sure reliability must be rejected as a necessary condition for a methodology to deliver warranted belief. Surely, one might ask, is it not clear that past experience does justify a high degree of belief in the proposition that the sun will rise tomorrow? This is, of course, intended as a rhetorical question. But I choose to take it as a serious query, and to answer the query in the negative. This is to exercise the second option and to accept Humean skepticism. The sting of this skepticism can be partly drawn by noting that it is consistent with this skepticism to have $\Pr(Qb_{n+1}/Qb_o \& Qb_1 \& \ldots Qb_n) \to 1$ as $n \to \infty$; indeed, this convergence to certainty about the next instance must hold if $'(\forall x)Qx$ is given a non-zero prior.[8] So in every world where the sun always rises from this day forward, anyone who uses such a $\Pr$ will correctly come to believe with increasing certainty that the sun will rise on the morrow after a long enough unbroken string of sunrises is experienced.

The morals from the Hume example can now be transferred to the case of theoretical hypotheses. It was seen in section 4 that for hypotheses couched in terms of a language richer than the language of the evidence $\mathfrak{E}$, EI_3 will be rampant with respect to the set of W of all logically possible worlds; and EI_3 can also be expected to be rampant for most rich sets of possible worlds. So to achieve almost sure subjective reliability, $\Pr$ must exhibit dogmatism of one of two forms. Either it can prejudge the issue by assigning to one of the elements of $\mathfrak{H}$ a prior probability of 1 and to the others prior probabilities of 0. Or it can give all the elements of $\mathfrak{H}$ a fighting chance by assigning each a non-zero prior; but at the same time it must assign measure one to a subset of worlds within which the $\mathfrak{H}$ are ED_3. I take it that this form of dogmatism is no more tolerable than it was in the Hume example. In the Hume example the unattractiveness of this second form of dogmatism could be made vivid by showing that it entails an *a priori* probabilistic certainty in the truth of physical determinism. In the case of theoretical hypotheses the corresponding form of dogmatism takes the form of a kind of conceptual determinism in that it entails an *a priori* probabilistic certainty that if any two worlds agree on the truth values of all of the empirical evidence statements $\mathfrak{E}$, then they also agree on the truth values of all the theoretical hypotheses in $\mathfrak{H}$. This *a priori* certainty in conceptual determinism has even less to recommend it than the *a priori* certainty in physical determinism since in the latter case one can imagine various mechanisms that would implement the determinism. Thinking of a mechanism to implement the conceptual determinism is not so easy.

7. HIDDEN DOGMATISM

The upshot of the discussion so far seems to be a plus for epistemic anti-realism with respect to theoretical hypotheses. Insofar as long-run reliability is required of a method of inquiry that is to yield warranted belief, then there is an asymmetry between observational and theoretical hypotheses, or more generally between hypotheses stated in the language of $\mathfrak{E}$ and hypotheses stated in a richer language. Namely, there are interesting species of the former hypotheses where warranted belief is possible, but because of the rampant empirical indistinguishability among the latter hypotheses we cannot achieve warranted belief without paying the price for one or another form of dogmatism, a price I urged was too high.

The would-be scientific realist, or anyone who wants to deny the asymmetry between warranted inference from the observed to unobserved observables vs. warranted inference to unobservables, can respond by noting that the success in the former case depends upon a hidden dogmatism. In Ex. 2 any observational hypothesis, even a multiply quantified one, will be ED_3 from its negation for all W. But as noted ED_3 though necessary for reliability is not sufficient. There is a method that will correctly conjecture that $(\forall x)(\exists y)Rxy$ is true for every $w \in W$ where it is true; namely, conjecture that it is true iff every object examined so far is R-related to some object in the sample. But neither this nor any other method will correctly conjecture that $(\forall x)(\exists y)Rxy$ is false for every $w \in W$ where it is false. Likewise there is a method that will correctly conjecture that $(\exists x)(\forall y)\neg Rxy$ is false for every $w \in W$ where it is false; but there is no method that will correctly conjecture that it is true for every $w \in W$ where it is true. But it was asserted that a countably additive $\Pr$ will be subjectively reliable for $\{(\forall x)(\exists y)Rxy, (\exists x)(\forall y)\neg Rxy\}$ with respect to $\mathfrak{E}$ and a subset $W' \subset W$ of measure one. Intuitively, this means that with respect to W' $(\forall x)(\exists y)Rxy$ is being treated as equivalent to $(\exists x)(\forall y)R'xy$ for some 'R''. In other words, the agent is probabilistically *a priori* certain that the quantifiers permute. And that is surely a form of dogmatism.[9]

I agree that there is a kind of dogmatism here. But it is not on a par with the forms of dogmatism discussed in the preceding section. Those dogmatisms resulted from the choice of particular probability functions with peculiar features. Here the dogmatism results not from the choice of any particular probability function but is rather a general consequence of the property of countable additivity. It is Bayesianism in general that stands convicted of dogmatism and not any particular Bayesian agent. Since Bayesianism seems to be the only methodology capable of offering a unified account of the great range of scientific inferences, and since countable additivity is needed to derive key reliability results, the common dogmatism simply has to be tolerated.

8. LOWERING THE SIGHTS

Some of the concerns of traditional epistemology, especially those relating to skepticism, can be seen as deriving from a demand for a method of inquiry that is reliable over all possible worlds, or at least over a far-flung set of worlds that contains such "far out" possibilities as evil demons, brains in a vat, and the like. The demand, if rigorously enforced, generates the familiar forms of philosophical skepticism. While some forms of skepticism surfaced in the examples discussed above, they were never of the philosophical variety. That philosophical skepticism did not rear its ugly head is an outcome that did not emerge in any natural way from the analysis but was, so to speak, put in by hand. For example, in Ex. 2 and Ex. 3 it was simply stipulated that the a_i are denoting, and it was tacitly assumed that they denote objects of the intended type and not pseudo-objects or illusions. That piece of artifice seems allowable in the present context since the goal is to assess the epistemological component of scientific realism, and the discussion of this issue typically starts from the assumption that philosophical doubts about the existence of ordinary objects have been resolved or at least set aside.

Nevertheless, once one starts down the road of requiring reliability only for a W that is a proper subset of all logically possible worlds, it is natural to investigate the concept of warranted belief that results from shrinking W so that none of the elements represents more than a small departure from the actual world. Carrying out that exercise for the "S knows that p" formula leads to Robert Nozick's analysis of knowledge.[10]

The Nozick-style necessary and sufficient conditions for S to know that p are that (1) p is true, (2) S believes p via a method M, (3) if p were not true and S were to use M to arrive at a belief whether (or not) p, then S would not believe via M that p, and (4) if p were true and S were to use M to arrive at a belief whether (or not) p, then S would believe via M that p. On Nozick's reading of subjunctive conditionals, the combination of (3) and (4) requires that M is reliable—"tracks the truth" in Nozick's jargon—about $\mathfrak{H} = \{p, \neg p\}$ for a set of worlds that forms a near neighborhood of the actual world.

A standard worry here is not that "nearness" or "similarity" of worlds is a vague notion, resulting in a neighborhood with ragged boundaries; rather the worry is that there is no single, definite notion of similarity, vague or not, but a myriad of such notions, leading to a panoply of near neighborhoods. Perhaps contextual factors single out the relevant dimension of similarity or relevant weightings for different dimensions, but if so it is not obvious how those factors operate in the context of the realism–anti-realism debate.

Even if a definite similarity relation is fixed, there remains the worry that not much succor is to be derived for the epistemological component of scientific realism. Restricting the demand for reliable tracking of truth to a near neighborhood of worlds may help to overcome some forms of philosophical skepticism, but it is far from evident that it helps to overcome skepticism about theoretical entities. The succor would have to come from the finding that although ED_3 for the theories at issue fails for $\mathfrak{E}$ and W, it nonetheless holds for an an appropriate neighborhood $N \subset W$ of the actual world. But as explained in section 4, when EI_3 exists for theoretical hypotheses (or more generally for hypotheses stated in a language that outruns the language of $\mathfrak{E}$), it tends to be rampant; and without relying on a question-begging notion of similarity of worlds, there is no reason to think that EI_3 rears its head in far out worlds but vanishes for near worlds.[11]

9. ARE THERE INTERESTING CASES OF EMPIRICALLY INDISTINGUISHABLE THEORIES?

The discussion so far would have the flavor of a scholastic exercise if one could not produce, or at least argue for the existence of, theories that are interestingly different from and empirically indistinguishable from successful scientific theories. Here I find the philosophical literature disappointing. Quine (1975) suggested the following procedure for generating examples of underdetermination. Start with some standard theory T of physics. Choose two kind terms from the theoretical vocabulary of T (say, 'electron' and 'molecule') and systematically intersubstitute the one term for the other to produce T'. T and T' would seem to be EI_3. I find it hard to get excited about such examples, but for those who have a lower threshold of philosophical excitement, Horwich (1982) provides a good way of undercutting the threatened underdetermination.

We get closer to an interesting form of underdetermination by means of an example discussed by Laudan and Leplin (1991, 457–58). Let TN stand for Newton's theory of mechanics and gravitation, R the postulate that the center of mass of the universe is at rest with respect to absolute space, and V the postulate that the center of mass is moving with velocity v relative to absolute space. Then $TN + R$ and $TN + V$ will be empirically indistinguishable in any interesting sense with respect to the intended possible worlds W and any evidence about the relative motions of bodies and their absolute accelerations. But again it is hard to get excited about this example. Since $TN + R$ and $TN + V$ involve exactly the same ontology and ideology for space, time, and motion, no very interesting form of underdetermination is in the offing.

I claim here that what we have is a shortcoming of the philosophical literature and not a failure of the underdetermination thesis. For I claim that there do exist examples of rival empirically indistinguishable theories that posit interestingly different theoretical structures. For instance, TN (*sans* absolute space) can be opposed by a theory which eschews gravitational force in favor of a non-flat affine connection and which predicts exactly the same particle orbits as TN for gravitationally interacting particles (see Havas 1964).

A different type of example flows from the fact that an oracle does not whisper the truth values of the elements of $\mathfrak{E}$ in our ears; rather we have to try to access these truth values by causally interacting with the world. Relativity theory tells us that the data available to an observer through such interactions are restricted to events that are swept out by the observer's past light cone.[12] Thus, for a given observer, questions about empirical indistinguishability are properly understood not in terms of $\mathfrak{E}$ but in terms of the subset $\hat{\mathfrak{E}}$ of propositions about events in the observer's past light cone. As a result, even idealized observers who live forever may be unable to empirically distinguish hypotheses about global topological features of some of the cosmological models allowed by Einstein's field equations for gravitation (see Glymour 1977 and Malament 1977).

Other examples could be given, but I trust that enough has been said to remove the worry that the underdetermination thesis in the form $EI_3 \Rightarrow U$ is vacuously true because there are no concrete examples of interestingly different but empirically indistinguishable rivals to successful scientific theories. And the production of a few concrete examples is enough to generate the worry that only a lack of imagination on our part prevents us from seeing comparable examples of underdetermination all over the map.

10. LAUDAN AND LEPLIN ON UNDERDETERMINATION

Laudan and Leplin (1991) launch a two-pronged attack on the underdetermination thesis. First they attack the widespread notion that all or many empirically successful theories have empirically equivalent counterparts. Second, they
maintain that even if there are interesting examples of empirically equivalent
theories, it does not follow that each of the equivalent theories is equally well
(or ill) supported by the evidence. The discussion of the preceding sections
provides the context for assessing the second prong.

The main argument that Laudan and Leplin offer in support of the second
prong consists of (a) the assertion that proponents of the underdetermination
thesis rely on the idea that a theory receives empirical support only through
its empirical consequences, and (b) the counterclaim that a theory can derive
support from pieces of evidence that are not consequences of the theory. To
(b) they might have added (c) the observation that two theories need not be
supported to the same degree by their common consequences.

It should be clear by now that worries about underdetermination need not
derive from the crude views about evidential support—essentially, hypothetico-
deductivism—contained in (a). Once this straw man is removed, the force of the
(correct) claim (b) and the observation (c) cannot be discerned until a positive
account is given of empirical support. Laudan and Leplin decline to give one.
Since Bayesianism is the only live contender on the current scene, I propose
to use it in assessing (b) and (c).

If C is a logical consequence of T_1 and of T_2, then an application of
Bayes's theorem shows that $\Pr(T_{1,2}/C) = \Pr(T_{1,2})/\Pr(C)$. Thus, although
Bayes's theorem endorses (c), it also reveals that the only way the posterior
probabilities of the theories, conditional on the common consequence, can differ
is for the prior probabilities of the theories to differ. Since priors are notoriously
subjective, such differences are not plausible sources for the good reasons for
belief that are likely to cut much philosophical ice in the realism debate. When
C is not a consequence of either T_1 or of T_2, Bayes's theorem shows that
$\Pr(T_1/C)$ and $\Pr(T_2/C)$ can differ not only because the priors of T_1 and T_2
differ but also because the likelihoods $\Pr(C/T_1)$ and $\Pr(C/T_2)$ differ. But
if neither C nor $\neg C$ is a consequence of either theory and if it is not the case
that T_1 and T_2 are statistical theories about a chance set-up while C is a report
of the outcomes of random trials, then the likelihoods will be just as much up
for grabs as the priors.

The Bayesian literature contains two responses to these subjectivity problems. The first is to supplement the core principles of Bayesian personalism—
the probability axioms and the rule of conditionalization—with additional principles that constrain assignments of values to the priors and likelihoods. These
supplementary principles are at best controversial and at worst unworkable
because they lead to conflicting results depending upon how the problem is
formulated. The second response is to acknowledge rather than to quash the
fact that different scientists start with differing opinions and to try to show that

objectivity emerges with the "washing out" of differing priors and likelihoods as a consequence of repeated conditionalizations on pieces of empirical evidence. This second approach reconnects with the previous discussion because most of the merger of opinion theorems for non-statistical hypotheses are parasitic on the convergence to certainty results that are at the heart of the reliability of Bayesian methodology. And in any case merged subjective opinion is no more interesting than individual subjective opinion unless it bears some systematic connection to the truth. It is precisely here that the empirical indistinguishability of theories does damage to the epistemic component of realism, or so I have argued.

I turn now to the first prong of Laudan and Leplin's attack, which is supposed to "cast doubt on empirical equivalence in general, as a relation among scientific theories" (1991, 451). They propose to sow doubt by appealing to three theses.

Thesis 1: The variability of the range of the observable (VRO). Any circumscription of the range of observable phenomena is relative to the state of scientific knowledge and the technological resources available for observation and detection. (451)

Thesis 2: The need for auxiliaries in prediction (NAP). Theoretical hypotheses typically require supplementation by auxiliary or collateral information for the derivation of observable consequences. (452)

Thesis 3: The instability of auxiliary assumptions (IAA). Auxiliary assumptions once sufficiently secure to be used as premises frequently come subsequently to be rejected, and new auxiliaries permitting the derivation of additional observational consequences frequently become available. (452)

Reflecting on the implications of these theses, Laudan and Leplin conclude that the empirical equivalence of hypotheses is not a purely formal matter but must be "relativized to a particular state of science" (454).

I would begin by responding that empirical equivalence or indistinguishability *is* a matter of the formal relationships among $\mathfrak{H}$, $\mathfrak{E}$, and W; but that if what is counted as the empirical evidence $\mathfrak{E}$ or the hypotheses $\mathfrak{H}$ is changed then, of course, the empirical equivalence or indistinguishability classes change. Although in part this response is only terminological it points to a difference in attitude concerning the importance of the empirical indistinguishability of hypotheses. Suppose that at some stage of inquiry $\mathfrak{E}$ adequately captures what was then meant by empirical evidence, and suppose further that relative to this $\mathfrak{E}$ and the appropriate W the hypotheses $\mathfrak{H}$ are EI_3. Then if the arguments of the preceding sections are correct, observational and experimental results couched in terms of $\mathfrak{E}$ cannot provide satisfactory reasons for believing one of the $\mathfrak{H}$'s over the others. Now suppose that at some later stage of inquiry

ℭ expands to ℭ′. Relative to W and the new ℭ′ the 𝔥 may be ED_3, and the new empirical evidence gathered from the expanded evidence matrix may provide a firm basis for believing one of the 𝔥's. But this shift in no way undermines the importance of underdetermination considerations. For it is cold comfort to tell the scientists who were in the former epistemic context that if their situation had been different then they would have been able to gather evidence that would decide among the 𝔥's

Perhaps the suggestion is that the notion of what counts as observational and more broadly as empirical evidence is so fluid and the notion of empirical indistinguishability is so sensitive to what belongs to ℭ that it is easy to overcome underdetermination. This is not what Laudan and Leplin say, but my guess is that is what they want us to conclude from the VRO thesis. But for the concrete examples of underdetermination given in section 9, this suggestion strikes me as flatly false. I find it difficult to imagine any plausible development in the concept of observability that would change the indistinguishability of these cases to distinguishability. In the cosmological example, an observer can take as evidence *any* proposition about events swept out by her past light cone, whether or not the proposition corresponds to a state of affairs that can be detected by the unaided eye, and still the alternative hypotheses about the global topological structure of space-time remain indistinguishable.

In sum, the doubts that Laudan and Leplin try to sow by appeal to VRO do not seem to me very troubling, and in the remainder of this section I will set them aside to deal with NAP and IAA. My first remark about the latter theses is that Laudan and Leplin's concern about auxiliary hypotheses stems in large part from their fixation on an imagined opponent who adheres to hypothetico-deductivism. But I have argued that the crucial form of distinguishability is not the one that flows from hypothetico-deductivism—namely, ED_2—but rather ED_3. And ED_3 can hold even though the 𝔥's do not have different empirical consequences. In Ex. 2 of section 5 the 𝔥's do not have any non-trivial consequences, and yet they are ED_3.

My second remark about NAP and IAA is that interesting examples of underdetermination tend to be robust with respect to different auxiliary hypotheses. Consider again the different versions of Newtonian gravitational theory mentioned in section 9, and more especially their implications for planetary orbits. Presumably the relevant sorts of auxiliary hypotheses concern possible extra-solar system sources of perturbation. But whatever hypothesis is made about such sources, the two theories in question will yield exactly the same predictions for planetary orbits.

Even if differences in auxiliary hypotheses do make for differences in empirical distinguishability, I fail to see how these differences undercut the force of underdetermination considerations. Suppose that the members of $𝔥 = \{H_1, H_2, \ldots\}$ are EI_3 with respect to ℭ and W, and further that there is no subset of W of Pr induced measure one for which ED_3 holds. Thus insofar as reliability is necessary for an inductive method to produce warranted belief, empirical evidence from ℭ cannot offer grounds for warranted belief in the

$\mathfrak{H}$'s. How is this negative result to be impugned by noting that if auxiliary A_i is added to H_i to form $H_i' = H_i \& A_i$, then the $\mathfrak{H}' = \{H_1', H_2', \ldots\}$ may be ED$_3$ with respect to $\mathfrak{E}$ and W? It isn't. The result is sidestepped if the focus is shifted from $\mathfrak{H}$ to $\mathfrak{H}'$, but that is changing the subject since what counts as the hypothesis has been changed. The other way to sidestep the no-go result without changing $\mathfrak{H}$ or $\mathfrak{E}$ is to change the starting presuppositions. For the Bayesian this would amount to changing to a new probability function Pr' which is equal to the old Pr conditional on one of the auxiliaries (or perhaps a conjunction of auxiliaries, if they are consistent). If the $\mathfrak{H}$ are ED$_3$ with respect to $\mathfrak{E}$ and a subset $W' \subset W$ of measure one in the Pr' induced measure, then the no-go result is bypassed. But for the auxiliaries to play this role they must go beyond the empirical evidence $\mathfrak{E}$, and thus their epistemic status will be just as open to question as that of the $\mathfrak{H}$. Adopting Pr' then amounts to another form of dogmatism that is not supported by any amount of empirical evidence $\mathfrak{E}$, for if evidence from $\mathfrak{E}$ allowed reliable Pr-convergence to certainty on the auxiliaries, then the $\mathfrak{H}$ would not be EI$_3$ with respect to a subset $W' \subset W$ of Pr-induced positive measure, which is contrary to assumption.

I conclude that Laudan and Leplin's invocation of VRO, NAP, and IAA does not draw the sting of underdetermination.

11. CONCLUSION

The considerations advanced here support a position that is similar to but not identical with van Fraassen's constructive empiricism. This position, which I dub *gentle empiricism*, does not enjoin us from believing in the truth of scientific theories. My preferred version of inductive methodology is Bayesianism, and the good Bayesianism is free to assign degrees of belief, high or low, to the latest version of quark theory or super string theory no less than to statements about observables. Indeed, a Bayesian could avoid assigning degrees of belief to theoretical statements only by restricting the domain of her degree of belief function to statements about observables.[13] Such a restriction is highly undesirable when it comes to evaluating ampliative inferences, even from observables to observables, that cannot be captured in terms of straightforward inductive generalizations. To borrow an example from Putnam (1963), prior to the first atomic explosion, induction by enumeration would have suggested that slamming together two rocks of a certain kind (that can be specified in terms of observable characteristics) would not produce a novel effect. But because of other empirical evidence in support of nuclear theory, scientists were confident that a novel effect would result if the masses of the rocks were increased beyond a critical value. So gentle empiricism does not say, "Don't have high degrees of belief in theories," much less "Don't even assign degrees of belief to theories," but only "When there are competing empirically indistinguishable theories, recognize that your degree of belief is an expression of personal opinion that may not bear any systematic relation to the truth." This gentle empiricism also endorses a milder asymmetry between the status of inferences to unobserved observables vs. inferences to unobservables. For reliability-generated skepticism about the latter spills over into the former. And in some cases the asymmetry hinges on "measure one" considerations of a subjective and narcissistic nature.

What is left open is the reach of the underdetermination thesis. I have argued, first, that empirical indistinguishability in the sense of EI_3 generates underdetermination and, second, that there are interesting examples of EI_3 theories. I did not tackle the questions of whether there are mechanical algorithms for generating EI_3 alternatives or whether many successful scientific theories do in fact possess interestingly different EI_3 rivals. To my mind these are the most important remaining problems in assessing the threat of underdetermination.

In closing I would like to offer a few general comments on the realism–anti-realism debate. In a series of stimulating articles, Arthur Fine (1984a, 1984b, 1986) has opined that the debate has entered a degenerative phase. His diagnosis is that the traditional positions are bankrupt, and he proposes that they be abandoned in favor of the "natural ontological attitude" (NOA). The ethos of NOA is to take science on its own terms and to eschew philosophical theories of science—theories of truth, theories of evidence, and theories of demarcation. While I share Fine's perception of a degenerating debate, my diagnosis and prescriptions are quite different from his. Rather than marching bravely into some post-realist–anti-realist regime, I think that we ought to attend more closely to issues that have been neglected of late by philosophers more interested in reaching for profundity. The toy examples studied above admittedly give but a pale imitation of real science. But nonetheless they do seem to me to help to define important issues about the reliability of inductive methodology, issues which have linkages to venerable problems about skepticism and to contemporary problems of scientific realism. If the NOAer says that issues about the reliability of inferences to theoretical entities cannot be legitimately raised, or that such issues are not worthy of attention, or that philosophers cannot aspire to construct theories of such matters—if that is the attitude of the NOAer, then NOAers are NOAernothings.[14]

NOTES

1. For expositions of Bayesian personalism and applications to issues of confirmation of scientific hypotheses, see Howson and Urbach (1989) and Earman (1992).

2. I conjecture that if there is a Pr which is completely .5 reliable, then there will be a Pr' which is completely reliable.

3. Here I am following the learning paradigm discussed in Kelly and Glymour (1989).

4. This follows from some simple results of the probability calculus; see Earman (1992).

5. A proof is given in Kelly and Glymour (1989).

6. For details, see Gaifman and Snir (1982).

7. See Theorem 2.1 of Gaifman and Snir (1982).

8. $Pr((\forall x)Px) > 0$ is a sufficient but not necessary condition for the probability of the next instance to go to 1; see Earman (1992).

9. For a precise statement and proof of this result, see Kelly (1992).

10. See Nozick (1987). Critical essays on Nozick's theory of knowledge are to be found in Luper-Foy (1987).

11. Externalist accounts of knowledge seem to me to provide the best approach to traditional problems about "S knows that p." But these accounts are doubly irrelevant to the philosophy of science. Here the main concern is rarely whether or not a scientist 'knows' that some theory is true but rather whether or not she is justified in believing it. And because science is a community enterprise the only forms of justification that are scientifically relevant are those which are stateable and open to public scrutiny.

12. If γ is the world line of an observer, then the relevant region of spacetime is $C^-(\gamma) \equiv \cup_{p \in \gamma} C^-(p)$, where $C^-(p)$ is the set of all space-time points q such that there is a future-directed causal curve from q to p.

13. An alternative way of expressing agnosticism about theoretical claims is to assign them vague or interval-valued probabilities; see van Fraassen (1989).

14. I am indebted to Clark Glymour, Laura Ruetsche, and Bas van Fraassen for helpful comments on earlier drafts of this paper.

REFERENCES

Earman, J. 1986. *A Primer on Determinism*. Dordrecht.

Earman, J. 1992. *Bayes or Bust? A Critical Examination of Bayesian Confirmation Theory*. Cambridge, Mass.

Earman, J., C. Glymour and J. Stachel. Eds. 1977. *Foundations of Space-Time Theories, Minnesota Studies in the Philosophy of Science*, Vol. 8. Minneapolis.

Fine, A. 1984a. "The Natural Ontological Attitude." In *Scientific Realism*, edited by J. Leplin. Berkeley.

Fine, A. 1984b. "And Not Anti-Realism Either." *Noûs* 18: 51–65.

Fine, A. 1986. "Unnatural Attitudes: Realist and Instrumentalist Attachments to Science." *Mind* 95: 149–79.

Gaifman, H., and M. Snir. 1982. "Probabilities on Rich Languages." *Journal of Symbolic Logic* 47: 495–548.

Glymour, C. 1977. "Indistinguishable Space-Times and the Fundamental Group." In Earman et al. (1977).

Havas, P. 1964. "Four-Dimensional Formulations of Newtonian Mechanics and Their Relation to the Special and General Theory of Relativity." *Reviews of Modern Physics* 36: 938–65.

Horwich, P. 1982. "How to Choose Between Empirically Indistinguishable Theories." *Journal of Philosophy* 79: 61–77.

Howson, C. and P. Urbach. 1989. *Scientific Reasoning: The Bayesian Approach*. La Salle, Ill.

Kelly, K. 1992. "Reliable Inquiry: A Learning Theoretic Perspective on Scientific Method," unpublished manuscript.

Kelly, K., and C. Glymour. 1989. "Convergence to the Truth and Nothing but the Truth." *Philosophy of Science* 56: 185–220.

Kripke, S. 1982. *Wittgenstein on Rules and Private Language*. Oxford.

Laudan, L., and J. Leplin. 1991. "Empirical Equivalence and Underdetermination." *Journal of Philosophy* 88: 449–72.

Luper-Foy, S. Ed. 1987. *The Possibility of Knowledge*. Totowa, N.J.

Malament, D. 1977. "Indistinguishable Space-Times: Comments on Glymour's Paper." In Earman et al. (1977).

Nozick, R. 1987. "Knowledge and Skepticism." In Luper-Foy (1987).

Putnam, H. 1963. "Degree of Confirmation and Inductive Logic." In *The Philosophy of Rudolf Carnap*, edited by P. A. Schilpp. La Salle, Ill.

Quine, W. V. O. 1975. "On Empirically Equivalent Systems of the World." *Erkenntnis* 9: 313–28.

van Fraassen, B. C. 1980. *The Scientific Image*. Oxford.

van Fraassen, B. C. 1985. "Empiricism in the Philosophy of Science." In *Images of Science*, edited by P. M. Churchland and C. A. Hooker. Chicago.

van Fraassen, B. C. 1989. *Laws and Symmetry*. Oxford.

Epistemology for Empiricists

ELLIOTT SOBER

In 1963, Paul Feyerabend published a paper called "How to Be a Good Empiricist." Feyerabend took empiricism to be the reigning orthodoxy of his time, but argued that it had lapsed into an anti-empirical dogmatism. Feyerabend tried to show empiricists how to do better. Had Feyerabend not beat me to it, I would have used the title he chose for his paper as the title for the one you now see before you. How one is to be a good empiricist is still a problem worth posing. Not that the ideas I will advance overlap very much with the ones he discussed. Questions about how theoretical terms change meaning when old theories give way to new ones and about whether science is cumulative are important, but they will not be important to me in what follows. And, of course, another difference between my project and Feyerabend's is that empiricism can no longer be called a reigning orthodoxy. Feyerabend's paper appeared during a time in which logical empiricism was giving way to scientific realism. My hunch is that if there is now a dominant ideology, it is realism, not empiricism.

Empiricism is nonetheless still a doctrine worth reckoning with, mainly due to the formulation and defense that Bas van Fraassen (1980, 1985, 1989) has more recently provided. Van Fraassen criticized scientific realism; realists replied in kind. In my opinion, some of these criticisms of van Fraassen's views are fundamentally sound. The question I wish to address is whether anything of value remains of empiricism. Empiricism is not now a reigning orthodoxy, but it still is worth asking whether empiricists can do better.

Empiricism is a thesis about the importance that experience should have in shaping our beliefs about the world. It is a normative claim—a thesis of epistemology, not psychology. The exact formulation the doctrine should receive is a problem to which I will return. For now, I will begin with van Fraassen's statement of the difference between his own version of empiricism—*constructive empiricism*—and scientific realism. According to van Fraassen (1980, 8), realism maintains that

Science aims to give us, in its theories, a literally true story of what the world is like; and acceptance of a scientific theory involves the belief that it is true.

Constructive empiricism, on the other hand, says that

Science aims to give us theories which are empirically adequate; and acceptance of a theory involves as belief only that it is empirically adequate. (van Fraassen 1980, 12)

Fundamentally, the conflict concerns the rational attitude we should have towards theories in science.

What does it mean to say that a theory is empirically adequate? The rough idea is that what the theory says about observable objects is true. Many theories talk about both observable and unobservable entities. According to van Fraassen, if such theories pass reasonable standards of adequacy, we should conclude that what they say about observables is true, but we should remain agnostic with respect to what they say about unobservables. The most we can assert, then, is that such theories are empirically adequate; we cannot say that they are true. Because true theories are empirically adequate, but not conversely, constructive empiricists are more circumspect than realists in the way they assign truth values.

Before taking my discussion of empiricism and realism any farther, several points need to be clarified concerning van Fraassen's statement of the problem and the gloss I have given of what he says. I have described both philosophical doctrines as *ought* statements. But van Fraassen's formulations are in the indicative mood—they describe what the aim of science *is*. This suggests that realism and empiricism are not epistemological theses at all (at least not if epistemological claims are normative claims concerning the regulation of belief).

Reflection on how van Fraassen understands theses about the "aim" of science reveals that his indicative formulations in fact express normative claims. His contrast between realism and empiricism is intended to offer two rival theories about what is required for one to play "the game of science." The empiricist says that the most that is *required* in science is belief in the empirical adequacy of theories; the realist says that belief in the truth of theories is sometimes *required* as well. Claims about the aims of science are, for van Fraassen, theses about the norms that constitute scientific activity.[1]

So there is no conflict between the indicative formulations of realism and empiricism that van Fraassen provides and my claim that both are normative theses in epistemology. But there is a further question, this one concerning the *kind* of normative claim that is at issue. Van Fraassen (1985, 252) describes empiricism and realism as views about what is *permissible*, not about what is epistemologically *obligatory* (see also van Fraassen 1989, 171–72). On this reading, van Fraassen's empiricism does not say that the realist is *irrational* in believing various theories true, only that such belief is not rationally required.

According to this formulation, the empiricist says only that one *may* remain agnostic concerning theories about unobservables, not that one *must* do so.

Whether epistemological norms should be thought of as norms of obligation or norms of permission is an interesting question, but it is not a question I will discuss in what follows. My reason for skirting this issue is that the problems I wish to address are not affected by which formulations one considers. For ease of exposition, I will describe empiricism as *requiring* agnosticism, and realism as *requiring* assent, with respect to claims about unobservables.[2] As will become clear shortly, these formulations are only initial points of departure.

One standard criticism of constructive empiricism is that it attributes to the distinction between observables and unobservables an epistemological significance it does not possess. According to van Fraassen, the moons of Jupiter are observable; they are observable because a properly situated astronaut could see them without the aid of instrumentation. In contrast, the AIDS virus is not observable, because detecting its presence and properties requires the use of instruments.

Although moons and viruses differ in the respect just mentioned (let us grant), they are similar in another respect. Neither has *actually* been observed. According to van Fraassen's preferred usage, we earthlings have not in fact *observed* the moons; rather, we have indirectly *detected* their presence by using instrumentation—a telescope. A parallel remark applies to viruses. We have not directly observed them either; rather, we have detected their presence with the help of an instrument—a microscope. The moons are observable, but unobserved; viruses, on the other hand, are both unobserved and unobservable.

Van Fraassen (1980, 16) is right to construe observability as a dispositional property:

> X is observable if there are circumstances which are such that, if X is present to us under those circumstances, then we observe it.

He also is right to say that empirical science will help us understand what makes some objects, but not others, observable.[3] Nonetheless it is entirely unclear why this difference between moons and viruses should matter to us epistemologically. Perhaps there are specific reasons why we should be more circumspect about some particular hypothesis about the AIDS virus than we are about some particular hypothesis concerning Jupiter's moons. But the mere fact that Jupiter's moons are observable while individual viruses are not is epistemically irrelevant.

This argument against van Fraassen has become standard. I think it is fundamentally correct. The question I wish to ask is whether some version of empiricism can survive its corrosive effect. What, if anything, is epistemologically special about observation? To begin answering this question, I want to locate the objection to van Fraassen's version of empiricism within a larger epistemological context. There is a general lesson to be drawn from this criticism that illuminates how the concept of evidence should be understood.

1. ACTUALISM

Consider two diagnostic problems that physicians confront. They want to gather evidence concerning whether a patient has diabetes; they also want to gather evidence about whether a patient has small pox.[4] In each case, a laboratory test is performed, and the outcome of the test is interpreted. For simplicity, let us suppose that each test produces one of two results, which we designate "positive" and "negative." Each test has two error characteristics (a,b and c,d, respectively), representing the chance of a false positive and the chance of a false negative. These are described in table 1.

Table 1

	Test for Diabetes	
	positive	negative
S has diabetes	$1-a$	a
S does not have diabetes	b	$1-b$
	Test for Small Pox	
	positive	negative
S has small pox	$1-c$	c
S does not have small pox	d	$1-d$

The entries in these tables denote the probability of a test result, conditional on the patient's situation; they do not describe the probability of a disease, conditional on a test result.

Let us suppose that these tests are *equally reliable*, though each is less than perfectly so. They have identical and nonzero error probabilities: $a = c \neq 0$ and $b = d \neq 0$. This assumption entails that the amount of information delivered by each test will be the same. If a positive result on the first test strongly favors the hypothesis that the patient has diabetes, then a positive result on the second test strongly favors the hypothesis that the patient has small pox. The tests are on an epistemic par.

Now let us consider a third test. This is a test for diabetes *and it is infallible* (table 2).[5]

Table 2

	An Infallible Test for Diabetes	
	positive	negative
S has diabetes	1	0
S does not have diabetes	0	1

This test is infallible because if you have diabetes you are certain to get a positive result and if you do not have diabetes you are certain to get a negative result. If we were to run this infallible test, we would gain more information about whether the patient has diabetes than either of the first two tests could provide. Suppose there is no similarly infallible test for small pox.

Now let's run the first two tests, but not the third. How should the outcomes of these two tests be interpreted? Suppose each delivers a positive outcome. Should we be more confident that the patient has diabetes than we should that the patient has small pox *on the grounds that there exists an infallible test for the former, but none for the latter?* I would say *no*. We should be guided by the evidence we *actually* possess; the fact that we *could* possess greater certainty about diabetes than we *could* about small pox is irrelevant. We could have run the third test, but in point of fact we did not; the mere possibility of running the third test is irrelevant. I will call the idea underlying this judgment the *principle of actualism*. It says that we should form our judgments about hypotheses based on the evidence we actually possess; possible but nonactual evidence does not count.[6]

Actualism is an empty idea until additional epistemic principles are specified. No doubt the modal fact that we *could* be certain about whether a patient has diabetes if we were to run the third test is grounded in some fact about what is *actually* the case. The point is that the mere fact that such a test procedure is available does not help us tell whether a patient has diabetes. Explaining why this is true requires both a substantive theory of evidence and substantive facts about the world.[7]

Actualism is an intuitive principle, on which both realists and empiricists *should* be able to agree. It says nothing about whether our knowledge of electrons and genes is less secure than our knowledge of dogs and tables. However, the point of interest is that van Fraassen's formulation of empiricism violates actualism. Let us return to the comparison of moons and viruses to see why.

We detect some fact about Jupiter's moons by looking through a telescope; we detect some fact about a sample of the AIDS virus by looking through a microscope. These two test problems may differ in many ways that may be relevant to the degree of certainty we are entitled to have about each of the matters under test. However, one thing that is epistemically irrelevant is a certain counterfactual possibility: if we were to get in a space ship and journey to Jupiter, we could see its moons without instrumentation, but no such journey would allow us to see an individual virus without the help of instruments.[8] The reason this difference is irrelevant is given by the principle of actualism.[9]

Empiricism is a thesis about the importance of observation. However, the way to develop the thesis is to show that *what we actually observe* has some special epistemic status. It is not the distinction between *observable* and *unobservable* that is fundamental, but the difference between *observed* and *unobserved*. Empiricists should attempt to show how *actual* observation provides more certainty than other supposed routes to knowledge. Let us bear this lesson in mind; we will return to it later.

2. ACCEPTANCE

I now want to consider a second, and less noticed, property of van Fraassen's formulations of empiricism and realism. It concerns the word *acceptance*.[10] According to van Fraassen (1980), science is in the business of accepting hypotheses; the dispute between realism and empiricism concerns what sorts of hypotheses scientists should accept.

At least since discussion of the lottery paradox (Kyburg 1970), the dichotomy between accepting and not accepting a hypothesis has struck many philosophers of science as too coarse-grained. Instead of deciding between accepting and not accepting, rational agents might be described as assigning *degrees of belief* to various hypotheses, or as deciding how *well supported* various hypotheses are.

A wholesale rejection of the concept of acceptance is not forced on one by the lottery paradox, but it is a plausible and common diagnosis of what the paradox shows. Imagine a fair lottery with a thousand tickets. Exactly one ticket will win and each ticket has the same probability of winning. Consider the statement "ticket #1 will not win." Based on the information given, this statement has a probability of 0.999. If our rule of acceptance is to accept a hypothesis when it is very probable (suppose the threshold for this is specified at 0.95), then we should accept this hypothesis about ticket #1. However, the same line of reasoning applies to each of the tickets. But if we accept *each* such hypothesis, we will have contradicted the starting assumption that some ticket will win.

One might interpret this problem as showing that rules for acceptance must be more complicated than the one described here. Instead of jettisoning the concept of acceptance, maybe we should try to fine-tune it. However, it is hard to see, on reflection, that a cut-off exists that separates what we believe *tout court* from what we do not. Rather, statements tend to grade off from the very plausible, to the moderately plausible, to the moderately implausible, and so on. Let us assume, then, that epistemology should have no truck with the concept of acceptance. How, then, are we to reformulate van Fraassen's opposition between empiricism and realism?[11]

One possibility, to which van Fraassen (1980) alludes, is Bayesianism. Hypotheses are assigned probabilities; as new evidence rolls in, these probabilities are updated by appeal to Bayes's Theorem. When we acquire the evidence E, we assign to the hypothesis H a posterior probability $P(H/E)$ in accordance with the following formula:

$$P(H/E) = P(E/H)P(H)/P(E).$$

If we adopt this Bayesian format, empiricism and realism must now be understood as advancing principles about how probabilities are to be assigned and updated.

Van Fraassen's empiricism must delimit an epistemic difference concerning how hypotheses about observables and hypotheses about unobservables should be treated. What might that difference come to in a Bayesian framework? Bayesians say that a hypothesis is confirmed by a piece of evidence when acquiring the evidence makes the hypothesis more probable than it was before. Perhaps empiricism should then be formulated as the claim that *hypotheses about unobservables are not confirmable; if* H *is about unobservables, then* $P(H/E) \not> P(H)$ This proposal articulates empiricist suspicions about hypotheses that talk about things we cannot observe; they are epistemically inaccessible in the precise sense that their probabilities cannot be boosted by observations.

It does not take much to see that this proposal is implausible. Rearranging Bayes's theorem, we obtain the following equality:

$$P(H/E)/P(H) = P(E/H)/P(E).$$

E confirms H precisely when the left-hand side is greater than unity. However, it is quite obvious that this quantity *can* be greater than unity, even for hypotheses that are about unobservables. For consider the right-hand side. Suppose that H deductively entails E. Then $P(E/H) = 1$. If E was not certain before the evidence was gathered, then $P(E) < 1$, which means that $P(E/H)/P(E) > 1$. When hypotheses about unobservables deductively entail true predictions that were not known with certainty beforehand, those hypotheses are confirmed.[12] Empiricism cannot deny this.

So the simple fact of the matter is that, within a Bayesian format, the probabilities of hypotheses about unobservables can go up and down, just like the probabilities of hypotheses about observables.[13] But perhaps there is some other asymmetry between observables and unobservables that can be described within a Bayesian framework. Rather than concentrate on *change* in probability, let us focus on the absolute values those probabilities attain. Perhaps hypotheses strictly about unobservables must always be less probable than hypotheses strictly about observables. This is the thesis that *for any evidence* E, $P(H_1/E) > P(H_2/E)$ if H_1 *is strictly about observables and* H_2 *is strictly about unobservables.*

This also is a proposal that cannot pass muster. Let H_1 be a hypothesis about observables on which we have no evidence whatever; let H_2 be a hypothesis about unobservables on which we have scads of positive evidence. Surely we are right to be more certain about whether someone has the AIDS virus in his blood, based on the results of laboratory tests, than we are about whether a *Brontosaurus* was standing exactly where the White House now is precisely 179 million years ago today.

So we cannot say that *all* statements about observables are more probable than *all* statements about unobservables. However, there is a much attenuated formulation of this asymmetry that cannot be denied. It is the one that van Fraassen (1980, 68–69) notes in the following passage:

> . . . we can distinguish between two epistemic attitudes we can take up toward a theory. We can assert it to be true . . . , and call for belief; or we can simply assert its empirical adequacy. . . . In either case we stick our necks out; empirical adequacy goes far beyond what we can know at any given time. . . . Nevertheless there is a difference; the assertion of empirical adequacy is a great deal weaker than the assertion of truth. . . .

The claim, as I understand it, is this: for any theory T that is about unobservables, and any evidence E, $\mathrm{P}(T$ is true$/E)$ $<$ $\mathrm{P}(T$ is empirically adequate$/E)$. Indeed, this is a theorem of probability theory: no matter what the evidence is, if one hypothesis entails another but not conversely, the first cannot be more probable than the second. The probability of "T is true" cannot exceed the probability of "T is empirically adequate."

This is true enough, but it is too slender a basis on which to rest the version of empiricism that van Fraassen defends. A realist can grant the mathematical point, but still maintain that the posterior probabilities we assign to theories about unobservables can be driven quite high. If "T is empirically adequate" has a probability of 0.95, then "T is true" cannot be assigned a greater value. But 0.94 is plenty big enough. The simple mathematical constraint on probability assignments is not enough to represent van Fraassen's claim that we should be agnostic as to the truth value of theories about unobservables.

A further reason can be given for being skeptical about this argument of van Fraassen's. Granted, "T is empirically adequate" is logically weaker than "T is true," but there are statements that are weaker still. "T is empirically adequate concerning events that occur on Mondays, Wednesdays, and Fridays" is even weaker, but surely it is arbitrary to say that nothing stronger than Monday-Wednesday-Friday claims about empirical adequacy is defensible. Van Fraassen's *argument from weakness* is defective unless it can be supplemented with an explanation of why reason may venture to a certain point, but no further.[14]

I so far have explored some Bayesian formulations of the epistemology that underlies constructive empiricism. The fact that these are unworkable does not show that no Bayesian approach can be made to work. Indeed, van Fraassen elaborates a more sophisticated proposal in his subsequent book, *Laws and Symmetries*. His idea is to represent the agnosticism that constructive empiricism counsels by using Kyburg's and Levi's ideas about interval probabilities. Van Fraassen's (1989, 193–94) proposal is that an agent is thoroughly agnostic about a hypothesis H precisely when he or she says that the probability of H is between 0 and 1 (inclusive).

Van Fraassen thinks this representation of agnosticism entails that thorough agnostics cannot learn from experience:

> What is the effect of new evidence? If hypothesis H implies E, then the vagueness of H can cover at most the interval $[0, \mathrm{P}(E)]$. So if E then becomes certain, that upper limit disappears. For the most thorough agnostic concerning H is vague on its probability from zero to the

probability of its consequences, and remains so when he conditionalizes on any evidence. (van Fraassen 1989, 194)

I want to question the adequacy of this representation of agnosticism as well as its suitability as a vehicle for expressing empiricist views concerning hypotheses about unobservables.

From Bayes's Theorem we know that if $P(E/H) = 1$, then $P(H/E) = P(H)/P(E)$. This is why the prior probability $P(H)$ must fall in the interval $[0, P(E)]$. As van Fraassen notes, once one learns that E is true (where E was not certain beforehand), the upper bound on the probability of H increases; the interval associated with $P(H/E)$ is $[0, 1]$. So learning that E is true does not leave one's attitude to H unchanged. Suppose I learn that smoking may be more dangerous than I had thought. I might say, "I used to think that my chance of cancer was somewhere between 0.001 and 0.01, but now I realize that my chance may be as high as 0.25." Increasing the upper bound reflects learning from experience, not the maintenance of one's previous epistemic state.[15]

If learning that E is true is genuinely to leave one's attitudes to H unchanged, then the interval associated with $P(H)$ must be the same as the interval associated with $P(H/E)$. However, when H entails E, this will be true only if $P(E) = 1$, which in turn is true only if E is entailed by both H and its negation. Observations are rarely related to theories in this way. My conclusion is that the assignment of interval probabilities is not an adequate method for codifying the agnosticism that figures in van Fraassen's constructive empiricism.

If Bayesianism does not provide a hospitable framework for van Fraassen's version of empiricism, it is well to consider other ways of quantifying how evidence bears on hypotheses. Many empirical scientists are reluctant to assign probabilities to the hypotheses they consider; at least such assignments do not figure in their "public" pronouncements about theories. However, this does not stop them from evaluating hypotheses by seeing how those hypotheses probabilify the observations they have made. That is, although scientists rarely discuss probabilities of the form $P(H/E)$, they often discuss probabilities of the form $P(E/H)$. The latter sometimes go by the technical term *likelihood*. The likelihood of a hypothesis is the probability it confers on the data, not the probability it possesses in the light of the data.

The likelihood concept is useful when it comes to *comparing* rival hypotheses. One evaluates competing hypotheses by seeing how each probabilifies the data. The basic idea is called the *Likelihood Principle* (Edwards 1972):

O favors H_1 over H_2 if and only if $P(O/H_1) > P(O/H_2)$.

If H_1 says that the data were to be expected while H_2 says that it is almost a miracle that the observations came out as they did, this means that the observations strongly favor H_1 over H_2.

How might the Likelihood Principle be used to carve out a difference between observable and unobservable entities? A scientist will test hypotheses about observables by finding observations that discriminate between them; for example, we might test claims about the dietary habits of different dinosaurs by looking at such things as their fossil remains. If hypotheses about *un*observables were not amenable to this approach, this would mark an important sense in which they are epistemically inaccessible. The idea is that *when H_1 and H_2 are both about unobservables, every observation O is such that* $P(O/H_1) = P(O/H_2)$. This expresses the thought that observations cannot distinguish between hypotheses that are about things we cannot observe.

Of course, this suggestion is absurd. As van Fraassen clearly realizes, competing hypotheses about unobservables make different predictions about what we can observe. By seeing which predictions come true, we are able to make judgments about which hypotheses the observations favor.

Not only does the Likelihood Principle apply both to hypotheses about observables and to hypotheses about unobservables. In a certain sense, it entails that observables and unobservables are on an epistemic par. If the observations we have made discriminate between "H_1 is empirically adequate" and "H_2 is empirically adequate," those observations must also discriminate between "H_1 is true" and "H_2 is true." Van Fraassen's constructive empiricism says that it is sensible to judge a theory empirically adequate, but metaphysical to say that the theory is true. The Likelihood Principle entails that this is an untenable dualism (Sober 1990a).

I have suggested that empiricism and realism should be formulated in terms of some matter-of-degree concept such as *degree of belief*, not in terms of the absolute concept of *acceptance*. The reason I have given for this suggestion has nothing special to do with realism and empiricism, but simply reflects a widely held interpretation of the lottery paradox.[16] However, there is an additional argument for this reformulation, one that is internal to the dispute between realism and empiricism.

Van Fraassen (1980, 16) agrees with earlier realists (e.g., Maxwell 1962) that the distinction between observable and unobservable entities is a matter of degree. However, unlike some of his realist opponents, van Fraassen insists that the absence of a precise boundary does not obliterate the distinction or its importance. At one end of the continuum there are objects that are clearly observable, while at the other there are objects that are clearly unobservable. In between is a gray area. I agree with *part* of what van Fraassen is saying here; we cannot reject the use of a distinction—in science, in philosophy, or in ordinary life—just on the ground that it is vague.

Nonetheless, there is a further issue, this one specific to how van Fraassen formulates empiricism. If our epistemic attitudes to a proposition are supposed to depend on whether the proposition is about unobservable entities, what attitude should we take to propositions that are about objects that fall in the gray area? Van Fraassen could elect to say nothing about these borderline cases. But surely empiricism can be given a less arbitrary formulation. If observability is a matter of degree, and if epistemic attitudes are to be keyed to observability, then those epistemic attitudes should themselves be formulated as matters of degree. A more natural formulation of empiricism might say something like this: *how certain we are entitled to be about a given proposition will depend on how observable the objects are that the proposition is about.* This is an internal reason for empiricists to shy away from the dichotomous category of acceptance.

3. SOURCE VERSUS SUBSTANCE

Van Fraassen's constructive empiricism is first and foremost a position in epistemology. Its fundamental claims address the question of how rational acceptance should be understood. Yet, when we try to equip that doctrine with an epistemology in which the all-or-nothing category of *acceptance* is replaced by the more nuanced idea of *degree of evidential support,* we run into trouble. It is entirely unclear how the distinction between observables and unobservables could be grafted onto a theory of the evidence relation that is probabilistic in character.

It is useful at this point to back up one step and try to see *why* van Fraassen's formulation of empiricism goes astray. I think a hint is provided in the following passage:

> . . . for James and Reichenbach, the core doctrine of empiricism is that experience is the sole source of information about the world and that its limits are very strict. . . . I explicate the general limits as follows: *experience can give us information only about what is both observable and actual.* (van Fraassen 1985, 253)

It is important to recognize a gap between the "core doctrine" and van Fraassen's proposed explication. The formulation he attributes to James and Reichenbach concerns the *sources* of evidence that are available to us; in contrast, van Fraassen's own proposal concerns the *content* of the hypotheses to which we have epistemic access. However, the claim about sources and the claim about substance are quite different. Even if evidence comes to us only through observation, that does not mean that our evidence bears only on what we believe about observables.

Perhaps it can be shown that the claim about *sources* leads inevitably to the claim about *substance.* But this is not at all obvious. For example, suppose that for some reason I am unable to look inside the gas tank of my car to see how much gas it contains. Perhaps the only access I have to claims about the contents of the gas tank derives from looking at the gas meter on the dashboard, noticing when the car stops moving (such events are curiously correlated with the needle's pointing to the letter "*E*"), and so on. My *sources* of information are limited to what I can observe, and I am supposing that I cannot observe what is in the gas tank. But surely it does not follow that the *content* of what

I believe should be limited to claims about what the meter says, about when meter readings are correlated with the car's grinding to a halt, and so on.[17]

It is interesting that immediately after giving the above quoted explication of what empiricism says, van Fraassen elaborates on it as follows:

> We may be rational in our opinions about other matters—Augustine's "faith in things unseen," which he rightly said, pervades even our everyday opinions about everyday matters—but any defense of such opinions must be by appeal to information about matters falling within the limits of the deliverances of experience. (van Fraassen 1985, 253)

This more modest formulation says that our evidence is limited to the deliverances of experience; however, it allows that experience can bear on hypotheses that describe "things unseen."

One salient feature of van Fraassen's approach is that he criticizes earlier formulations of empiricism in which scientific theories are construed nonliterally. The philosophical positions he criticizes include proposed phenomenalistic reductions, in which a scientific theory apparently about electrons is said to be translatable into a theory about experiences. Van Fraassen wants to emancipate empiricism from this semantic formulation. Indeed he does so, but a residue of semantics remains. According to van Fraassen, the epistemic attitude we are supposed to have towards a theory depends on what the theory is about; we may assign truth values to statements strictly about observables, but should remain agnostic with respect to statements that are about things we cannot observe. Perhaps the difficulties facing van Fraassen's formulation can be overcome if we formulate empiricism as a claim about *sources* of information, and not, in the first instance, as a claim concerning the *subject matter* of hypotheses about which we can hope to get evidence.[18]

Even after the claim about sources is separated from the claim about subject matter, the empiricist claim that experience is our sole source of evidence requires refinement. It should not entail that reason plays no role in evaluating hypotheses. Suppose we are comparing various hypotheses that seem *prima facie* to be possible explanations for some body of data; we then discover that one of those hypotheses contains a logical inconsistency. Surely this discovery can be grounds for rejecting the hypothesis. Here it is reason, not sense experience, that drives our decision. In the light of this, what claim can empiricism propose about the privileged status of experience?

This question suggests that one needs to think of stronger and weaker forms of empiricism; these differ with respect to how determining a role they attribute to experience. The most extreme formulation will deny that *a priori* considerations play any role at all in hypothesis evaluation. If deductive logic is relevant to hypothesis evaluation, then this sort of empiricism will claim that the credentials of logic derive entirely from sense experience. A less extreme formulation will grant that deductive logic is *a priori*, but will draw the line there. Further along the continuum is the view that deductive logic and Bayesian coherence are the sole *a priori* constraints on what we ought to believe. If we move a good deal farther in this direction, we come to a point of view in which a slew of supposedly *a priori* principles are thought to constrain which theories we may accept. Kantian theses about synthetic *a priori* knowledge exemplify this type of position; coherent theorizing must assume that every event has a cause, that space is Euclidean, and the like. Here one has moved along the continuum sufficiently far to say that the doctrine at hand is no longer an instance of empiricism.

To decide where along this continuum the truth is to be found, one must address the issue of whether the "superempirical" virtues of a theory—its simplicity, elegance, generality, degree of unification—play a role in determining how we should evaluate theories. If the simplicity of a theory has nothing to do with how well supported it is by available observations, then an empiricist should say that simplicity is not a sign of truth—it is irrelevant to assessing the theory's plausibility. Although few present-day realists would endorse the Kantian principles alluded to above, many uphold the relevance of such superempirical virtues. Accordingly, this is a very important issue for the dispute between realism and empiricism.[19]

The intuitive idea that the empiricist must try to explicate is that theory evaluation should be driven by observations, not by *a priori* preconceptions. If we observe many emeralds and find that all have been green, the empiricist has no trouble explaining why we should prefer the hypothesis that all emeralds are green over the hypothesis that only 87 percent of them are. Logic may be *a priori*, but the use of logic in this instance allows the observations to decide which hypothesis is more plausible. A more difficult problem for the empiricist is to explain why we should prefer "all emeralds are green" over the claim that "emeralds are green until the year 2100, but thereafter are blue." Can observations be cited that help one discriminate between these two hypotheses, or is some irreducible appeal to simplicity the only thing that stands between us and skepticism? If the latter, empiricists will be driven to skepticism; the question will then be whether we should follow them there, or view this consequence of their position as a *reductio*.

So the dispute about empiricism should not be understood as a clash between reason and experience. The question is not "should we reason or should we observe?" Rather, the problem concerns what reason dictates as policies of belief change. If hypothesis evaluation should be driven by observations, the empiricist wins; but if legitimate discriminations among self-consistent hypotheses far outrun what is mandated by the observations we have available, then empiricism is mistaken. Again, let me emphasize that this is not a dispute between clearcut dichotomies, but concerns matters of emphasis.

4. THREE DESIDERATA AND A PROPOSAL

I have suggested three *desiderata* concerning how empiricism should be formulated. First, the doctrine should claim that observations are special *sources* of information concerning which theories are true; it should not claim that observables are the sole *subject matter* about which we can have reasonable opinions. Second, the doctrine should not be expressed in terms of the dichotomous concept of acceptance, but should be formulated in terms of some matter-of-degree concept like evidential support. And finally, the doctrine should conform to the principle of actualism; the significance attributed to the distinction between observable and unobservable entities should attach instead to the distinction between what is observed and what is not.

One natural way to satisfy these *desiderata* is by viewing the Likelihood Principle as the sole vehicle by which competing hypotheses can be evaluated. If one theory is claimed to be more plausible than another, this cannot be due to some irreducible fact about how simple or unified the two theories are; comparisons of plausibility must be based on observations that discriminate among the competitors.

The Likelihood Principle imposes no limitation on the subject matters or vocabularies of competing hypotheses. If two hypotheses about electrons make different predictions about what we can observe, then observations will allow us to say which is more plausible. In this respect, hypotheses about electrons are on the same footing as hypotheses about the moons of Jupiter. If the data we actually possess discriminate among the competitors, we can make a scientific assessment; if not, not.[20]

It follows from this epistemology that certain kinds of skeptical challenges cannot be answered. If the evil demon hypothesis is formulated in such a way that no possible experience can discriminate between it and the "normal" hypotheses we believe, then science is in no position to say which of them is true. Philosophers may talk of *ad hocness*, simplicity, and explanatoriness all they wish, but they should realize that they are not invoking the epistemological standards of science, but ones of their own devising. Science is not in the business of attempting to discriminate between predictively equivalent hypotheses. The scientific method provides no handy solution to such problems; rather, such problems are passed over as unscientific (Sober 1990a).

The name I give to this approach to scientific inference is *contrastive empiricism* (Sober 1990a; 1993). It involves a compromise between realism and empiricism. It draws from realism the idea that hypotheses about unobservables are just as much within the scope of scientific evaluation as hypotheses about observables. From empiricism, it draws the idea that theory evaluation must be driven by observations; appeal to the so-called "superempirical" virtues is idle.[21]

5. OBSERVATION: BETWEEN SCYLLA AND CHARYBDIS

I have already pointed out that van Fraassen (1980) grants that the distinction between observable and unobservable is a matter of degree, though this does not stop him from imposing on it a dichotomous epistemology, in which some statements, but not others, may be regarded as true. However, it is only fair to note that this fixation on dichotomous epistemological categories also is prominent in the writings of van Fraassen's realist opponents. For example, after emphasizing the highly fallible nature of scientific theorizing, Churchland (1985, 36–37) asks:

> Why, then, am I still a scientific realist? Because these reasons fail to discriminate between the integrity of observables and the integrity of unobservables. If anything is compromised by these considerations, it is the integrity of theories generally. That is, of *cognition* generally. Since our observational concepts are just as theory-laden as any others, and since the integrity of these concepts is just as contingent on the integrity of the theories that embed them, our observational ontology is rendered *exactly as dubious* as our nonobservational ontology.

According to Churchland, observables and unobservables are *in the same boat*, because *cognition in general* is fallible and theory-laden.

It is a startling thesis that our confidence in the existence of dogs and tables is no more secure than our confidence in the existence of oncogenes and leptons. Granted, all cognition is fallible and theory-laden. But is all cognition fallible and theory-laden *to the same degree*? This monism is no less bizarre than van Fraassen's dualism; scientists would no doubt scratch their heads in wonderment at the claim that everything is just as dubious as everything else.

Perhaps Churchland's choice of words in this passage overreaches what he wants to assert. He advances the *positive* claim that statements about observables and statements about unobservables are equally dubious. But maybe all he had in mind was a *negative* thesis; perhaps his point was just to reject the idea that every statement about unobservables is more dubious than every statement about observables. As noted in Section 3, I reject this idea as well. However, much is left unsettled by this very modest point. Granted, there exists at least one statement about unobservables that is more certain than at least one statement about observables. Still, it does not follow that the observable-unobservable distinction is epistemologically irrelevant. There is at least one smoker who has better health than at least one nonsmoker; however, it would be a mistake to conclude from this that the smoking-nonsmoking distinction is irrelevant to health.

So the question that needs to be addressed is whether there is any matter-of-degree epistemic difference that is associated with the distinction between observables and unobservables, given that the latter distinction may itself also be a matter of degree. I think the answer to this question is *yes*. I will argue for this thesis in two steps. First, I will map the distinction between observable and unobservable onto the distinction between *direct* and *indirect*; then I will provide a likelihood representation of the idea that we have more decisive evidence about matters with which we have direct contact than we have concerning matters that are known only by indirection.

I begin with what I regard as a truism: *when something is unobservable, the only way we can find out about it is by finding out about something that is observable*. Electrons are too small to see. So we use instruments. We usually encounter few problems in seeing what state a measuring device is in; we then infer from the state of the measuring device what state the electrons occupy in our experiment. This intuitive idea may be summarized as follows: *our knowledge of unobservables is necessarily indirect.* Here I use "unobservable" in a wider sense than van Fraassen does—we cannot observe flesh-and-blood dinosaurs because of our temporal location, the moons of Jupiter because of our spatial location, and present-day viruses because of our size. Since we are large earthlings who live in modern times, we can know about each only indirectly.

Things from which we are not cut off in this way can be known without a pathway through unobservables. We can know what state the voltage meter is in—what number the pointer points to—without knowing the state of the electrons in the experiment. And we can know what a fossil looks like without already knowing what the dinosaur was like from which the fossil derives. Knowledge of unobservables is necessarily mediated by knowledge of observables, but knowledge of observables need not pass through knowledge of unobservables. Here is an asymmetry between observables and unobservables that is worth pondering.[22]

The asymmetry I am proposing does not entail that our knowledge of observables is infallible or theory-neutral. Granted, we can be wrong when we judge what state a measuring device is in; and we can tell what state the device is in only by exploiting background information. When I say that unobservables are knowable only indirectly, while observables are knowable directly, I am positing a difference in degree. Maybe it is true that *all* perceptual judgments about physical objects are generated by inferential processes (Marr 1982). *Direct* does not mean totally *non*inferential, but that inferences are present to a lesser extent. Perhaps, when we tell what state an electron occupies by looking at a measuring device, there are *two* layers of inference. First, we figure out what state the device is in; then, on that basis, we infer what state the electron occupies.

There is a second misinterpretation that also should be avoided. I am not saying that every belief that is about an unobservable must involve "more" inference than every belief that is about an observable. I do not know how to compare such quantities, especially when they attach to hypotheses that are about quite different subject matters. Doubtless, it is possible to imagine an inference about electrons that involves three "steps" and an inference about the birds living in Antarctica that involves four. In any case, there is no need to propose a *total ordering* of *all* propositions in the one category and *all* propositions in the other. Rather, my proposal posits a *partial ordering;* it is, so to speak, "chain internal." When we infer something about an unobservable u, based on beliefs we have concerning an observable o, u will be known more indirectly than o is; the inferences grounding our belief about o will form a subpart of the inferences grounding our belief concerning u.

I hope it also is clear that my thesis requires no precise boundary between observable and unobservable, or between what is observed and what is inferred from observation. Van Fraassen says we do not "observe" when we use instrumentation; others have adopted a more liberal terminology. For me, it makes no difference whether or where a line is drawn. Whether we "observe" electrons in a cloud chamber or merely "detect" them, the point is that our belief about the electrons is based on a judgment we form about the state of the cloud chamber screen.

Having argued that the distinction between observable and unobservable maps onto the distinction between direct and indirect, I now need to explain why this fact is epistemologically significant. Why should extra inferences, so to speak, have the effect of weakening the testimony of experience? Even if I decide what state the electron occupies by looking at a measuring device, why can't my experience provide more evidence about the electron than it does about the state of the measuring device?

The reason indirection affects strength of evidence can be understood within the framework of the Likelihood Principle. Consider a causal chain, in which an effect E traces back to a (relatively) direct cause (D) and thence to a more indirect cause (I):

$$I \longrightarrow D \longrightarrow E$$

Let us suppose that this causal chain has the *Markov* property: the intermediate link D "screens off" E from I. To see what this means, imagine that each of the three nodes can be in one of two states, which I will denote as "0" and "1". When E is in state 0, I will express this fact by saying that "$E = 0$," and similarly for other nodes and other states. Screening off means that

$$P(E = i/D = j) = P(E = i/D = j \ \& \ I = k) \neq P(E = i/I = k),$$
for $i, j, k = 0, 1$.

The idea is that if one knows the state of D, one's ability to predict the state of E remains the same, whether or not one also knows the state of I.

Suppose we know that $E = 0$. On this basis, we wish to infer what state D is in and also what state I occupies. In each case, we will use the Likelihood Principle to interpret the data. We will compute the likelihood ratio $P(E = 0/D = 0)/P(E = 0/D = 1)$. This will tell us whether and to what degree the observation that $E = 0$ favors $D = 0$ over $D = 1$. Likewise, we will compute $P(E = 0/I = 0)/P(E = 0/I = 1)$, the point this time being to see how strongly the observation favors one hypothesis about the indirect cause over the other. In particular, we will be interested in the relationship between these two likelihood ratios. When does the state of E provide more information about the state of D than it does about the state of I?

The answer is *always* (so long as the screening-off relation obtains and the probabilities are strictly between 0 and 1): E tells us more about D than E tells us about I. More specifically, the claim is:

If $P(E = 0/D = 0)/P(E = 0/D = 1) > 1$, then
$$P(E = 0/D = 0)/P(E = 0/D = 1) >$$
$$P(E = 0/I = 0)/P(E = 0/I = 1) \text{ and}$$
$$P(E = 0/D = 0)/P(E = 0/D = 1) >$$
$$P(E = 0/I = 1)/P(E = 0/I = 0).$$

The antecedent of this conditional means that $E = 0$ favors $D = 0$ over $D = 1$; the consequent means that $E = 0$ favors $D = 0$ over $D = 1$ more than $E = 0$ favors either state of I over the other. The same point holds for the other state that E might occupy, namely $E = 1$; this observation also provides more information about D than it does about I. So no matter which state E happens to occupy, observing that state tells us more about the direct cause than it does about the indirect cause. A proof is provided in the Appendix.

Consider an example of this three-step process. The state of an electron in some apparatus causes a measuring device to go into some particular state; the state of this measuring device causes a visual image to form in the mind of the experimenter. The chain goes from electron to meter reading to sensation. The sensory image is evidence about the state of the meter, in that the sensory image favors one hypothesis about the meter over another; and the state of the meter is evidence about the state of the electron, in that the state of the meter favors one hypothesis about the electron over another. As it happens, the favoring relation is *transitive;* it follows that the sensory state favors one hypothesis about the electron over the other (see the Appendix). But more to the point, the sensory state provides better evidence about the state of the meter than it does about the state of the electron.[23]

Perhaps the belief that *the meter reads 9.3* results from an inference of the same general type as the inference that generates the belief that *the electron is in state F*. Both involve "abduction," which is to say that in both cases data are interpreted by appeal to the Likelihood Principle. However, this *qualitative* parity between the two hypotheses does not entail that they are on a *quantitative* par. And indeed they are not. The hypothesis that is more directly related to sensation is the one about which sensation has more to say. *Experience teaches us the most about matters that are closest to experience.*[24]

Empiricism says that belief revision ought to be driven by observation. In privileging observation, the claim is not that observations are known with certainty nor that they can be known without using an interpretive theory. Nor is it plausible to claim that statements about observables are knowable, whereas other statements are not. These *absolute* categories have no place in a probabilistic epistemology. Observation is special because of its *relative* status. Our knowledge concerning "things unseen" is mediated by our knowledge of the things we see. This asymmetry has epistemological consequences, which empiricism must seek to describe without exaggeration and without the imposition of false dichotomies.

APPENDIX

In the Markov chain from I to D to E, we define the following conditional probabilities:

$$P(E = 0/D = 0) = a \qquad P(D = 0/I = 0) = c$$
$$P(E = 0/D = 1) = b \qquad P(D = 0/I = 1) = d,$$

all of which we assume to be strictly between 0 and 1. Then

$$P(E = 0/D = 0)/P(E = 0/D = 1) > P(E = 0/I = 0)/P(E = 0/I = 1)$$

if and only if

$$a/b > [ca + (1 - c)b]/[da + (1 - d)b].$$

This latter quantity simplifies to

$$(a - b)[ad + b(1 - c)] > 0,$$

which must be true if $a > b$. A symmetrical argument goes through for the observation that $E = 1$.

Another consequence of this format is that the favoring relation is transitive:

$$\text{If } P(E = 0/D = 0)/P(E = 0/D = 1) > 1 \text{ and}$$

$$P(D = 0/I = 0)/P(D = 0/I = 1) > 1,$$

$$\text{then } P(E = 0/I = 0)/P(E = 0/I = 1) > 1.$$

In other words: if $a > b$ and $c > d$, then

$$ca + (1 - c)b > da + (1 - d)b.$$

The consequent of this conditional simplifies to $(a - b)(c - d) > 0$.

The argument developed in this Appendix in terms of likelihood ratios is specifically for the case in which nodes each have two possible states. When this is generalized to the case of n states, the natural measure to use is R. A. Fisher's idea of *mutual information*. The mutual information between two random variables X and Y is defined as

$$M(X, Y) = \sum_x \sum_y P(X = x \ \& \ Y = y)$$

$$\log[P(X = x \ \& \ Y = y)/P(X = x)P(Y = y)].$$

For the Markov chain from I to D to E, Van Rijsbergen (1983) and Forster (unpublished) have each shown that $M(E, D) > M(E, I)$. See Sober and Barrett (1992) for discussion.

NOTES

I thank Martin Barrett, Ellery Eells, Malcolm Forster, Alan Musgrave, Alan Sidelle, Bas Van Fraassen, and Leora Weitzman for useful comments on earlier drafts of this essay.

1. Van Fraassen does not deny that scientists do other things besides make judgments about empirical adequacy. Sometimes they apply for grants, stab each other in the back, and work long hours. But these are not, for him, *requirements* for one to play the game of science. Scientists who do not participate in these activities may still be "scientific."

Likewise, van Fraassen does not deny that scientists sometimes *ought* to do things besides make judgments about empirical adequacy. Scientists have moral obligations, like the rest of us. But here again, van Fraassen does not regard the obeying of moral norms as requirements for one to play the game of science. Immoral scientists can still be "scientific."

I thank van Fraassen (personal communication) for helping me clarify the issues involved here, though I do not claim that he will agree with what I say.

2. Readers are invited to replace "requires" with "permits" when constructive empiricism is under discussion to check my claim that the arguments I will advance do not turn on which of these formulations is used.

3. Van Fraassen (i) assigns to observability a crucial role in his version of empiricism, (ii) defines observability in terms of a conditional that presumably must be interpreted nontruth functionally, and (iii) says that elucidating the property of observability is a scientific problem. Nonetheless, van Fraassen (1980, 115–16) also maintains that counterfactual conditionals are not part of the content of any scientific theory, since such conditionals, he says, do not describe objective facts about nature.

4. It won't matter for the purposes of this example whether "S has diabetes" and "S has small pox" are "strictly about observables." If the reader believes that they are not, and feels that this affects the point, he or she should reformulate the argument such that the hypotheses under test concern the number of moons of Jupiter and the number of moons of Mars.

5. Whether the third test is *perfectly* infallible is irrelevant to the point I wish to make; merely imagine that its error probabilities are *lower* than those for the first two tests.

6. Actualism plays a role in a number of epistemic problems—for example, in the optional stopping problem in statistics (Hacking 1965, 107-9) and in the task of explaining why the asymptotic convergence of an inference method is inessential for the method to be justified when applied to finite data (Sober 1988). Eells (1993) formulates a similar thesis: "we have to work with what we have to work with."

7. A scenario can be invented in which the existence of the third test procedure bears on how the results of the first two tests should be interpreted. Merely imagine that Clyde, a mad scientist, would be inclined to invent a procedure of the third kind only if he were inclined to change the error characteristics of the first test. I take it that the mere *possibility* of such scenarios is consistent with the principle of *actualism*.

8. Could we directly perceive the AIDS virus if we were to shrink to the right microscopic size? Even if we could, van Fraassen thinks this would not show that the virus is observable. Van Fraassen's (1980, 17) notion of observability rests on the idea that our location in space is accidental, while our approximate size is essential to us, *qua* human beings. It is hard to resist chiding van Fraassen—an avowed foe of real modality—for these essentialist expostulations. One is reminded of Quine's question about the bicycling mathematician: Is he essentially rational and accidentally bipedal, or accidentally rational and essentially bipedal? Surely empiricism ought not to depend on deciding whether our location is accidental and our size essential, or the other way around.

9. Musgrave (1985, 206) puts the point very well: "it is a curious sort of empiricism which sets aside the weight of *available* evidence on the ground that a casual observer might one day see his mouse or yeti, while the scientist can never see (but can only detect) his electrons."

10. Van Fraassen uses the term "acceptance" so that empiricists and realists have different views about what it means. For empiricists, accepting a theory means thinking that *it is empirically adequate*; for realists, acceptance means believing that *the theory is true*. Notice that in both cases, acceptance involves believing-true some proposition or other (the issue being what the proposition ought to be). I will depart from van Fraassen's usage in what follows. For me, accepting a statement simply means believing that it is true. Realists say that available evidence sometimes warrants accepting theories about unobservables; constructive empiricists deny this. This terminological shift leaves the substantive issues unchanged.

11. This problem is obviously not news to van Fraassen (1980); right after his formulation of realism as a doctrine about acceptance (p. 8), and in his accompanying note 3 (p. 216),

van Fraassen says that a realist might adopt a Bayesian framework in which hypotheses are assigned degrees of belief. Van Fraassen remarks that this would be an unusual stance for a realist, but one that is not ruled out by any realist principle. However, the present point is not that realists *could* jettison acceptance, but that both realists and empiricists *should* abandon that category.

I also should mention that van Fraassen takes up the problem of providing a probabilistic representation of his version of empiricism in his book *Laws and Symmetries* (Van Fraassen 1989). I will discuss this proposal in due course.

12. It might be objected, on Duhemian grounds, that hypotheses about unobservables never, *on their own*, entail observational predictions. This point, I reply, can be accommodated. Instead of saying that H entails O, let's say that $H \& A$ entails O (here A provides the requisite "bridge" principles). Then the probability of $H \& A$ is sensitive to the observations, even though $H \& A$ is not strictly about observables.

13. Skyrms (1984) explores a formulation of empiricism in which empirical propositions are the ones whose probabilities can be raised or lowered by observations. It is notable that Skyrms's development of this idea fails to coincide with important elements in van Fraassen's position.

14. Giere (1985, 83) provides a nice version of this response to van Fraassen's argument by locating constructive empiricism in a continuum of positions, ranging from solipsism of the present moment at one extreme to a form of modal-realism at the other.

15. If second-order probabilities were incorporated into this framework, one could go farther. For some second-order distributions (e. g., a uniform distribution), one could say that increasing the upper bound on the first-order probability would entail that the first-order probability has increased. This would further undermine van Fraassen's proposal for how agnosticism should be represented.

16. Using a dichotomous category of acceptance is harmless so long as what one says can be reconciled with the fact that the dichotomy rests, somewhat arbitrarily and contextually, on the matter-of-degree concept of plausibility. In this respect, these epistemic concepts are related to each other the way the concept of *being tall* is related to the concept of *height in inches*. The problem exposed by the lottery paradox is that empiricism and realism should be subjected to a kind of consistency check. Can their pronouncements be represented within an epistemology in which plausibility is construed as a matter of degree?

17. It is worth considering what the word "information" means in the quotation from van Fraassen that is under discussion. I suggest that this concept should be understood within the framework of the Likelihood Principle (see Section 5 and the Appendix). An experience gives us information about a set of competing hypotheses precisely when those hypotheses (perhaps in conjunction with a plausible background theory) confer different probabilities on the experience's occurring. Of course, this explication conflicts with the dictates of constructive empiricism. Van Fraassen needs to explain what alternative explication of "information" he wishes to endorse.

18. The challenge to empiricism, then, is to show that some thesis about substance *follows* from the thesis about sources. I try to defend such an inference in Section 5.

19. Sober (1988; 1990b), and Forster and Sober (forthcoming) develop proposals for bringing simplicity considerations within the purview of a broadly empiricist framework.

20. When prior probabilities are based on a model of a chance process, there is no problem in using them as well as likelihoods to evaluate the overall plausibility of hypotheses. In addition, it is not to be denied that background assumptions shape our evaluation of hypotheses in ways that are additional to the testimony of the data presently at hand. These background assumptions trace their credibility back to likelihood considerations, relative to earlier data sets.

21. If a connection between simplicity and likelihood could be established (in which case simplicity would not be a *super*empirical virtue of theories), then contrastive empiricism would grant that simpler theories are more plausible. There may be *local* contexts in which

this connection can be established. For example, in Sober (1988) I explore the extent to which the principle of parsimony used in phylogenetic inference can be understood within a likelihood framework. And Forster and Sober (forthcoming) provide a framework of considerable generality within which the simplicity of a curve is linked with its predictive accuracy. However, I doubt that there is any global and presuppositionless connection between simplicity and evidential warrant (Sober 1990b).

22. This asymmetry is important in understanding the role that observations play in resolving theoretical disputes. However theory-laden the observations may be, it usually is possible to reach agreement about what the observations are without already agreeing about which theory under test is true (Sober 1993).

23. Although I described the chain from I to D to E as a *causal* chain, the causal idea is quite irrelevant to the above result. What is essential is that the chain have the Markov property. See Sober and Barrett (1992) for further discussion.

When the likelihood ratios discussed above have values different from unity, the states of the different nodes are *correlated*. The likelihood conception of how observation or detection ought to be understood accords well with some of the main ideas behind the treatment in Wilson (1985).

24. To clarify what this Likelihood Result entails, it is useful for a moment to lapse into a Bayesian idiom. If $P(H)$ is the prior probability of H and if $P(H/E)$ is the posterior, then $P(H/E)/P(H)$ measures the degree to which E confirms H. It is quite possible that H should be quite certain and that E should confirm it only slightly or not at all; simply imagine that $P(H/E) = P(H) = 0.95$. Likewise, it is possible that E confirms H a lot, even though H is quite improbable in the light of H; simply imagine that $P(H) = 0.000001$ and that $P(H/E) = 0.001$. In short, degree of confirmation is a *diachronic* concept, measuring *changes* in probability, whereas posterior probability is, so to speak, *synchronic*.

Now let us return to the chain from I to D to E and consider the posterior probabilities that propositions about D and I have in the light of the observation that $E = 0$. By Bayes's Theorem we can deduce

$$\frac{P(I = 0/E = 0)}{P(I = 1/E = 0)} = \left[\frac{P(E = 0/I = 0)}{P(E = 0/I = 1)}\right]\left[\frac{P(I = 0)}{P(I = 1)}\right]$$

and

$$\frac{P(D = 0/E = 0)}{P(D = 1/E = 0)} = \left[\frac{P(E = 0/D = 0)}{P(E = 0/D = 1)}\right]\left[\frac{P(D = 0)}{P(D = 1)}\right].$$

The likelihood result developed in the text describes a relationship between the first ratio on the right side of the first equation and the first ratio on the right side of the second. However, this result does not tell us how to evaluate the two left-hand ratios. The result says nothing about how the priors are related, and so does not determine the relation of the posterior probabilities (Sober and Barrett 1992).

This is why the proper motto is that *experience teaches us the most about matters that are closest to experience*, not that *experience allows us to be most certain about matters that are closest to experience*. This latter thesis would be obtainable from the former one, once the likelihood result is supplemented with a thesis about priors. However, I am not Bayesian enough to defend any such further thesis.

REFERENCES

Churchland, P. 1985. "The Ontological Status of Observables: In Praise of the Superempirical Virtues." In *Images of Science: Essays on Realism and Empiricism*, edited by P. Churchland and C. Hooker, 35–47. Chicago.

Edwards, A. 1972. *Likelihood*. Cambridge.

Eells, E. 1993. "Probability, Inference, and Decision." In *Foundations of Philosophy of Science*, edited by J. Fetzer, 192–208. New York.

Feyerabend, P. 1963. "How to be a Good Empiricist—a Plea for Tolerance in Matters Epistemological." In *The Delaware Seminar II*, edited by B. Baumrin, 3–39. Newark, Delaware.

Forster, M. Unpublished. "In Defence of a Causal Theory of Mental Representation."

Forster, M. and E. Sober. Forthcoming. "How to Tell When Simpler, More Unified, or Less *Ad Hoc* Theories will Provide More Accurate Predictions."

Giere, R. 1985. "Constructive Realism." In *Images of Science: Essays on Realism and Empiricism*, edited by P. Churchland and C. Hooker, 75–98. Chicago.

Hacking, I. 1965. *The Logic of Statistical Inference*. Cambridge.

Kyburg, H. 1970. "Conjunctivitis." In *Epistemology and Inference*. Minneapolis.

Marr, D. 1982. *Vision*. San Francisco.

Maxwell, G. 1962. "The Ontological Status of Theoretical Entities." In *Minnesota Studies in the Philosophy of Science*, edited by H. Feigl and G. Maxwell, 3–27. Minneapolis.

Musgrave, A. 1985. "Realism vs. Constructive Empiricism." In *Images of Science: Essays on Realism and Empiricism*, edited by P. Churchland and C. Hooker, 197–221. Chicago.

Skyrms, B. 1984. *Pragmatics and Empiricism*. New Haven.

Sober, E. 1988. *Reconstructing the Past: Parsimony, Evolution and Inference*. Cambridge, Mass.

Sober, E. 1990a. "Contrastive Empiricism." In *Scientific Theories: Minnesota Studies in the Philosophy of Science*, vol. 14, edited by W. Savage, 392–412. Minneapolis.

Sober, E. 1990b. "Let's Razor Ockham's Razor." In *Explanation and Its Limits*. Royal Institute of Philosophy Supplementary Volume 27, edited by D. Knowles, 73–94, Cambridge.

Sober, E. 1993. "Mathematics and Indispensability." *Philosophical Review*.

Sober, E. and M. Barrett. 1992. "Conjunctive Forks and Temporally Asymmetric Inference." *Australasian Journal of Philosophy* 70: 1–23.

van Fraassen, Bas. 1980. The *Scientific Image*. Oxford.

van Fraassen, Bas. 1985. "Empiricism and the Philosophy of Science." In *Images of Science: Essays on Realism and Empiricism*, edited by P. Churchland and C. Hooker, 245–308. Chicago.

van Fraassen, Bas. 1989. *Laws and Symmetry*. Oxford.

Van Rijsbergen, K. 1983. "A Discrimination Gain Hypothesis." In *Proceedings of the 6th International ACM Conference on Information Retrieval*, 101–104, London.

Wilson, M. 1985. "What Can Theory Tell Us About Observation?" In *Images of Science: Essays on Realism and Empiricism*. edited by P. Churchland and C. Hooker, 222–44. Chicago.

MIDWEST STUDIES IN PHILOSOPHY, XVIII (1993)

Wittgensteinian Bayesianism

PAUL HORWICH

B elief is not an all-or-nothing matter. Rather, there are various *degrees* of conviction which may be represented by numbers between zero and one. Were we ideally rational, our full beliefs (of degree one) would comply with the laws of deductive logic; they would be consistent and closed under logical implication. And similarly, our *degrees* of belief should conform to the probability calculus.[1] This enrichment of epistemology—provided by the addition of degrees of belief and an appreciation of their probabilistic 'logic'— fosters progress with respect to many problems in the philosophy of science.

These statements form the core of a program, which I will call "therapeutic Bayesianism," whose primary goal is the solution of various puzzles and paradoxes that come from reflecting on scientific methods. Its creed is that many of these problems are the product of oversimplification, and that the above-mentioned elementary probabilistic model of degrees of belief often contains just the right balance of accuracy and simplicity to enable us to command a clear view of the issues and see where we were going wrong.[2] This somewhat Wittgensteinian goal and creed distinguishes therapeutic Bayesianism from more systematic enterprises in which probabilistic degrees of belief play a prominent role: for example, Bayesian decision theory, Bayesian statistics, Bayesian psychology, Bayesian semantics, and Bayesian history of science. It is especially important to appreciate the difference between the problem-solving orientation of therapeutic Bayesianism—that of exploiting a simple, idealized model in order to help illuminate notorious philosophical perplexities—and the quite distinct project of providing a perfectly true and complete (descriptive or normative) *theory* of scientific practice. The latter task might well involve the postulation of belief-gradations, and might also be done in the name of philosophy of science. However, its aims are quite different; and one must beware of judging one project by adequacy conditions appropriate to the other.[3]

Therapeutic Bayesianism is not self-evidently beneficial, but it does have some prima facie plausibility. Moreover, this plausibility is enhanced by substantial accomplishments, and, as we shall see, a great deal of the criticism it has received is misdirected—commonly for the reason just indicated. In this paper I would like to try to make a case for the program by discussing it from three, progressively abstract, points of view: substantial, foundational, and metaphilosophical. More specifically, there will follow sections on: (I) "The fruitfulness of therapeutic Bayesianism," in which I will sketch treatments of the 'raven' paradox and the question of diverse data and mention various other applications; (II) "Probabilistic foundations," in which the propriety of certain idealizations will be defended—particularly the representation of belief by numbers, the adoption of probabilistic canons of reason governing such beliefs, the definition of confirmation as increase in rational degree of belief, and the idea that induction may be codified in a confirmation function; and (III) "Misplaced scientism," in which I criticize a metaphilosophical perspective that does not properly distinguish science from the philosophy of science, and which overvalues the use of symbolic apparatus. Along the way, I shall respond to some criticisms of therapeutic Bayesianism that have recently been advanced.

I. THE FRUITFULNESS OF
THERAPEUTIC BAYESIANISM

A good illustration of therapeutic Bayesianism at work is its way of treating the notorious 'raven paradox'. It is plausible to suppose that any hypothesis of the form 'All Fs are G' would be supported by the observation of an F that is also G. But if this is generally true, then the discovery of a non-black non-raven (e.g., a white shoe) confirms that all non-black things are non-ravens; and thereby confirms the logically equivalent hypothesis, 'All ravens are black'—a seemingly bizarre conclusion. This is 'the paradox of confirmation'. The Bayesian approach to this problem is to argue that observing a known raven to be black will *substantially* confirm "All ravens are black," whereas observing that a known non-black thing is not a raven will confirm it only *negligibly*—the difference being explained, roughly speaking, by the fact that, given our background beliefs about the chances of coming across ravens and black things, the first of these observations is more surprising, more of a test of the hypothesis, and therefore more evidentially powerful, than the second. Thus, the paradoxical flavor of our conclusion comes from the not unnatural confusion of negligible support with no support at all—a confusion sustained by inattention to degrees of belief and their bearing on confirmation.

A formal version of this analysis proceeds from the following premises:

(a) That the amount of support for hypothesis H provided by evidence E is the factor by which the rational degree of belief in H is enhanced by the discovery of E—which is indicated by the ratio of subjective probabilities, $\mathrm{P}(H/E)/\mathrm{P}(H)$, for a rational person.

(b) That a rational person's degrees of belief will ideally conform to the probability calculus; and, in particular, will obey Bayes's Theorem:

$$\frac{P(H/E)}{P(H)} = \frac{P(E/H)}{P(E)}$$

(To appreciate the intuitive plausibility of this theorem, note that it derives from the fact that the conditional probability of H given E is equal to the probability of the conjunction of H and E, divided by the probability of E: i.e., $P(H/E) = P(H\&E)/P(E)$. See figure 1.

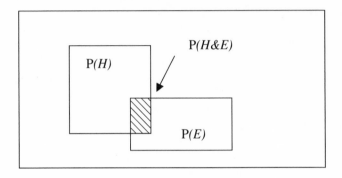

Therefore, since $P(H\&E) = P(E\&H)$, we obtain $P(H/E)P(E) = P(E/H)P(H)$, and hence Bayes's Theorem).

(c) That our degree of belief (prior to the investigation, and given the known scarcity of ravens) that a randomly selected non-black thing would turn out to be a non-raven is high.

(d) That our prior degree of belief (prior to the investigation, and given the known abundance of non-black things) that a randomly selected raven would turn out not to be black is substantial.

Now let us compare the support for the hypothesis, H, that all ravens are black, provided, first, by the discovery concerning a known raven that it is black (which is symbolized as $R^* B$), and, second, by the discovery that a known non-black thing is not a raven ($-B^* - R^*$). Applying premise (a) and then (b), we find:

$$\frac{\text{Support provided}}{\text{by } (R^* B)} = \frac{P(H/R^* B)}{p(H)} = \frac{P(R^* B/H)}{P(R^* B)}$$

$$\frac{\text{Support provided}}{\text{by } (-B^* - R)} = \frac{P(H/ - B^* - R)}{p(H)} = \frac{P(-B^* - R/H)}{P(-B^* - R)}$$

But our hypothesis *entails* that any known raven would be black and any known non-black thing would not be a raven; therefore, $P(R^* B/H) = 1$ and $P(-B^* - R/H) = 1$; Therefore

$$\text{Support from raven found to be black} = \frac{P(H/R^* B)}{P(H)} = \frac{1}{P(R^* B)} = \text{1/prior degree of belief that a known raven would be black}$$

$$\text{Support from non-black thing found not to be a raven} = \frac{P(H/-B^* - R)}{P(H)} = \frac{1}{P(-B^* - R)} = \text{1/prior degree of belief that a known non-black thing would not be a raven}$$

Now one may assume (premise (c)) that a normal investigator of the hypothesis has prior background knowledge about the rough distribution of ravens and black things in his vicinity, and that this will lead him to expect that there is a very good chance that a randomly selected non-black thing will turn out not to be a raven. Thus $P(-B^* - R)$ is very nearly 1; and the amount of support for the hypothesis provided by observing that a non-black thing is not a raven is very little.

On the other hand, we would expect the background of investigation to dictate, in addition, (premise (d)) that the likelihood of a randomly selected raven being black is not especially high. After all, as far as we know at the outset of the research, there are many colors that the raven could perfectly well have. Thus $P(R^* B)$ is a good deal less than one. Therefore, the amount of support provided by observing that a known raven is black is substantial.

One might object to this reasoning that the final assumption is false, since the *objective* chances of finding that a raven is black are actually extremely high. However, this objection is based on a slip which is easy to identify. It confuses "probability" in the sense of *subjective degree of belief* and "probability" in the sense of *relative frequency*. All the probabilities mentioned in the argument are rational subjective probabilities, and it is under that construal that we may reasonably assume that $P(R^* B)$ is not near to 1. The feeling that this assumption is wrong derives from incorrectly reading $P(R^* B)$ as a relative frequency assertion. In that sense, since in fact almost all ravens are black, the probability that a randomly selected raven will be black is indeed very great. But this fact has no bearing on the argument.[4]

A similar objection is to deny that there could be any difference in evidential import between identifying a known raven as black and identifying a known black thing as a raven. Howson and Urbach,[5] for example, maintain that the only difference between these two data is the time order in which the elements of the observed fact are established. They think that in each case what is eventually known is the same, so there can be no variation in confirmation power between the two discoveries. Imagine, however, that an ornithologist instructs her assistant to go and find a black raven and bring it back to the lab for inspection. Surely, that inspection would count for nothing. And there is no paradox here, even though we might loosely speak of 'seeing a black raven' in all three cases. For a more precise characterization of the evidence shows that what is discovered in each case is not really the same. That a randomly selected raven turns out to be black, that a randomly selected black thing turns out to be a raven and that a randomly selected black raven turns out to be a black raven, are very different pieces of information, and it should not be surprising that they confirm our hypothesis to different degrees.

Therapeutic Bayesianism handles other issues in the philosophy of science similarly, putting a lot of weight on premises (a) and (b). By combining the idea of confirmation as enhancement of rational degree of belief, with the principle that rational degrees of belief should satisfy the probability calculus, we get a way of treating those problems that hinge upon considerations having to do with *degree of support*. Therefore the method has a wide scope. In particular, one can expect to shed light on why 'surprising' predictions have relatively great confirmation power, what is wrong with ad hoc hypotheses, whether prediction has more evidential value than mere accommodation of data, why a broad spectrum of facts can confirm a theory more than a narrow data set, why we base our judgments on as much data as possible, how statistical hypotheses can be testable despite their unfalsifiability, what is peculiar about 'grue-like' hypotheses, and various other problems.

These issues are unified by their involvement with the notion of 'varying evidential quality'; and this is why traditional epistemology, with its fixation on all-or-nothing belief, is not able to resolve them. It is only to be expected that the introduction of degrees of belief, together with an understanding of the rational constraints to which they are subject, would open the way to progress. Of course, there is not the space here to fully substantiate this thesis by describing all these applications of therapeutic Bayesianism. Let me, however, give one further illustration of the approach.

How is it that a broad spectrum of different kinds of fact, when entailed by a hypothesis, will confirm it to a greater degree than a uniform, repetitive set of data? It is natural to answer as follows. To the extent that our observations cover a broad range of phenomena, they are capable of falsifying a large number of alternative hypotheses, which then bequeath substantial credibility to those hypotheses that survive. Now, this solution does not quite work. For a narrow data set can preclude just as *many* hypotheses as a diverse data set. Nevertheless, we can repair the solution by noting that there is a significant difference in the *kinds* of hypotheses that are excluded by the two sets of facts. We should notice that the diverse data tend to exclude more of the *simple* hypotheses than do the narrow data. Given a representation of simplicity in terms of high prior probability,[6] this suggests that diverse data tend to rule out more high-probability alternatives than narrow data. But if so, then a hypothesis that survives relatively diverse observations becomes more probable than one that is left in the running by a narrow set of data. (See figure 2).

In particular, the data points E(narrow) exclude (given experimental error) just as many alternatives to the line H(straight) as does E(diverse). Nonetheless E(diverse) confirms H(straight) more strongly than E(narrow) does, because E(diverse) is better than E(narrow) at excluding simple alternatives to H(straight)—for example, gradual curves—which have an initially high

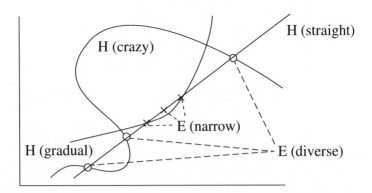

probability. Thus H(gradual) is ruled out by E(diverse) but not by E(narrow). On the other hand, the sort of hypothesis, like H(crazy), prohibited by E(narrow) yet not by E(diverse), is not very probable anyway; so excluding it does not greatly benefit those hypotheses that survive. Thus, with the help of a probabilistic representation of simplicity, we can begin to account for our methodological intuitions concerning diverse data.[7]

II. PROBABILISTIC FOUNDATIONS

In the last section I have tried to indicate something of the fruitfulness of therapeutic Bayesianism. Let me now consider various foundational questions that might be thought to cast doubt on the project:

(1) Do people actually have numerical degrees of belief?

(2) If so, can it be shown that *rational* degrees of belief conform to the probability calculus?

(3) Is it correct to identify degree of confirmation with rational enhancement of subjective probability?

(4) Are there objective facts of confirmation?

(5) Does reason require *merely* that one's beliefs conform to the probability calculus? Or is it the case (as Carnap thought) that a rational system of beliefs is subject to several further constraints?

(6) If further constraints are needed, then what are they?

On the first question, perhaps we should be agnostic. The successes of therapeutic Bayesianism will reinforce the evident fact that its basic principles are at least *roughly* correct. Thus we know that there are belief gradations of some sort, that there are rational constraints governing them (prohibiting, for example, a high degree of confidence in two contradictory propositions), and that confirmation is not wholly unrelated to increasing belief. Moreover, the Bayesian representation of these ideas has a great deal of plausibility. Consider a spectrum of situations in which we know that the propensities (as manifested by relative frequencies) of certain events are $x_1, x_2, \ldots, x_N$. For each such case there is a corresponding epistemic attitude—a degree of confidence—that the next trial will produce an event of the designated type. Presumably the appropriate attitude will vary with the relative frequency. Specifically, since the frequencies range over numbers between zero and one, so will the degrees of confidence.

However, despite the attractiveness of such considerations, one must of course acknowledge that the Bayesian framework might be wrong. The crude ideas that it represents should not be controversial. However, it is quite possible that the Bayesian articulation of those ideas is not absolutely right; and that, in particular, the assumption of precise-valued, numerical degrees of belief is incorrect.

Even so, such a model could be an excellent idealization, sufficing perfectly well for the primary purposes of therapeutic Bayesianism: namely, to dispel confusion, solve problems, and thereby improve our understanding of the scientific method. For the paradoxes are caused by forgetting the *crude* facts (that there are gradations of belief, etc.) or by failing to recognize their significance. And so the solutions will involve noticing that those rough ideas have been overlooked and coming to appreciate how they bear on the problems. This sort of treatment will not depend essentially on any particular theoretical refinements. The function of the Bayesian framework is merely to cast the crude, uncontroversial ideas into a form where their impact on our problems can have maximal clarity and force.

What then is the import of studies that cast doubt on the existence of numerical degrees of conviction and which develop more complex and allegedly more realistic conceptions of belief? Let me stress that this work falls well outside the focus of therapeutic Bayesianism, for there is no reason to believe that such improvements will help to solve the standard problems in the philosophy of science. Perhaps these developments are important in psychology, statistics, semantics, or decision theory; perhaps they will become important to philosophy when we have progressed enough in our understanding of science so that the *details* of an inductive logic become items of reasonable concern. But at this juncture, confusion is rampant, the traditional problems are still very much with us, and it seems rather unlikely that the slight gains in accuracy to be derived from a more realistic theory of belief would be worth the price—in terms of loss of simplicity—that we would have to pay for it.

Our treatment of the 'raven paradox' is a case in point. The problem was solved by exposing a certain misconception (that a non-black, non-raven would be irrelevant to our hypothesis), and by explaining why we are so tempted by that misconception: namely, that in forgetting about degrees of belief, we lose sight of the distinction between very slight confirmation and no confirmation at all. The simple Bayesian model of belief provides a sufficiently perspicuous representation of the situation to enable us to put this in a clear way. Further accuracy regarding the nature of belief would distract us from the main point, ruin the argument, and not help us to understand the basis of the paradox.

Let me give another example. The conflict between realism and instrumentalism with respect to the acceptance of scientific theories is fueled by a shared tendency to think in terms of all-or-nothing belief. The instrumentalist argues, in light of previous scientific revolutions, that it is foolishly optimistic to expect that our current theories are true and will not eventually be refuted. Whereas the realist complains that it is a distortion of science to distinguish rigidly between credible observation reports and incredible theoretical claims. However, once we see that the issue is not 'To believe or not to believe?' but rather, 'To what degree shall we believe?' then there is room for reconciliation. The crucial move is the elimination of the shared misconception. There is no reason to think that a fancy model of belief, even more accurate than the Bayesian idealization, would be any further help with the problem.

I do not mean to be suggesting that it is not worthwhile to investigate more sophisticated models of belief. On the contrary, I can readily imagine research programs—e.g., Bayesian psychology, or attempts to give a perfectly accurate description of scientific practice—in which this would be crucial. My point is that there is another enterprise—the one I am calling "therapeutic Bayesianism"—whose focus is on solving the traditional methodological puzzles and paradoxes, and for which the introduction of such complex models is likely to do more harm than good.

Suppose, then, that we do have numerical degrees of belief. Is there any way of justifying the Bayesian assumption that, to be rational, these degrees of belief must conform to the probability calculus? Although there are indeed various lines of reasoning which purport to establish this thesis, none is compelling. The best known of them is the 'dutch book' argument[8] and it goes roughly as follows. Defining a person's degree of belief in a proposition as a function of the odds at which he is prepared to bet on its truth, it can be proved that if his degrees of belief do not satisfy the probability calculus then he will be prepared to accept a collection of bets which is guaranteed to lead to a loss. Therefore, since it would surely be irrational for him knowingly to put himself in such a no-win situation, it would be irrational to have a system of degrees of belief that violates the probability calculus. QED. However, the definition of 'degree of belief' that is employed in this argument presupposes that people maximize their expected utility. And there is a lot of room for skepticism about that assumption (and about the preference axioms to which it is equivalent). So the 'dutch book' argument is far from airtight. Worse still, there is positive reason to think that its conclusion is false, for it requires logical omniscience. The probability of any logical truth is 1 and of any contradiction is 0. Yet it is surely quite rational to be less than perfectly confident in the truth of *some* logical truths—those that are especially hard to prove—and quite rational to give non-zero degrees of belief to contradictions that are hard to recognize as such.

The proper response to these difficulties is to repeat that the picture of rational degrees of belief obeying the probability calculus should be regarded as an *idealization* of the real normative facts. It is uncontroversial that one ought to be certain of elementary logical truths, and that one ought not be confident of the truth of obviously incompatible hypotheses. The probabilistic model of belief provides a sharp, perspicuous way of capturing these trivialities, and to the extent that it goes beyond them it need not be construed realistically.

A similar answer may be given to the third question concerning the definition of confirmation. In our discussion of the raven paradox we defined "the degree by which E confirms H" as "the ratio, $P(H/E)/P(H)$, for a rational person." Evidently, this explication has at least *some* prima facie plausibility, and it certainly helps us to give a neat, compelling solution to the problem. Nonetheless, it is often argued that this particular explication is 'wrong'—yielding counterintuitive consequences—and that there are better definitions of confirmation which should be used instead.[9]

However, these criticisms have little relevance to the project of therapeutic Bayesianism. No doubt our explication leads to some strange-sounding consequences. No doubt it is strictly speaking false that the ordinary meaning of "confirms" is given by our explication. No doubt there are definitions (perhaps involving non-probabilistic notions) that come closer to what we ordinarily mean. But the object of therapeutic Bayesianism is not to give a theory of science. We are not trying to find the most accurate analyses of our concepts, but rather to use explications that are at least roughly right, and which are conducive to simple, convincing dissolutions of philosophical problems. Since we assume that these problems are the product of confusion, it is desirable to look for ways of clarifying the issues, which have the proper blend of accuracy and simplicity. Of course it is possible to *over*simplify. But one can conclude that this has happened only after finding that the admittedly idealized models do not in fact help to solve our problems.[10]

On the fourth question—Are there objective facts of confirmation?—it seems evident that judgments of credibility and confirmation do purport to capture objective normative facts. They do not state what any individual's degrees of belief *actually* are, but rather they say something about what one's degrees of belief *ought* to be, or how they *ought* to change given the circumstances. Thus we should acknowledge non-subjective facts regarding confirmation.

A natural way of capturing this idea, due to Carnap, is to suppose that an attribution of probability to a hypothesis reflects the belief in an objective, logical fact about the degree to which one statement—a summary of the available evidence—probabilifies another statement—the hypothesis in question. Such logical facts might be codified in a confirmation function, $c(p/q) = x$, which would specify explicitly the degree, x, to which q confirms p, and would specify implicitly the degree to which one should believe p if the total evidence is q. Carnap says, for example:

> Probability-1 is the degree of confirmation of a hypothesis h with respect to an evidence statement e, e.g., an observation report. This is a logical, semantical concept. A sentence about this concept is based,

not on observation of facts, but on logical analysis; if it is true, it is L-true (analytic). . . . Probability-2 [relative frequency] is obviously an objective concept. It is important to recognize that probability-1 is likewise objective.

Let h be the sentence 'there will be rain tomorrow' and j the sentence 'there will be rain and wind tomorrow'. Suppose someone makes the statement in deductive logic: 'h follows logically from j'. . . . The statement 'the probability-1 of h on the evidence e is 1/5' has the same general character as the former statement. . . . Both statements express a purely logical relation between two statements. The difference between the two statements is merely this: while the first states a complete logical implication, the second states, so to speak, a partial logical implication; hence, while the first belongs to deductive logic, the second belongs to inductive logic.[11]

Thus, Carnap held that certain facts about confirmation are analytic and *objective*, and thought of inductive probability as a *partial* version of the logical relation of entailment.[12]

On the fifth question—Does reason impose constraints on belief *over and above* the requirement of conformity with the probability calculus?—there are grounds for sympathy with Carnap's view that it does. For it is hard to see how the probabilistic constraint alone can account for our intuitions about the relative plausibility of competing hypotheses that equally well fit the current data. In particular, it is hard to see how it can solve the 'grue' problem.[13]

Suppose that such further constraints are indeed required. Still, to take up the sixth question, it is no trivial matter to say what they are. Carnap tried out various constraints and employed them to derive confirmation functions for certain extremely simple formal languages. Unfortunately, these functions have the counterintuitive property that laws of nature are never able to acquire more than a negligible probability. And this shows a deficiency in Carnap's constructions: either the languages are too simple, or the constraints are wrong. However, one can certainly not conclude that anyone who endorses a Carnapian conception of logical probability *must* hold that general laws never attain a non-negligible probability. This is a non sequitur, arising from a failure to distinguish between the general conception of logical probability and the admittedly inadequate prototypes with which Carnap experimented.[14]

For the treatment of various problems it is helpful to suppose that inductive reasoning is represented by a specific (but unspecified) real-valued Carnapian confirmation function, c, allowing general laws to achieve a non-negligible credibility. In light of our responses to questions (1), (2), and (3), we see that it can be no objection to this procedure that our inductive practice is not in fact precisely described by a single c-function. For, once again, the intention is not to get at the exact truth, but merely to employ a useful idealization. Nor—as we have just said—is it fair to complain that *some* c-functions—

those Carnap toyed with—always give zero probability to general laws. For we can suppose that 'the right c-function' is one of those that do *not* have that counterintuitive feature.

III. MISPLACED SCIENTISM

Much criticism of therapeutic Bayesianism arises from a conflation of philosophy and science. More exactly, it derives from a failure to recognize the legitimacy (even the existence) of non-scientific philosophical projects—those prompted, not by a desire to expose the whole truth regarding some domain, but by an interest in the resolution of paradoxes. Let me elaborate.

What one might call '*theory*-oriented philosophy of science' aims for a systematic account of the scientific method. The criteria of success are just those that pertain to theory construction *within* particular sciences: namely, empirical adequacy, scope, depth, simplicity, internal consistency, and coherence with the rest of our knowledge. More specifically, a perfect theory of the scientific method would be expected to conform with specific intuitions about the way that good science is done, to cover all aspects of methodology in detail, to expose fundamental principles enabling the complex, superficial aspects of scientific practice to be unified and explained, and to respect results in psychology and sociology. Thus it seems appropriate to regard theory-oriented philosophy of science as itself a department of science—a branch of naturalized epistemology. This characterization neither ignores nor denies that scientific methods are normative. A description of science will contain a codification of the basic norms which are implicit in the evaluation of theories. Moreover, it is quite possible that an identification of basic normative principles will result in the exposure of cases in which science is being done badly. One might thereby effect an improvement in the conduct of some science. Indeed, this may well be a motive for engaging in theory-oriented philosophy of science.

In contrast, the approach portrayed in this paper, '*problem*-oriented philosophy of science', has very different goals, methods, and adequacy conditions. It aims at the resolution of deep puzzles and paradoxes that arise from reflection upon science. It includes in its domain, for example, the problem of induction, the paradox of confirmation, the question of total evidence, and the issue of prediction versus accommodation. The problems here are not simply to fill various undesirable gaps in our knowledge about science. Characteristically, they are conceptual tensions, contradictions, absurd conclusions—that is to say, symptoms of confusion. We have somehow gone astray, and the task is to understand how this has happened and to get a clear view of the issue so that our misguided ways of thinking will be exposed and no longer seem so attractive.

These two approaches to the philosophy of science do not compete with one another. They are distinct projects with distinct objectives—not *wholly* unrelated to one another, but by no means simply parts of the same enterprise. Thus, it is not the case that the sort of full understanding provided by a successful theory-oriented philosophy of science would automatically solve the puzzles that form the domain of problem-oriented philosophy of science. For the resolution of a paradox requires a great deal more than just locating the wrong move in a fallacious argument. It is crucial to a proper resolution that one comes to see why that fallacy was natural. And it is important that one obtains a new perspective on the issue—a point of view from which the old and troublesome habits of thought no longer seem plausible. These elements of the solution do not simply fall out of a complete theory of science. (Similarly, the explanation of conjuring tricks does not follow from physics.) Thus, theory-oriented philosophy of science is not simply a more thorough, systematic, and ambitious project than problem-oriented philosophy of science.

Neither is it necessary, in order to solve problems, that one be in possession of an adequate theory of science. For confusions can be identified, understood, and removed without a theory of any particular depth or generality. Granted, assumptions about methodology will often be involved in the diagnosis and treatment of a problem, and if these were *wildly* false then it is unlikely that the discussion would be helpful. However, there is no reason why such assumptions should be *true* as long as their replacement with the truth would not undermine the solution that is based on them. Indeed it is quite possible that the perfect theory of science would be a very bad tool for solving problems. For the truth may be so complicated that it cannot provide the sort of simple and relevant perspective that is needed.

If the practice of conceptual troubleshooting is confused, as it often is, with the scientific search for a theory of science, then therapeutic Bayesianism will be wrongly subjected to all of the methodological requirements that are properly applied only in science. Let me describe some of the bad effects of this confusion.

One consequence, discussed above, of not seeing the distinctive aim of therapeutic Bayesianism is a tendency to misjudge the function of various helpful idealizations. Thus one commonly finds objections to the use of precise-valued degrees of belief, to the assumption that these should conform to the probability calculus, to the adoption of a particularly simple explication of confirmation, and to the idea that our inductive practice may be represented by a single Carnapian confirmation function. Doubtless some of these assumptions are, strictly speaking, false. (Just as it is false that a gas is made of point masses.) And in a different kind of study—one aimed at truth—it would be very important to discuss more realistic models. However, for the purposes of therapeutic Bayesianism it is important to use the simplest roughly accurate models of degrees of belief and of confirmation that will help to clarify the issues, and it is sufficient to proceed on the basis of their intuitive plausibility and to justify these models in retrospect in terms of their utility.

Secondly, a scientific understanding of confirmation aims for the truth, the *whole* truth, and nothing but the truth. Consequently, those wedded to this conception of the philosophy of science will find fault with studies that do not discuss every significant aspect of the phenomenon of confirmation. Consider, for example, *prior probability assignment*, the procedures for deciding, before data have been gathered, the various 'intrinsic' plausibilities of hypotheses; *belief-kinematics*, the way that systems of belief change over time in the light of new discoveries; or *direct inference*, the impact on our degrees of belief of a knowledge of empirical probabilities. These are fascinating topics, and a good theory of science must deal with them. But there is no reason why a paradox-oriented Bayesian program should incorporate a complete, systematic account of all such elements of methodology.[15]

In the third place, scientism in philosophy engenders a 'hyperformalist' fixation on symbolic technique—an overvaluation of logico-mathematical machinery. Among the symptoms of this hyperformalist state are: (a) a blindness to the possibility of philosophical problems distinct from the scientific and mathematical issues that arise in statistics, decision theory, sociology of science, etc., further questions being dismissed as 'merely verbal';[16] (b) a dissatisfaction with informal discussions and conclusions; (c) an exaggerated concern with formal rigor for its own sake; and (d) an obsession with the elimination of any potential ambiguity or vagueness, leading to the feeling that the English language is too confusing and vague a medium for intellectual progress, and that it should, wherever possible, be replaced with mathematics or logic.

Thus, even if an approach employs formal techniques, as therapeutic Bayesianism clearly does, it may still be subjected to hyperformalist criticism. I think this is an unhealthy point of view—in philosophy generally, and particularly in the philosophy of science, where it is especially common. No doubt there are occasions when clarity is gained and confusion allayed with the help of formal apparatus. This, I believe, is one of the morals of Bayesianism's success. However, one can withdraw too quickly into the secure, regulated territory of a formal system. It is certainly a tempting relief from the frustrating vagaries of philosophy to be able to obtain definite, proven results and get clear answers to clear questions. But, unless we are very careful, these answers and results might have little to do with the problems that have traditionally motivated philosophy of science. Our methodological puzzles arise when we reflect informally about scientific practice; and they can be solved only with an appreciation of the misconceptions and confusions to which we are prone and an understanding of the ways in which they are fostered by the rich conceptual resources put at our disposal by natural language. It seems to me that only when that sort of understanding is eventually attained will we know what we are looking for in a fully fledged inductive logic; and then, perhaps, be in a better position to devise one. But this level of understanding will not be achieved by trying to express as many questions as possible within a formal system, proving some theorems, and dismissing the residue as intractable and uninteresting. At its worst, such scientistic hyperformalism betrays a lack of concern for truly philosophical problems. If "merely verbal" issues are any that do not make a scientific difference, and if only scientific problems are worth worrying about, then philosophy is truly an endangered enterprise.

I hope to have clarified what I believe is a valuable approach to the philosophy of science, and to have shown that many of the complaints about it derive from scientistic hyperformalism and are therefore misconceived. The goal is not a theory of science but the unravelling of puzzles surrounding our ideas about surprising data, prediction versus accommodation, ad hoc postulates, statistical hypotheses, our thirst for new data, the tenability of realism, and other aspects of methodology. And given some of the successes of therapeutic Bayesianism there is reason to have a fair amount of confidence in its basic principles.

Thus, the notion of rational degrees of belief conforming to the probability calculus has an important role in the philosophy of science. It would no doubt be easier to think in terms of all-or-nothing belief, but that oversimplification is part of what engendered our methodological puzzles in the first place. On the other hand, there are more complex and realistic conceptions of belief, but the cause of clarity is not served by using them. Therapeutic Bayesianism appears to offer the ideal compromise between accuracy and simplicity, enabling us to represent the issues starkly without neglecting the essential ingredients or clouding them with unnecessary details.[17]

NOTES

1. The axioms of elementary probability theory are as follows: (1) probabilities are less than or equal to one; (2) the probability of a necessary truth is equal to one; (3) if two statements are jointly impossible, then the probability that at least one of them is true is equal to the sum of their individual probabilities; and (4) the conditional probability of p given q equals the probability of the conjunction of p and q divided by the probability of q.

2. This project is attempted in my *Probability and Evidence* (Cambridge, 1982), henceforth abbreviated as *P&E*. The metaphilosophical outlook is inspired by Wittgenstein's *Philosophical Investigations*, paragraphs 88–133.

3. Bayesian programs of various kinds have been developed in the work of Rudolf Carnap, David Christensen, R. T. Cox, Bruno de Finetti, Ron Giere, I. J. Good, John Earman, Ellery Eells, Hartry Field, Allan Franklin, Ian Hacking, Mary Hesse, Jaakko Hintikka, Colin Howson, E. T. Jaynes, Richard Jeffrey, Harold Jeffreys, Mark Kaplan, J. M. Keynes, Henry Kyburg, Isaac Levi, Patrick Maher, Roger Rosencrantz, Wesley Salmon, L. J. Savage, Teddy Seidenfeld, Abner Shimony, Brian Skyrms, Patrick Suppes, Peter Urbach, Bas van Fraassen and others. Much of this work (especially the studies by Good, Hesse, Howson & Urbach, and Earman) contains contributions to therapeutic Bayesianism. However, I cannot attribute to these philosophers the project that I have in mind by that label, since their work is oriented towards the discovery of a 'theory of science', and thus reflects a metaphilosophical point of view that is quite distinct from that of the program which I am calling "therapeutic Bayesianism."

4. Stephen Spielman's objection is based on the mistake described here: the identification of the probabilities with objective proportions. (See his review of *P&E*, *Journal of Philosophy* [March 1984]: 168-73. Page references for Spielman are to this work).

To keep things relatively simple I have assumed that there are just *two* observations in question: namely, the discovery regarding a randomly selected raven that it is black and the discovery regarding a randomly selected non-black thing that it is not a raven. If we consider instead the discovery that a known black thing is a raven, or various other ways of seeing black ravens and non-black non-ravens, then the existence of confirmation depends on the presence of special additional background assumptions (e.g., that ravens are quite likely all to have the same color). Nonetheless a similar contrast between the degrees of confirmation provided by black ravens and non-black non-ravens may be established. In *P&E* I suggest that these other ways of seeing black ravens would provide *no* confirmation of the hypothesis. This is misleading. Sometimes our background theories include a belief in the projectibility of the generalization in question, and in that case all the ways of observing an instance of it will normally provide confirmation.

My treatment of the paradox is, in a couple of respects, different from Patrick Suppes's analysis ("A Bayesian Approach to the Paradoxes of Confirmation," in *Aspects of Inductive Logic*, edited by J. Hintikka and P. Suppes, [Amsterdam 1966]). In the first place, he does not distinguish between the discovery that a randomly selected object is a black raven and the discovery that a randomly selected raven is black; whereas it is a significant feature of my account that in certain circumstances only the latter datum would confirm the hypothesis. And secondly, he does not obtain his results from the basic principles of Bayesianism—the thesis that degrees of belief should conform to the probability calculus; rather, he starts with the assumption that surprising observations have greater confirmation power; and this, though correct, is much better derived than simply presupposed.

5. C. Howson and P. Urbach, *Scientific Reasoning: The Bayesian Approach* (La Salle, Ill., 1989).

6. An argument for associating simplicity with high prior probability is given in *P&E.*, 70–71.

7. Teddy Seidenfeld maintains that this account goes in the "wrong direction." But he gives no grounds for that claim other than to note the above-mentioned deficiencies in our understanding of simplicity—our inability to solve either the descriptive or the normative problems surrounding it. And it seems to me that his observation is irrelevant in the absence of any reason to believe either (a) that we can get a satisfactory explication of simplicity in terms of evidential diversity, or (b) that the Bayesian account would not withstand a better grasp of simplicity. (See his review of *P&E, Philosophical Review*, [July 1984].)

8. For a good assessment of this argument and various others see John Earman's *Bayes or Bust* (Cambridge, Mass., 1992). Bruno de Finetti perhaps deserves the credit for first having argued that degrees of belief *ought* to be 'coherent', i.e., conform to the probability calculus—though they *need* not be coherent if the believer is irrational ("Foresight: Its Logical Laws, Its Subjective Sources," translated in *Studies in Subjective Probability*, edited by H. E. Kyburg, Jr., and H. E. Smokler [New York, 1964]). I hesitate to credit Frank Ramsey's earlier paper (in *Foundations: Essays in Philosophy, Logic, Mathematics and Economics*, edited by D. H. Mellor [Atlantic Highlands, N.J., 1977]) with this result, since he defines "degrees of belief" in such a way that they *must* conform to the probability calculus. On Ramsey's account there is no room for the existence of someone who has degrees of belief that are not coherent.

9. For example, I. J. Good (in the *British Journal for the Philosophy of Science* 19 [1968]: 123–43) advocates:

$$\text{Weight of evidence concerning } H \text{ provided by } E = \log \frac{P(E/H)}{P(E/-H)}$$

And Seidenfeld (op. cit.), noting that on our account E might confirm both H_1 and H_2 yet disconfirm the conjunction $(H_1 \ \& \ H_2)$, suggests that confirmation cannot be defined in terms of probability alone.

10. A further complaint is that our definition of confirmation seems to go badly wrong when we apply it to measure the evidential value of *already known* data. For in that case $P(E) = 1$, therefore $P(H/E) = P(H)$. This problem for Bayesians was first posed by Clark Glymour (see his *Theory and Evidence* [Princeton, 1980]). It has been forcefully reiterated by James Woodward (in his review of *P&E, Erkenntnis* 23 (1985): 213–19) and treated thoroughly by John Earman (in *Bayes or Bust*). In order to deal with it we should

remember that the idea of the definition is to compare the credibility of a hypothesis, H, given the knowledge that E is true, with its credibility in the absence of such knowledge. Thus we should take the prior probability to be that which H would have had if the truth of E had not been discovered. Then, in order to assess E's confirmation power, we should consider what the absolute subjective probability of E would have been in that counterfactual situation, and also what the conditional probability of E given H would have been. Then we can employ Bayes's Theorem to calculate the factor by which the prior probability of H would have been increased. Doubtless, there is substantial indeterminacy in the assessment of these counterfactual probabilities. But this is no objection, since we generally have no reason to expect the magnitude of E's confirmation-power to be an especially determinate matter.

11. Rudolf Carnap, *Logical Foundations of Probability* (Chicago, 1962), 19, 31.

12. According to Spielman, this construal of Carnap is a "distorted caricature" (170), for "any careful reading of LFP [Logical Foundations of Probability] would show that Carnap never talks about 'objective relations of probabilification' or 'objective' relations of partial entailment" (170). Here I am at a loss to explain how Spielman could have arrived at his interpretation, and I can only refer the reader back to Carnap's work.

13. For further discussion of this point see *P&E*, 32–36 and 74–81 and Earman's *Bayes or Bust*, chapter 6.

14. Spielman (171) falls into this error, complaining that one cannot endorse logical probability and yet still assume that laws can have a non-negligible credibility.

15. Thus Woodward writes: "The principal defect of *Probability and Evidence* is its unsystematic character. Horwich does not give us a fully worked out general theory of confirmation but rather a series of essays which offer solutions to various particular puzzles, where the interconnections among these solutions are by no means always clear" (214).

16. This is starkly revealed in Seidenfeld's dismissal of therapeutic Bayesianism on the grounds that it is no substitute for a combination of excellent, but highly technical, foundational studies in decision theory and statistical inference by Jeffrey, Fishburn, and Lindley—works that hardly touch upon the traditional philosophical puzzles that form the domain of therapeutic Bayesianism. In a similar vein, Spielman is bothered by the "fail(ure) to see that the only difference between an 'objectivist' account [of the 'grue' problem] and a personalist account would be verbal: an objectivist would say that we *ought* to assign a much higher probability to H_1 than to H_2, and a subjectivist says that this is what intelligent informed people in fact do" (170). Spielman thinks the issue between them is 'merely verbal'.

17. I have greatly benefited from James Woodward's thorough and perceptive criticism. I would also like to thank Ned Block, Susan Brison, Josh Cohen, Marcus Giaquinto, Mark Kaplan, and Judith Thomson for helping me to improve earlier drafts of this paper.

Carnapian Inductive Logic
for a Value Continuum

BRIAN SKYRMS

One of the unfinished problems of Carnap's program for inductive logic was the construction of useful confirmation functions for the case where outcomes can take on a continuum of possible values. This is, as Carnap (1971) clearly realized,[1] requisite to application of inductive logic to a wide range of real scientific problems. By 1973 the problem had been completely solved in a way quite consonant with Carnapian techniques by Ferguson (1973) and Blackwell and MacQueen (1973).[2] Carnap never learned of the solution, although Ferguson was at his own university, because Ferguson, Blackwell, and MacQueen—mathematical statisticians—were working in a different tradition and a different intellectual community. It is ironic that Carnap—a life-long advocate of the unity of science—should be deprived in this way of knowledge of the progress of inductive logic.

It is not Carnap alone who was so deprived. Although these methods are important tools of non-parametric Bayesian statistics, most philosophers have no knowledge of them. Indeed, the current general feeling among philosophers of science seems to be that Carnap's program for the development of inductive logic is dead. I hope that this essay makes some small contribution to bridging the communications gap that appears to exist.

In fact Ferguson gives just the natural generalization of Carnap's inductive rules to the case of a value continuum. In order to set the stage so that the smoothness of the generalization is apparent, I will first describe Carnap's continua of inductive methods.

1. CARNAP'S CONTINUA

Suppose that we have an exhaustive family of k mutually exclusive categories, and a sample of size N of which n are of category F. Carnap (1950) originally proposed the following inductive rule, C^*, to give the probability that a new sampled individual, a, would be in F on the basis of the given sample evidence, e.

$$pr(Fa|e) = \frac{1 + n}{k + N}$$ C*

On the basis of no sample evidence, each category gets equal probability of $1/k$. As the sample grows larger, the effect of the initial equiprobable assignment shrinks and the probability attaching to a category approaches the empirical average in the sample, n/N. Soon Carnap (1952) shifted from this method to a class of inductive methods of which it is a member, the λ-continuum of inductive methods.

$$pr(Fa|e) = \frac{\lambda + n}{\lambda k + N}$$ λ-continuum

Here, again, we have initial equiprobability of categories and predominance of the empirical average in the limit with the parameter, $\lambda(\lambda > 0)$, controlling the rate at which the sample evidence swamps the prior probabilities. In his posthumous (1980) paper, Carnap introduced the more general λ-γ continuum:

$$pr(F_i a|e) = \frac{\lambda \gamma_i + n}{\lambda + N} \left(\sum_i \gamma_i = 1 \right)$$ λ-γ continuum.

The new parameters, $\gamma_i > 0$, allow unequal *a priori* probabilities for different categories. For Carnap these are intended to reflect different "logical widths" of the categories. The parameter, λ, again determines how quickly the empirical average swamps the prior probability of an outcome.

(Note that one could equivalently formulate Carnap's λ-γ rule this way: Take any k positive numbers, $b_1 \ldots b_k$, and let the rule be:

$$pr(F_i a|e) = \frac{b_i + n}{\sum_j b_j + N}$$ λ-γ continuum

where $\lambda = \sum_j b_j$ and $\gamma_i = b_i/\lambda$.)

We can think about the problem addressed by Carnap's inductive rules in the following way. The experiment is represented by a discrete random variable, taking as possible numerical values the integers 1 through k, according to whether the experimental result is $F_1 \ldots F_k$. This experimental result generates a measurable space, $S = \langle W, A \rangle$, where the points in W are the k possible outcomes and the propositions (measurable sets) in A are gotten by closing the atoms under Boolean combination.

Induction takes place when the experiment is iterated and a general analysis does not place any finite upper bound on the number of possible iterations. Thus we are led to consider an infinite sequence of such random variables, indexed by the positive integers. The relevant measurable space is a product of an infinite number of copies of the one-shot probability space, $S_1 \times S_2 \times S_3 \times \ldots$. The points in the infinite product space are infinite sequences, $w_1, w_2, \ldots$, of integers from 1 through k. In this setting we can give a fairly general definition of an inductive rule.

> An *inductive rule*, R, takes as input any finite initial sequence of results, $\langle w_1, \ldots, w_j \rangle$ and any proposition about the experimental outcome the next time around, a_{j+1} in A_{j+1} and outputs a numerical prediction in [0,1] such that: (i) for any finite outcome sequence, $\langle w_1, \ldots, w_j \rangle$, the rule gives a probability, $R(\langle w_1 \ldots w_j \rangle, \cdot)$ on the space for the next moment of time, $\langle W_{j+1}, A_{j+1} \rangle$ and (ii) For every proposition about the next outcome, a_{j+1}, $R(\cdot, a_{j+1})$ is measurable on the space of histories up until then, $S_1 \times \ldots \times S_j$.

> Note on (i): This includes empty sequences, so the rule must specify a prior probability on S_1.

> Note on (ii): This technical regularity requirement is automatically satisfied in the finite case considered in this section, but is required for more general applications.

Then an *inductive rule* determines a unique probability on the infinite product space, $S_1 \times S_2 \times \ldots$, according to which the probabilities given by the inductive rules are conditional probabilities (Neveu 1965, chap. 5). Thus, the properties of an inductive rule can be studied at either the "operational" level of a predictive rule or at the "metaphysical" level of a prior probability on an infinite product space.

The rules in Carnap's λ-γ continuum are all inductive rules in the foregoing sense. For each of these rules the corresponding probability on the infinite product space is *symmetric* or *exchangeable:* that is, invariant under finite permutations of outcomes. By de Finetti's theorem, any such probability has a unique representation as a mixture (or average) of probabilities which make the outcome random variables independent and identically distributed. That is to say that the Carnapian inductive logician behaves as if she were multinomial sampling. The examples of throwing a die with unknown bias or sampling with replacement from an urn of unknown composition are canonical.

2. THE CLASSICAL BAYESIAN PARAMETRIC VIEW OF CARNAP'S λ-γ CONTINUUM

One way of thinking about Carnap's rules is to take this representation at face value. There is a fixed statistical model of the chance mechanism with subjective prior uncertainty as to the true values of the parameters of the model. There is a chance setup (the die and the throwing mechanism) generating a sequence of independent and identically distributed random variables, but the distribution of the random variables (the bias of the die) is initially unknown. Inductive inference depends on this prior uncertainty *via* Bayes's theorem. The goal of inductive inference is to learn the true bias of the coin. This is the *parametric Bayesian* conceptualization of the problem.

From this point of view, Carnap's postulate that inductive methods should satisfy *the Reichenbach Axiom*—that the probabilities given by the inductive rule converge to the empirical average as the number of trials goes to infinity—is well motivated. By the strong law of large numbers, this is just the requirement that with chance equal to one, the inductive rule learns in the limit the true bias of the die, no matter what the true bias is. That is to say that Carnap's methods correspond to priors on the bias which are *Bayesian Consistent* with respect to the multinomial statistical model.

Not every consistent prior for multinomial sampling generates an inductive rule which is a member of Carnap's λ-γ continuum. Carnap's rules correspond to the class of *natural conjugate priors* for multinomial sampling: the Dirichlet priors. As the predictive rules which are generated by Dirichlet priors, Carnap's λ-γ methods are well known to Bayesian statisticians.[3]

3. THE SUBJECTIVE BAYESIAN VIEW OF CARNAP'S λ-γ CONTINUUM

For pure subjective Bayesians, talk of objective chances is strictly meaningless and the multinomial statistical model discussed in the last section is only an artifact of the de Finetti representation. Exchangeability is consistent with degrees of belief which give rise to inductive rules outside Carnap's continuum. Is there any subjective characterization of the λ-γ continuum?

There is. It is due to the Cambridge logician W. E. Johnson, whose pathbreaking work in inductive logic should be more widely known. Johnson introduced the concept of exchangeability or symmetry in 1924 (before de Finetti). We defined exchangeability in terms of the infinite product space induced by an inductive rule as invariance under finite permutations of trials. There is an equivalent formulation in terms of the inductive rule itself. That is that vector of frequency counts of the possible outcomes is sufficient to determine the probabilities for the next trial: $R(\langle w_1 \ldots w_j \rangle, \cdot) = R(\langle w_i' \ldots w_j' \rangle, \cdot)$ if the outcome sequences $\langle w_1 \ldots w_j \rangle$ and $\langle w_1' \ldots w_j' \rangle$ have the same frequency counts. For example, in rolling a die, the outcome sequences 1234561 and 1135264 would lead to the same probabilities for the eighth trial; order is presumed not to be relevant.

Johnson was led to postulate a stronger kind of sufficiency. That is that:

(i) The probability of an outcome on the next trial should only depend on the frequency that *it* has occurred in the preceding trials (for each fixed number of trials) and not on the relative frequencies of the other outcomes.

(ii) The dependence should be the same for all categories.

In the example of the die, (i) says that the probability of a one on the eighth trial given initial outcome sequences 1213654 and 1155555 should be the same;

(ii) adds that this probability should be equal to the probability of a 2 on the eighth trial given the initial sequence 1253266. Taken together (i) and (ii) are Johnson's *sufficientness postulate*.

If we have (1) exchangeability but not independence, (2) sufficientness in Johnson's sense, (3) the number of categories is at least three, and (4) the relevant conditional probabilities are all well defined, then we get Carnap's 1952 λ-continuum of inductive methods. If Johnson's (ii) is dropped from the foregoing, we get Carnap's λ-γ continuum.[4] Thus, from a purely subjective point of view, Carnap's continua correspond to strong symmetries in a predictor's degrees of belief.

4. BLACKWELL-MACQUEEN INDUCTIVE RULES

Carnap was interested in developing inductive logic to the point where it could make contact with mathematical physics. He saw the first important step to be to generalize his methods so that they could deal with the case where the outcome of an experiment could take on a continuum of values. We will accordingly change our canonical example from that of a die to that of a wheel of fortune with unit circumference. Repeated spins of the wheel produce as (ideal) outcomes, real numbers in the interval [0,1). We equip this outcome space with the metric corresponding to the shortest distance measured around the circumference of the circle.[5] We will take as our problem the specification of an inductive rule in this setting. The techniques which solve this problem, however, will apply very generally.[6]

A useful way to think about the problem is to think about how a Carnapian in possession only of Carnap's λ-γ continuum for finite numbers of outcomes could approximate a solution. The natural thing to try might be to partition the unit interval into a finite number of subintervals and apply a method from the λ-γ continuum taking the elements of the partition as outcomes. Finer and finer partitions might be thought of as giving better and better approximations.

Recall the reformulation of Carnap's λ-γ continuum of section 1. Take any k positive numbers, $b_1 \ldots b_k$ and let the rule be:

$$pr(F_i a | e) = \frac{b_i + n}{\sum_j b_j + N} \qquad \lambda\text{-}\gamma \text{ continuum}$$

(where $\lambda = \sum_j b_j$ and $\gamma_i = b_i / \lambda$.) Thus, for a partition into k subintervals, we have a class of Carnapian inductive rules whose members are specified by k parameters, $b_i \ldots b_k$.

In the case of a continuum of possible outcomes, the appropriate parameter will be a non-negative bounded measure, α, defined on the Borel algebra of the unit interval. As a generalization of Carnap's condition that the $b_i s$ should be positive in the finite case, we will require that the measure, α, be absolutely continuous with respect to Lebesgue measure on the unit interval. (Thus, in particular, every interval of positive length gets positive measure.) For any Borel set of the unit interval, O, and evidence, e, consisting of n points in

O in N trials, we will take the inductive rule with parameter, α, to give the probability of an outcome in O in the next trial as

$$pr(O|e) = \frac{\alpha(O) + n}{\alpha([0,1)) + N} \qquad \text{Blackwell-MacQueen}$$

We will call this the Blackwell-MacQueen inductive rule, since Blackwell and MacQueen used this formula in generalizing Polya urn models for what philosophers know as Carnap's λ-γ inductive rules, to more general Polya urn models for non-parametric Bayesian inference.

It can be seen that the Blackwell-MacQueen rules are consistent with Carnap's λ-γ continuum in the following sense. For a Blackwell-MacQueen rule and a "coarse-graining" of outcomes according to which member of a finite partition they fall into, the induced inductive rule for the finite partition is a member of Carnap's λ-γ continuum. For a simple illustration, let the parameter, α, for the Blackwell-MacQueen rule, just be Lebesgue measure. In particular, if I is a subinterval of $[0,1)$, then $\alpha(I)$ is just the length of I. Partition the unit interval into k equal subintervals. Then for some fixed one of these subintervals, I, let n be the number of sample points in I and N be the total number of sample points. The Blackwell-MacQueen rule gives the probability that the next point will fall into I as

$$Pr(I|e) = \frac{\frac{1}{k} + n}{1 + N}$$

which is a member of Carnap's λ-continuum. (Notice that as the partitions get finer the values of λ get proportionally smaller to preserve consistency with the Blackwell-MacQueen rule.) It should be clear that this class of inductive rules is the natural generalization of Carnapian rules to problems where the outcomes can be represented as real numbers in the unit interval.

Blackwell-MacQueen rules are *Inductive Rules* in the sense made precise in section 1. Thus they induce a probability measure on the infinite product space. This probability measure makes the random variables which represent the experimental outcomes exchangeable. By de Finetti's theorem, it can be represented as a mixture of probabilities which make the trials independent and identically distributed (henceforth abbreviated as IID). From a classical Bayesian viewpoint, the mixing measure is the ignorance prior over the true chances governing the IID process. The IID probabilities correspond to a distribution on $[0,1)$. The prior corresponds to a distribution on the distributions on $[0,1)$.

5. THE CLASSICAL BAYESIAN NON-PARAMETRIC
VIEW: FERGUSON DISTRIBUTIONS

One might try to generalize Carnap's inductive methods in a different way—by working at the level of the de Finetti priors rather than at the level of inductive rules. As noted in section 2, the members of Carnap's λ-γ continuum are just those rules which arise from multinomial sampling with Dirichlet priors. The natural generalization of a Dirichlet prior is a Ferguson distribution (called by Ferguson and also known as the Dirichlet process). A *Ferguson distribution* with parameter α is a distribution which for every k member partition of the interval, $P_1 \ldots P_k$ is distributed as Dirichlet with parameters, $\alpha(P_1) \ldots \alpha(P_n)$.[7]

The good news for Carnap's program is that both roads lead to the same place. The de Finetti prior distribution corresponding to the probability on the infinite product space induced by a Blackwell-MacQueen inductive rule with parameter α is a Ferguson distribution with parameter α.[8]

Furthermore, just as the class of Dirichlet probabilities is closed under multinomial sampling, the class of Ferguson distributions is closed under (IID) sampling of points in [0,1). Consider a finite sample sequence consisting of data points $x_1, \ldots, x_n$ and let $\delta_1 \ldots \delta_n$ be probability measures giving mass one to the points $x1, \ldots, xn$, respectively. Let the prior be given by a Ferguson distribution with parameter α. Then conditioning on the data points takes one to a posterior which is a Ferguson distribution with parameter α', where:

$$\alpha' = \alpha + \sum_{i=1}^{n} \delta_i.$$

From the classical Bayesian viewpoint, Ferguson distributions are natural conjugate priors for this non-parametric sampling problem in just the same way that Dirichlet priors are natural conjugate priors for multinomial sampling.

Blackwell-MacQueen Inductive Rules satisfy a version of *Reichenbach's axiom*. As evidence accumulates, the probability for the next trial is a weighted average of the a priori probability and the empirical relative frequency probability, with all weight concentrating on the empirical relative frequency probability as the number of data points goes to infinity. Can we say in the classical Bayesian setting that Blackwell-MacQueen inductive rules will—with chance one—learn the true chances?

In this case "learning the true chances" is not just learning the values of a finite number of parameters as in the multinomial case, but rather learning the true chance probability on [0,1). Thus we need a sense of convergence for the space of all probability measures on [0,1). We will say that a sequence of probabilities measures, μ_n, converges to measure iff μ for every bounded continuous[9] function of [0,1), its expectation with respect to the measures μ_n converges to its expectation with respect to μ. (This is convergence in the weak* topology.)

A prior (or alternatively, the corresponding inductive rule) is Bayesian-consistent with respect to a chance probability μ, if with probability one in μ, the posterior under IID (in μ) sampling will converge weak* to μ. A prior is Bayesian-consistent if it is Bayesian-consistent for all possible chance probabilities (Freedman [1963], Fabius [1964], Diaconis and Freedman [1986]).

In the multinomial case, all the rules of Carnap's λ-γ continuum (alternatively, the corresponding Dirichlet priors) are Bayesian-consistent.

The Blackwell-MacQueen rules as defined in section 4 are also Bayesian-consistent in this sense.[10] It should be noted that this is a consequence of a restriction that I put on the parameter, α, of Blackwell-MacQueen inductive rules; that is that α be absolutely continuous with respect to Lebesgue measure on the unit interval. I take this to be a natural generalization of Carnap's requirement of regularity (or strict coherence) in the finite case.[11] And I take Bayesian consistency to be a way of capturing the basic motivation for Reichenbach's axiom.

In summary, if we adopt the classical Bayesian position which embraces objective chances, the status of the Blackwell-MacQueen inductive rules is quite analogous to that of the Carnap rules in the finite case. They arise from a class of conjugate priors for non-parametric problems—the Ferguson distributions. And they are Bayesian-consistent.

6. SUBJECTIVE BAYESIAN ANALYSIS OF BLACKWELL-MACQUEEN INDUCTIVE RULES

In this section we return to the subjective point of view. The notion of random variables which are independent and identically distributed according to the true unknown chances gives way to the subjective symmetry of exchangeability. In the case where the random variables take on a finite number of values, an additional symmetry assumption—W. E. Johnson's sufficientness postulate—(together with a few other technical assumptions) gets us Carnap's λ-γ continuum. In the case under consideration, where our random variables can take on a continuum of values in $[0,1)$, we have the analogous result. That is—roughly speaking—that Exchangeability + Sufficientness gives the Blackwell-MacQueen inductive rules.

Let us first review the case of random variables, taking a finite number of values discussed in section 3 in a little more detail. This discussion is based on Zabell (1982). Suppose that we have an infinite sequence of random variables, $X_1, X_2, \ldots$, each taking values in a finite set $O = \{1, \ldots, k\}$. And suppose that the number of possible outcome values, k, is at least three. We consider a number of conditions on our probabilities:

(1) *Exchangeability:* This guarantees by de Finetti's theorem that our (degree of belief) probability can be uniquely represented as a mixture of probabilities which make the outcomes independent and identically distributed. (2) *Non-Independence:* If our beliefs make the outcomes independent, then we will not learn from experience. (3) *Strict Coherence*: The probability of any finite outcome sequence is non-zero. This is a kind of open-mindedness condition. It guarantees that all the conditional probabilities in the next condition are well defined. (4) *Generalized Sufficientness*: Let n_i be the frequency count of outcomes in category i in the trials $X_1, \ldots X_n$. Then:

$$Pr(X_{N+1} = i | X_1, \ldots, X_N) = f_i(n_i).$$

That is, for each category, i, the probability of the next outcome being in that category is a function only of the frequency count for that category in the preceding sequence of observations. (Johnson assumed that these functions would be the same for all categories, but that is not assumed here.) Zabell shows that under assumptions 1–4, the predictive conditional probabilities

$$Pr(X_{N+1} = i | X_1, \ldots, X_N)$$

are just those given by the inductive rules of Carnap's λ-γ continuum.

Now consider the case of the wheel of fortune, where an infinite sequence of random variables takes on values in the interval $[0,1)$. We assume (1) *Exchangeability* and (2) *Non-independence* as before. As (3*) *Regularity* we assume that for any measurable set, B, which has non-zero Lebesgue measure and for any finite sequence of observations, $X_1, \ldots, X_n$:

$$Pr(X_{N+1} \in B | X_1, \ldots, X_N) > 0.$$

As (4*) *Generalized Sufficientness* we require that for any measurable set, B_i, and any finite sequence of observations, the probability that the next observation falls in B_i is a function only of the count, n_i, of previous observations that have fallen in B_i:

$$Pr(X_{N+1} \in B_i | X_1, \ldots, X_N) = f_i(n_i).$$

If 1, 2, 3*, and 4* are fulfilled, then our predictive conditional probabilities are Blackwell-MacQueen inductive rules.

This is almost immediate from the finite case. Here is a sketch of an argument. If 1, 2, 3* and 4* are fulfilled, then for any finite partition of the unit interval whose members have non-zero Lebesgue measure, 1, 2, 3, 4 hold. Then for any such partition with three or more members, Zabell's version of W. E. Johnson's result holds. Thus the probabilities of falling in members of the partition update according to Carnap λ-γ rules. This must also be the case for even partitions of only two members, which can be seen by subdividing them into partitions of four members. Then the de Finetti priors for all these partitions must be distributed as Dirichlet, and the de Finetti prior for $[0,1)$ must be a Ferguson distribution. Thus the inductive rules induced by the prior must be Blackwell-MacQueen inductive rules.

7. CONCLUSION

Carnap would not have thought that inductive logic for a value continuum ended with the class of Blackwell-MacQueen inductive rules but rather that it started there. These rules are just right for those settings which characterize them—where 1, 2, 3* and 4* hold.[12] But in some contexts they will not all hold.

Ferguson (1974) considers the 4* as a drawback of the use of Dirichlet Processes (Ferguson distributions):

> One would like to have a prior distribution for P with the property that if X is a sample from P and X=x, then the posterior guess at P gives more weight to values close to x than the prior guess at P does. For the Dirichlet process prior, the posterior guess at P gives more weight to the point x itself, but it treats all other points equally. In particular, the posterior guess at P actually gives less weight to points near x but not equal to x.[13]

Carnap makes exactly the same point in his last writings on inductive logic under the heading of the problem of *analogy by similarity*. Where it is desirable that sample x gives more weight to values close to x, W. E. Johnson's Sufficientness postulate must be given up. Perhaps the simplest way to do this in the finite case is to move from Dirichlet priors to mixtures of Dirichlet priors. These lead to "hyperCarnapian" inductive methods which can indeed exhibit "analogy by similarity" effects.[14] Carnap also discussed a different kind of analogy which his methods could not represent as the problem of *analogy by proximity*. This is the problem of taking into account temporal patterns in the data, for instance, in inference about Markov chains. In such contexts, exchangeability must give way to weaker subjective symmetries. Nevertheless, in the finite case, methods which are essentially Carnapian can incorporate such sensitivity to order.[15] Inductive logic for a value continuum should address these problems of analogy, both singly[16] and in combination. Progress in this area will be facilitated by a realization of the inseparability of philosophy of science and science itself.

NOTES

I would like to thank Persi Diaconis for introducing me to the Dirichlet process, Richard Jeffrey for discussion, and Sandy Zabell and Peter Vanderschraaf for comments on an earlier version of this paper.

1. Carnap (1971, 50 ff.).

2. Building on earlier work of Freedman (1963) and Fabius (1964).

3. See Good (1965), DeGroot (1970, chap. 5). Carnap's (1952) λ continuum is generated by the symmetric Dirichlet priors and his (1950) confirmation function, c*, by the uniform prior.

4. See Zabell (1982) for rigorous proofs and additional historical information.

5. Thus our outcome space is a complete separable metric space. For completeness, note that the Cauchy sequence 1/2, 3/4, 7/8, . . . converges to the point 0.

6. The methods to be described in the following generalize to any standard Borel space. Thus, for example, our experimental results could be n-dimensional vectors in R^n.

7. A Ferguson distribution is a distribution over random chance distributions for a random chance probability, p. The parameter, α, of the Ferguson distribution can be any finite, non-null measure on [0,1). Thus for any finite measurable partition, $\{P_1, \ldots, P_n\}$ of [0,1), there is a corresponding vector of numbers, $\langle \alpha_1, \ldots \alpha_n \rangle$ where $\alpha_i = \alpha(P_i)$. The requirement

is then that for any such partition, the random chance probability vector for members of the partition, $\langle p(P_i), \ldots, p(P_n) \rangle$, has a Dirichlet distribution with parameter $\langle \alpha_i, \ldots, \alpha_n \rangle$.

8. In fact, the paper of Blackwell and MacQueen was written to give a simple proof of the existence of Ferguson distributions.

9. We mean continuous with respect to the metric gotten by measuring around the circle here, so the continuous function must be bounded, but we state the definition in a more generally applicable form.

10. This follows from results of Fabius (1964) and Blackwell and MacQueen (1973) since Ferguson priors are special cases of the "tail-free" priors of Fabius. For an overview see Diaconis and Freedman (1986).

11. See Carnap (1971, section 7).

12. It is also worth noting that Blackwell-MacQueen rules are intimately related to Bayesian inductive rules for "sampling of species" problems, where the total number of categories is unknown. In this regard, see Zabell (1992) and Hoppe (1984).

13. Ferguson (1974, 622). Note that in Ferguson's characterization 3 on that page, Tl, T2, and T3 are eliminated by my non-independence and regularity conditions.

14. I. J. Good has long been an advocate of use of mixtures of Dirichlet priors. See Good (1965, 1979). For the use of mixtures of Dirichlet priors to implement Carnapian "analogy by similarity," see Skyrms (forthcoming).

15. See Skyrms (1991).

16. The use of mixtures of Dirichlet priors to construct inductive methods sensitive to "analogy by similarity" can be generalized to the case of a value continuum. Here mixtures of Ferguson distributions can be used to implement "analogy by similarity." The study of mixtures of Ferguson distributions was initiated by Antoniak (1974).

REFERENCES

Antoniak, C. 1974. "Mixtures of Dirichlet Processes with Applications to Bayesian Non-parametric Problems." *Annals of Statistics* 2: 1152–74.

Blackwell, D., and J. B. MacQueen, 1973. "Ferguson Distributions via Poyla Urn Schemes." *Annals of Statistics* 1: 353–55.

Carnap, R. 1950. *Logical Foundations of Probability*. Chicago. (Second edition 1962.)

Carnap, R. 1952. *The Continuum of Inductive Methods*. Chicago.

Carnap, R. 1971. "A Basic System of Inductive Logic, Part 1." In *Studies in Inductive Logic and Probability: Volume I*, edited by R. Carnap and R. Jeffrey, 35–165. Berkeley.

Carnap, R. 1980. "A Basic System of Inductive Logic, Part 2." In *Studies in Inductive Logic and Probability: Volume II*, edited by R. C. Jeffrey. Berkeley.

DeGroot, M. 1970. *Optimal Statistical Decisions*. New York.

Diaconis, P., and D. Freedman. 1986. "On the Consistency of Bayes Estimates." *Annals of Statistics* 14: 1–26.

Fabius, J. 1964. "Asymptotic Behavior of Bayes Estimates." *Annals of Mathematical Statistics* 35: 846–56.

Ferguson T. 1973. "A Bayesian Analysis of Some Non-Parametric Problems." *Annals of Statistics* 1: 209–30.

Ferguson, T. 1974. "Prior Distributions on Spaces of Probability Measures." *Annals of Statistics* 2: 615–29.

Freedman, D. 1963. "On the Asymptotic Behavior of Bayes' Estimates in the Discrete Case." *Annals of Mathematical Statistics* 24: 1386–1403.

Good, I. J. 1965. *The Estimation of Probabilities: An Essay on Modern Bayesian Methods*. Cambridge, Mass.

Good, I. J. 1979. "Some History of Hierarchical Bayesian Methodology," In *Bayesian Statistics* [Proceedings of the International Meeting on Bayesian Statistics 1979, Valencia, Spain], edited by J. M. Bernardo et al., 489–519. Valencia. Reprinted in *Good Thinking: The Foundations of Probability and its Applications*, 95–105. Minneapolis, 1983.

Hoppe, F. 1984. "Polya-like Urns and the Ewens Sampling Formula." *Journal of Mathematical Biology* 20: 91–94.

Johnson, W. E. 1924. *Logic, Part III: The Logical Foundations of Science*. Cambridge.

Johnson, W. E. 1932. "Probability: The Deductive and Inductive problems." *Mind* 49: 409–23.

Neveu, J. 1965. *Mathematical Foundations of the Theory of Probability*, translated by A. Feinstein. San Francisco.

Skyrms, B. 1991. "Carnapian Inductive Logic for Markov Chains." *Erkenntnis* 35: 439–60.

Skyrms, B. Forthcoming. "Analogy by Similarity in HyperCarnapian Inductive Logic." In *Philosophical Problems of the Internal and External Worlds*, edited by Earman et al. Pittsburgh.

Zabell, S. L. 1982. "W. E. Johnson's 'Sufficientness' Postulate." *Annals of Statistics* 10: 1091–99.

Zabell S. L. 1992. "Predicting the Unpredictable." *Synthese* 90: 205–32.

How to Defend a Theory Without Testing It: Niels Bohr and the "Logic of Pursuit"

PETER ACHINSTEIN

1. THE "LOGIC OF PURSUIT"

Sometimes a scientist presents a new theory without attempting to argue that it is true or even probable. Indeed, no tests of the theory may be reported or made. Or if they are, the results do not establish the theory or give it substantial support. Instead the aim is to show that the theory is *reasonable to pursue*; i.e., that it is reasonable to try to work out consequences of the theory, apply it to more complex systems, add or reformulate assumptions, devise possible ways to test it, and so forth. Doing this is important because when a theory is first proposed the scientific community needs to know whether research on it should be encouraged, and whether tests should be planned and conducted. Funding agencies with limited resources need to know whether proposals to investigate the theory should be supported. And scientists need to know whether it is reasonable to invest their own time and energy in its pursuit.

To show these things a theorist goes beyond simply stating basic assumptions of the theory. Use is made of some form of argumentation to arrive at the conclusion that the theory is a reasonable one to pursue. Some philosophers of science speak of this as a "logic of pursuit," others as a "logic of discovery." It was of particular concern to Charles Peirce at the turn of this century, to N.R. Hanson in the 1950s and 1960s, and to several philosophers since then influenced by Peirce and Hanson. Various claims have been made by one or more of these authors about the sort of reasoning scientists employ to show that a new theory is reasonable to pursue. Among the claims are these:

1. The reasoning is weaker than that which establishes a theory or shows it probable.[1]

Peirce writes:

Deduction proves that something *must* be; Induction shows that something *actually is* operative; Abduction [the sort of reasoning Peirce thinks is applicable here] merely suggests that something *may be*.[2]

Hanson quotes this passage approvingly.[3] John Stuart Mill discusses reasoning of a kind later stressed by Peirce and Hanson, and he rejects the idea that such reasoning establishes a theory or shows that it is probable. Nevertheless, Mill claims,

. . . it may be very useful by suggesting a line of investigation which may possibly terminate in obtaining real proof.[4]

It can tell us "what cause *may* have produced the effect."[5]

> 2. *The reasoning occurs before experimental tests have shown the theory to be probable or improbable.*

This is not a corollary of the first claim, which is concerned with the strength of the reasoning. The present claim indicates when the reasoning occurs. Peirce asserts that it occurs in the very act of discovery when the scientist's idea first emerges:

The abductive suggestion comes to us like a flash. It is an act of *insight*, although of extremely fallible insight. It is true that the different elements of the hypothesis were in our minds before; but it is the idea of putting together what we had never before dreamed of putting together which flashes the new suggestion before our contemplation.[6]

Despite this passage, Martin Curd interprets Peirce to be speaking of reasoning involving a "methodological appraisal of hypotheses after they have been generated but before they have been tested."[7] In what follows, I will suppose simply that at some point—whether during the thought process culminating in the discovery of the idea, or afterwards but before the theory has been experimentally tested and shown to be probable or improbable—scientists may engage in reasoning which concludes that the theory is reasonable to pursue. This is not to suggest that at this point in the theory's history there is a complete absence of experimental results. There may be results that render probable some of the assumptions being made. If so, however, they do not do so for the most central assumptions or for the theory as a whole.

> 3. *The reasoning is governed by a "logic."*

Exactly what this means is not made clear. It suggests reasoning subject to some universal principles whose validity or applicability does not vary from one scientist, theory, or context of reasoning, to another. It also suggests that the reasoning has some general form that can be represented schematically. The form suggested by Peirce—"retroduction" or "abduction" is this:

R_1: The surprising fact C is observed
But if A were true C would be a matter of course
Hence, there is reason to suspect that A is true. (5.189)

Peirce makes it clear that the second premise is concerned with the idea of explanation: if hypothesis *A* were true then fact *C* would be explained.[8]

Hanson on different occasions suggests different forms for retroduction, all of which, however, explicitly involve explanation. Here is one:

R_2: 1. Some surprising, astonishing phenomena $p1, p2, \ldots$ are encountered.
2. But $p1, p2, p3 \ldots$ would not be surprising or astonishing if *H* were true—they would follow as a matter of course from *H*; *H* would explain $p1, p2, p3, \ldots$.
3. Therefore there is good reason for elaborating *H*—for proposing it as a possible hypothesis from whose assumption $p1, p2, p3 \ldots$ might be explained.[9]

Note that the conclusion of Hanson's R_2 is somewhat different from that of Peirce's R_1.[10] Martin Curd, who defends the idea of a logic of discovery, distinguishes between a "logic of probability" and a "logic of pursuit."[11] The former governs reasoning to the conclusion that a theory is (highly) probable, the latter governs reasoning to the conclusion that a theory is worth pursuing, working on, or taking seriously. He construes Peirce to be proposing a "logic of pursuit" rather than a "logic of probability." This seems plausible since Peirce rejects the idea that retroductive reasoning renders an hypothesis (highly) probable, and he speaks of such reasoning in connection with the question of when an hypothesis warrants examination.[12]

4. *The reasoning may involve an appeal to certain general methodological criteria in addition to explanation or deduction of phenomena, e.g., "simplicity" and "testability." Or to put the same point differently, such general constraints on the explanation or derivation may be imposed.*

Peirce, e.g., notes that numerous hypotheses may explain the phenomena. To determine which of these hypotheses to consider Peirce invokes two rules. The first appeals to simplicity:

> In those subjects [physics and psychology], we may, with great confidence, follow the rule that one of all admissible hypotheses which seems the simplest to the human mind ought to be taken up for examination first. Perhaps we cannot do better than to extend this rule to all subjects where a very simple hypothesis is at all admissible.

Peirce's second rule invokes testability (in a strong sense which requires an actual means of testing):

> This remark at once suggests another rule, namely, that if there be any hypothesis which we happen to be well provided with means for testing, or which, for any reason, promises not to detain us long, unless it be true, that hypothesis ought to be taken up early for examination. (6.533)

Thus Peirce regards simplicity and testability, along with explanation of phenomena, as conditions that contribute to rendering an hypothesis reasonable to examine.

Other such criteria are frequently invoked by those who discuss explanatory reasoning. William Whewell, e.g., cites "consilience," the idea that the theory should "explain and determine cases of a *kind different* from those which were contemplated in the formation of our hypothesis."[13] Now Whewell thinks that if a theory is consilient, this will "impress us with the conviction that the truth of our hypothesis is certain."[14] That such a conviction is justified is emphatically denied by Mill, who argues that even if a theory explains all the relevant observed phenomena, where these are as different as you like from each other and from those that prompted the theory in the first place, you cannot infer that the theory is true or probable, but only that it may be true, or that it is reasonable to pursue. Accordingly, with respect to the kind of reasoning at issue in this paper, Mill is closer to Peirce and Hanson than Whewell is.[15]

   5. The reasoning may also involve an appeal to the practical costs of pursuit, such as time, money and energy.

Practical costs may make the pursuit of a given theory unreasonable even if that theory explains various phenomena and satisfies other general methodological criteria. Peirce, in particular, notes this practical consideration in a number of passages in which he discusses "the economy of research":

> Having a certain fund of energy, time, money, etc., all of which are merchantable articles to spend upon research, the question is how much is to be allowed to each investigation; and *for us* the value of that investigation is the amount of money it will pay us to spend upon it. (1.122)

> Proposals for hypotheses inundate us in an overwhelming flood, while the process of verification to which each one must be subjected before it can count as at all an item, even of likely knowledge, is so very costly in time, energy, and money . . . that Economy would override every other consideration even if there were any other serious considerations. In fact there are no others. For abduction commits us to nothing. It merely causes a hypothesis to be set down upon our docket of cases to be tried. (5.602)

> Indeed, the two general methodological criteria Peirce mentions in the previous point—simplicity and testability—might for him be justified solely by reference to the practical costs of pursuit: simple hypotheses and those we have actual means of testing require less energy, time, and money to pursue. Alternatively, for Peirce these methodological criteria, as well as others noted earlier, have some intrinsic value and are not to be justified solely in terms of the practical costs of pursuit. In either case, he seems to be defending an account of pursuit which considers not just intellectual benefits such as explanatory power, but practical costs as well.[16]

It is possible that Peirce reserved the terms "retroduction" and "abduction" exclusively for inferences of form R_1 in which only explanation is invoked. On this interpretation, to determine whether a theory is reasonable to pursue one considers retroduction *together with* other methodological criteria.[17] However, Peirce does say that "the rules of scientific abduction ought to be based exclusively upon the economy of research" (7.220, note 18). And he clearly believes that various criteria in addition to explanation are relevant for determining which theory should be examined. Accordingly, Peirce's schema R_1 might be formulated as follows, so as to take into account the various factors he introduces:

> R_3: T if true would correctly explain $O_1, \ldots, O_n$
> $\quad$ T has other general methodological virtues (e.g., simplicity and testability)
> $\quad$ Practical costs of pursuing T are reasonable
> $\quad$ Hence
> $\quad$ It is reasonable to pursue T

What if there are two theories T_1 and T_2 both of which satisfy R_3, but T_1 explains more phenomena, has more methodological virtues, and lower practical costs of pursuit, than T_2? And suppose that the costs of pursuing both theories together are too high. In view of this we might conclude, contrary to R_3, that it is not reasonable to pursue the inferior theory. To this a defender of R_3 can reply by saying that while it is reasonable to pursue each, it is more reasonable to pursue T_1 than T_2, or it is reasonable to pursue T_1 before T_2. What may not be reasonable is *to pursue T_2 at the cost of not pursuing T_1*. But from "it is reasonable to pursue T_2" it does not follow that the emphasized action is reasonable.

> 6. *Finally, the reasoning is different in kind from that employed to establish a theory or show that it is probable.*

For Peirce, retroduction is involved in devising a theory, induction in testing it. He classifies inductions into three sorts, all of which he regards as different from retroductions. In the present context he is concerned with "qualitative" induction:

> Induction consists in starting from a theory, deducing from it predictions of phenomena, and observing those phenomena in order to see *how exactly* they agree with the theory. The justification for believing that an experiential theory which has been subjected to a number of experimental tests will be in the near future sustained about as well by further such tests as it has hitherto been, is that by steadily pursuing that method we must in the long run find out how the matter really stands. (5.170)

Again:

> Induction is the experimental testing of a theory. . . . Abduction consists in studying facts and devising a theory to explain them. . . . Abductive

and Inductive reasoning are utterly irreducible, either to the other or to Deduction. . . . (1.145–1.146)

Induction tests a theory by deriving new predictions from it and then determining by experiment whether they hold. To the extent that they do one is justified in believing the theory, especially as such tests are multiplied. Abduction explains what is already observed. To the extent that it does, one is justified not in believing the theory, but in pursuing it, especially if it is simple and readily testable. And abduction and induction are not mutually reducible. These passages in Peirce are noted approvingly by Hanson, who also distinguishes hypothetico-deductive reasoning which can establish a theory from retroductive reasoning which cannot.[18]

These six claims represent what I take to be a standard position on reasoning used by scientists before testing a theory to show that the theory is reasonable to pursue. Whether it is reasonable to pursue a theory depends, of course, on the aims of the pursuit, which can vary considerably. One might pursue a theory solely for the purpose of criticizing it or solely to understand its historical significance. It should be obvious from the previous discussion that these are not the aims which defenders of a "logic of pursuit" have in mind. They are discussing pursuit from the perspective of developing a theory that has what they regard as *desirable scientific virtues*, such as truth, explanatory power, and simplicity. Accordingly, in what follows in speaking of arguments purporting to show that some theory is reasonable to pursue, these are the sorts of aims with respect to which reasonableness of pursuit is to be judged.

Is the theory of pursuit represented by the previous claims adequate? I propose to critically test it by examining reasoning employed by Niels Bohr during the second decade of this century in connection with his theory of atomic structure. Bohr, I will argue, was presenting reasons to pursue his theory which he did not regard as establishing it or showing it to be probable. Is his reasoning adequately reflected in the claims just noted? In sections 2–4 I will discuss Bohr's reasoning, and then in sections 5–7 consider to what extent it conforms to the previous theory of pursuit.

2. BOHR'S FIRST QUANTUM THEORY

In July, September, and November of 1913 Bohr published his ground-breaking trilogy "On the Constitution of Atoms and Molecules,"[19] which develops the first successful quantum theory of the atom. He begins with Rutherford's model of the atom, according to which atoms are composed of a positively charged nucleus, where most of the mass of the atom is concentrated, surrounded by a system of negative electrons subject to attractive forces from the nucleus. The existence of such relatively massive nuclei, Bohr notes, is demonstrated by experiments on alpha particle scattering conducted by Rutherford, Geiger, and Marsden.

However, Bohr continues, Rutherford's model has serious difficulties. Suppose we consider a simple system containing a small positively charged nucleus and an electron in some closed stationary orbit around it. From classical electrodynamics, an orbiting and thus accelerating electron, being charged, will radiate energy. This will produce a continual decrease in the radius of the orbit, and hence a continuous increase in the mechanical frequency of the electron, resulting in a continuous increase in radiation emitted from the atom. But, Bohr notes, this presents various problems. First, as the radius of the electron's orbit decreases the total energy of the atom, which is inversely proportional to the negative of the radius r, decreases. If r becomes very small so does the energy of the atom; and thus the energy emitted becomes very large. Bohr claims that a simple calculation shows that the energy emitted "will be enormously great compared with that radiated out by ordinary molecular processes" (p. 164).

Second, he notes that real atoms "in their permanent states seem to have absolutely fixed dimensions and frequencies," not changing ones which continuous radiation would require. Third, he points out that in the case of known molecular processes in which radiation is emitted, after a certain amount of energy is radiated the systems settle into a stable state of equilibrium in which the distances between the particles are the same order of magnitude as before the process.

To solve these problems Bohr proposes to introduce some radically new assumptions:

1. In an atomic system radiation is not emitted or absorbed in a continuous way (assumed in classical electrodynamics), but only when the atomic system passes from one "stationary" state to another.

2. The "stationary" states are governed by laws of classical mechanics; but these laws do not hold when the atom passes between different "stationary" states.

3. During the transition between "stationary" states the radiation emitted is homogeneous, i.e., it has a unique frequency and wavelength. Moreover, it is quantized in such a way as to satisfy the relationship $E = h\nu$, where E = total energy emitted, ν = frequency of the radiation, and h is Planck's constant.[20]

4. In an atom consisting of an electron rotating around a positive nucleus, the ratio between the total energy emitted during the formation of the configuration and the frequency of revolution of the electron is a whole multiple of $h/2$. (If the orbit is circular, this is equivalent to assuming that the electron's angular momentum is quantized and is a whole multiple of $h/2\pi$.)

These are not the only assumptions Bohr makes, but they are the most important and the most radically different from those of any previous theory.

With assumptions (1) and (2), Bohr by fiat avoids the problems in the Rutherford model associated with continuous energy radiation. But he is able to do much more than this with his assumptions, for example:

(a) Derive values for certain observed quantities associated with the atom.

From his quantization assumptions (3) and (4), together with experimentally determined values for the charge of the electron, the ratio of its charge to its mass, and Planck's constant, Bohr derives specific values for the following quantities: (i) the radius of an electron's orbit (an orbit produced when an electron is bound and assumed, for simplicity, to be circular); (ii) the frequency of revolution of the electron in this orbit; (iii) the ratio of the binding energy to the charge of the electron. He notes with satisfaction that these values have the same order of magnitude as previously accepted values, respectively, for linear dimensions of atoms, optical frequencies of radiation emitted by atoms, and ionization potentials (the voltage that will ionize the atom).

    (b) Derive laws governing various series of lines in the spectrum of hydrogen.

When light from hydrogen is analyzed using a spectroscope it is seen to consist of a series of sharp lines of definite wavelengths. In the visible region of the hydrogen spectrum these lines satisfy the Balmer formula

(1) $$\frac{1}{\lambda_n} = R\left(\frac{1}{4} - \frac{1}{n^2}\right)$$

where R is a constant, and $n = 3, 4, 5, 6, \ldots$ for each of the lines. From his assumptions (1)–(4) Bohr derives the formula

(2) $$\nu = \frac{2\pi^2 m e^4}{h^3}\left(\frac{1}{n_2^2} - \frac{1}{n_1^2}\right)$$

in which $\nu = $ the frequency of the radiation emitted by the hydrogen atom as it passes from stationary state n_1 to n_2; $m = $ the mass of the orbiting electron; $e = $ the charge of the electron; $h = $ Planck's constant. Equation (2) is similar in form to (1). (They can be given exactly the same form by relating wavelength λ to frequency ν by $\nu = c/\lambda$, since wavelength = speed/frequency, and the speed of light is a constant c.) Bohr notes that if $n_2 = 2$ and n_1 varies we derive the Balmer formula (1). If $n_2 = 3$ and n_1 varies we derive the Paschen series in the infrared region of the hydrogen spectrum. And if $n_2 = 1$ and $n_1 = 4, 5, \ldots$ we obtain a series in the ultraviolet, which, Bohr notes, was not at that time observed but might be expected. However, perhaps the most striking agreement with what had been empirically determined was the value for the constant R in equation (1), or more precisely R/c, viz. 3.290 x 10^{15}. Substituting empirically determined values for m, e, and h, Bohr obtains 3.10 x 10^{15} for the factor outside the parentheses in the theoretical equation (2). He regards this agreement between theory and observed values as inside the uncertainty due to experimental errors.

    (c) Determine configurations of electrons within atoms.

To the basic assumptions (1)–(4) Bohr adds several new ones which form the basis for claims about how electrons are distributed in atoms more complex than hydrogen and about the radii and frequency of orbit. For example, he assumes that all the electrons in an atom lie on a plane through the nucleus, that they are arranged at equal angular intervals in coaxial rings rotating around the nucleus (p. 189), and that the number of orbital electrons in an atom is equal to the number of the position of the corresponding element in the series of elements arranged by increasing atomic weight (p. 198). From these and other assumptions he determines values for electron radii, frequency, and binding energy as a function of those of the hydrogen electron in its ground state. And for each type of atom he constructs a table giving the number of electrons in each of its rings.

What did Bohr conclude from these initial successes? In the "Concluding Remarks" of his trilogy Bohr says that he has been attempting "to develop a theory of the constitution of atoms and molecules based on quantum ideas of Planck and the structure of the atom proposed by Rutherford" (p. 232). His claim is that with the assumptions he introduces "it is possible to account for the laws of Balmer and Rydberg connecting the frequency of the different lines in the line-spectrum of hydrogen." And he says that the theory he formulates about the constitution of atoms and molecules generally "on several points is shown to be in approximate agreement with experiments" (p. 233). But neither here nor anywhere else in the paper does he claim that these or other experiments establish the theory or even show it to be probable.

After completing this work, in a letter to Mosely (November 21, 1913) he acknowledges the speculative character of his theory and the need for experiments:

> For the present I have stopped speculating on atoms. I feel it is necessary to wait for experimental results. (p. 547)

Earlier, on February 7, 1913, Bohr wrote a letter to Hevesy outlining his theory and mentioning various phenomena that it explains. The fact that it does so, he writes, gives him

> hope to obtain a knowledge of the structure of the systems of electrons surrounding the nuclei in atoms and molecules, and thereby hope of a detailed understanding of what we may call the "chemical and physical" properties of matter. . . . I am sure that you will understand . . . that I don't speak of the result which I mean I can obtain by help of my poor means, but only of the point of view—and the hope to and believe in [sic] a future (perhaps very soon) enormous and unexpected?? development of our understanding—which I have been led to by considerations as those above.[21]

The initial explanatory success of his theory makes it reasonable to hope that knowledge of the structure of the atom will be obtained by pursuing the theory. It does not show that it has been.[22]

Despite the fact that Bohr regarded his theory as speculative, he clearly believed that in his trilogy he was offering a defense of it. He was not *merely* speculating. This defense, he thought, justified his "hope to obtain a knowledge of the systems of electrons surrounding the nuclei in atoms and molecules." The question to which I now turn is the nature of Bohr's defense.

3. BOHR'S DEFENSE

There are three parts of this defense that I will distinguish: (i) motivation, (ii) derivations and explanations, and (iii) non-retroductive arguments for individual assumptions. The first two will be discussed in the present section, the third in section 4.

Motivation. Part of Bohr's defense consists in "motivating" his theory, i.e., saying what general type of theory he seeks and why a theory of that type is of interest, needed, important, or useful. He begins the trilogy by indicating a basic assumption of Rutherford's theory of the atom, and the experiments upon which it is based. Almost immediately he notes the major difficulty with this theory, viz. the instability of the system of electrons. He also mentions a lacuna in the theory, viz. the fact that the charges and masses of the electrons and positive nucleus—the only quantities characterizing Rutherford's atom—do not afford a sufficient basis for determining the size of the atom. He then claims that Planck's radiation theory may provide a solution to these problems. That theory, which Bohr takes to be supported by a range of different phenomena, suggests the "inadequacy of classical electrodynamics in describing the behaviour of systems of atomic size," and the need "to introduce in the laws in question [governing electrons] a quantity foreign to the classical electrodynamics, i.e., Planck's constant" (p. 162).

In effect, then, Bohr defends his theory (in part) by claiming that

(a) It is of interest, important, useful to develop a theory of the following general type: one that will solve the instability and dimensionality problems in Rutherford's model of the atom; that will do so by applying (parts of) Planck's theory of radiation; and that will be applicable to atoms and molecules generally.

(b) The reason that a theory of this type is of interest is that developing models of the atom has come to be important to physicists since the discovery by Thomson in 1897 that electrons are constituents of atoms (Bohr mentions Thomson's own model); that central ideas of Rutherford's model (by contrast to Thomson's) are strongly supported by experiments on alpha particle scattering; that this model, which is based on assumptions of classical electrodynamics, does in fact have the problems he notes; and that Planck's theory, which abandons certain ideas of classical electrodynamics, is supported by a range of phenomena and applies to systems of atomic size.

(c) His theory can do the things noted in (a); i.e., it is a theory of this general type.

Claims (a), (b) and (c) comprise Bohr's "motivation." Claim (a) indicates a general type of theory to be sought without formulating all (or necessarily any) of the specific assumptions that will constitute that theory. It does so by indicating questions to be answered (How can stability be achieved in Rutherford's model? How can dimensions of the atom be determined from this model? How can the spectrum of hydrogen be derived?) It will also include *instructions* for answering such questions, i.e., conditions which answers to these questions should satisfy (e.g., that the answers be applicable to atoms and molecules generally, that they be quantitatively expressed, that conservation of energy be satisfied, etc.)[23]. In (b), claim (a) is defended in one or both of the following ways: by providing information about current physicists—showing that they have raised or, given their interests and standards, should raise such questions and seek answers subject to these instructions; and by providing information that lends empirical support to conditions presupposed by the questions or the instructions (e.g., by pointing out that experimental results with alpha particle scattering support a central assumption of Rutherford's model).

The defense of (c)—that Bohr's theory is a theory of the type described in (a)—comes next.

Derivations and Explanations. This part of Bohr's defense consists in producing the specific assumptions of his theory and showing how these can be used to answer questions he raises so that the conditions imposed are satisfied. In some cases this is done simply by noting the assumptions being made. (For example, the instability problem is solved by the assumption that radiation is emitted or absorbed only when the atom passes from one stationary state to another.) In most cases, however, questions are answered by producing demonstrations of what can be derived from the assumptions or explained in terms of them. To answer the dimensionality question Bohr derives formulas governing the radius of an electron's orbit and the frequency of its revolution. Introducing known values for some of the terms in these formulas, he computes values for the radius and frequency which are close to experimentally determined ones. In addition he derives formulas (including Balmer's) for the various series of lines in the hydrogen spectrum.

Bohr is aware of methodological criteria in addition to explanation and deduction of phenomena. He explicitly mentions simplicity and coherence. For example, he writes:

> The intention, however, has been to show that the sketched generalization of the theory of the stationary states possibly may afford a simple basis of representing a number of experimental facts which cannot be explained by the help of the ordinary electrodynamics. . . . (p. 179)

While he recognizes that the assumptions he introduces "conflict with the admirably coherent group of conceptions which have been rightly termed the classical theory of electrodynamics," he urges "that it may also be possible in the course of time to discover a certain coherence in the new ideas."[24]

However, probably the most important of the general constraints on explanation for Bohr is "consilience," the need to apply the theory to phenomena of different sorts, including ones different from those that prompted the theory in the first place. In an unpublished *Memorandum* written in June or July of 1912 for discussion with Rutherford, Bohr claims that his theory offers explanations of phenomena of quite different sorts including the periodicity of atomic volumes of the elements, Whiddington's law concerning the minimum velocity of an electron necessary to excite characteristic x-rays of an element, Bragg's law of the absorption of alpha rays by different elements, and the stability and heat of combination of some simple compounds. In the late fall or early winter of 1912, after the Rutherford *Memorandum*, Bohr turned to the question of atomic spectra (which at first he considered too complex to generate ideas about atomic structure).[25] After a colleague urged him to look at Balmer's formula, he realized how he could use the assumptions of his theory to derive this formula.

In the "motivation" Bohr says what he is *trying* to do, which can be expressed in terms of questions to be answered and instructions to be satisfied. In addition, he attempts to justify the questions and instructions. In "derivations and explanations" he provides specific answers to these questions usually by derivation from the assumptions, and he considers whether the instructions are satisfied. However, these are not the only ways he defends his theory.

4. NON-RETRODUCTIVE ARGUMENTS
FOR INDIVIDUAL ASSUMPTIONS

In addition to showing what can be derived and explained using his theory, Bohr defends a number of assumptions by other means. Since this aspect of Bohr's defense may be considered the most controversial, I will examine it in some detail.

One assumption he defends, which is incorporated into his third postulate, is that radiation is emitted by an atom in discrete quanta such that the amount of energy emitted is equal to a constant times the frequency of the radiation. This idea, Bohr emphasizes, is suggested by Planck's theory of radiation. The latter is concerned with thermal radiation produced by heating a black body (which emits heat radiation but reflects none incident upon it). In 1901 Planck published a derivation of his law relating the energy density (energy per wavelength) of black-body radiation to the temperature of the radiation. To derive the law Planck assumed that the walls of an enclosure whose surface is black contain electric oscillators of all frequencies which cause the emission and absorption of radiation. Experiments on electromagnetic waves by Hertz had confirmed the electromagnetic theory of light and suggested that other forms of radiation including thermal were also electromagnetic in origin.[26] Planck extended Hertz's idea that electric oscillators are responsible for optical radiation to thermal radiation.

In a later treatment in 1911 and 1912 Planck supposed that the electric oscillators as a result of their vibrations emit thermal radiation, *but not continuously:*

> ...we shall assume that the emission does not take place continuously, as does the absorption, but that it occurs only at certain definite times, suddenly, in pulses, and in particular we assume that an oscillator can emit energy only at the moment when its energy of vibration, U, is an integral multiple n of the quantum of energy, $e = h\nu$. Whether it then really emits or whether its energy of vibration increases further by absorption will be regarded as a matter of chance. . . .
>
> It will be assumed, however, that if emission does take place, the entire energy of vibration, U, is emitted, so that the vibration of the oscillator decreases to zero, and then increases again by further absorption of radiant energy.[27]

In his trilogy Bohr says that his aim is "to apply the main results obtained by Planck" to atoms of the sort Rutherford postulates. In his lecture "On the Spectrum of Hydrogen" (p. 296), as well as in the trilogy, he speaks of this as an *analogy*. Planck's electric oscillators become the electrons in Bohr's atom oscillating about the nucleus. The thermal radiation emitted from a black body becomes the optical radiation emitted from the atom. Planck's assumption that radiation from an electric oscillator in a black body does not take place continuously but only when the energy of vibration is an integral multiple of $h\nu$ is transformed into Bohr's crucial assumption (part of his (1) and (3)) that radiation from an electron in an atom does not take place continuously but only when the energy emitted is a multiple of $h\nu$.

There is another important transformation, which Bohr notes in his lecture "On the Spectrum of Hydrogen." The energy of the Planckian oscillator when it emits radiation is a whole multiple of $h\nu$. But the average value of the kinetic energy of the oscillator is equal to its potential energy. So for Planck the average value of the kinetic energy is a whole multiple of $h\nu/2$, or $h\omega/2$ (where $\omega =$ the mechanical frequency of the oscillator), since when energy is emitted Planck assumes that ν, the frequency of the radiation emitted, is equal to ω, the mechanical frequency of the oscillator. Accordingly, the ratio between the kinetic energy and the mechanical frequency of the oscillator is a whole multiple of $h/2$.

Bohr transforms this by focusing on a hydrogen atom consisting of a nucleus and an electron in an elliptical orbit around it. He considers the energy that must be added to the system to remove the electron an infinite distance from the nucleus; or equivalently, the energy required to capture an electron at an infinite distance from the nucleus (whose value will be that of the radiant energy emitted during the process). This can be shown to be equal to the average value of the kinetic energy of the electron during a single revolution. Since for a Planckian oscillator the average value of the kinetic energy of the oscillator is a whole multiple of $h\omega/2$, Bohr makes the corresponding assumption for the orbital electron. This yields his fundamental assumption (4) that in an atom consisting of an electron rotating around a positive nucleus the ratio between the total energy emitted during the formation of the configuration and the frequency of revolution of the electron is a whole multiple of $h/2$.

Bohr goes on to say that although "it was just in this way that I was originally led into these considerations [i.e., to his fourth assumption], it appears to me misleading to use this analogy as a foundation."[28] And indeed in his trilogy this particular defense of assumption (4) does not appear. There Bohr assumes that when an electron is captured from rest by a nucleus the frequency of the radiation emitted depends solely on the frequencies of revolution of the electron in the initial and final states. Since the frequency of revolution of the electron in a rest state is zero and the frequency of revolution in the final state is ω, Bohr takes a simple average, $\omega/2$, of these two to be the frequency ν of the radiation emitted during this transition. Since he assumes that the energy of this radiation is a whole multiple of $h\nu$, and therefore a whole multiple of $h\omega/2$, assumption (4) follows. However, Bohr offers no defense, even by analogy with Planck, for taking the frequency ν of the emitted radiation to be equal to a simple average $\omega/2$ of the initial and final frequencies of revolution.

How seriously does Bohr, or should we, take the analogy between a Planckian system of oscillators and a Bohr atom? Did Bohr regard this analogy as merely illustrative and not "foundational"? That is, did he think the analogy provided simply a useful device for presenting his assumptions and not an argument that supported them? In favor of this interpretation we might note several disanalogies between the two systems.

(i) Planck but not Bohr is dealing with systems producing thermal radiation. (ii) More importantly, the energy states in a Planckian oscillator are continuous (only the radiant energy emitted by the system is discontinuous). By contrast, in a Bohr atom the energy states of the electron are discontinuous (only certain ones are permitted, the so-called "stationary states"). (iii) In a Planckian system after the oscillator emits radiant energy it stops oscillating until there is new absorption. In a Bohr atom there are various energy states; i.e., after the electron emits radiation it does not stop oscillating (revolving). (iv) As Bohr himself points out, in the case of Planck's oscillator, the frequency of oscillation is independent of the amount of energy contained in the system. In Bohr's atom the frequency of revolution of the electron does depend on the energy of the system (p. 292). (v) Finally, Planck assumes that an oscillator can emit several energy quanta at once at the same frequency. Bohr at the outset of the trilogy makes the analogous assumption. However, a bit later in the work he recognizes the inadequacy of this assumption, "for as soon as one quantum is sent out the frequency is altered" (p. 172). So the new assumption, which is disanalogous to Planck's, is that in any radiation only a single energy-quantum $h\nu$ is emitted.

The passage from Bohr that perhaps most clearly indicates how he is using Planck's ideas in developing his own theory is this:

We stand here almost entirely on virgin ground, and upon introducing new assumptions we need only take care not to get into contradiction with experiment. . . . With this in mind let us first examine the experiments on temperature [black-body] radiation. The subject of direct observation is the distribution of radiant energy over oscillations of the various wave lengths. Even though we may assume that this energy comes from systems of oscillating particles, we know little or nothing about these systems. No one has ever seen a Planck's resonator [oscillator], nor indeed even measured its frequency of oscillation; we can observe only the period of oscillation of the radiation which is emitted. It is therefore very convenient that it is possible to show that to obtain the laws of temperature radiation it is not necessary to make any assumptions about the systems which emit the radiation except that the amount of energy emitted each time shall be equal to $h\nu$, where h is Planck's constant and ν is the frequency of the radiation. Indeed, it is possible to derive Planck's law of radiation from this assumption alone, as shown by Debye, who employed a method which is a combination of that of Planck and of Jeans. (p. 293)

If this passage (from his December 1913 address) can be taken to represent his thinking in the trilogy published a few months earlier, then the only part of the analogy with Planck that Bohr regards as support for his theory is this: as in the case of a Planckian system, energy from an atom is emitted discontinuously in such a way that the amount is equal to $h\nu$. Indeed, in the trilogy this is the only part of Planck that Bohr invokes.[29]

Accordingly, Bohr took Planck's ideas about discontinuous emissions and the quantization of energy in the case of black-body radiation as providing some defense for similar claims he makes about atomic radiation in assumptions (1) and (3). He was led to his assumption (4)—the ratio of energy to frequency of revolution is a whole multiple of $h/2$—by thinking about the corresponding situation in a Planckian oscillator. But in the latter part of 1913, when the trilogy was published, he did not regard this analogy as providing any defense of assumption (4). In the trilogy it is not even mentioned.

Now for Bohr's claims about "stationary states" in his assumptions. In "On the Spectrum of Hydrogen" he characterizes these as "waiting places between which occurs the emission of energy corresponding to the various spectral lines" (p. 293). Bohr's most general characterization of such states is that they are ones in which no energy is emitted (or absorbed). What does he assume about these states? On the basis of Rutherford's ideas he assumes that in an atom consisting of a nucleus and an electron the electron describes a stationary elliptical orbit about the nucleus. He then assumes that the frequency of revolution ω and the major axis $2a$ of the orbit will depend on the amount of energy in the system (which he takes to be the amount of energy W required to remove the electron to an infinite distance from the nucleus). These assumptions, together with the assumption $W = th\omega/2$ (which is equivalent to assumption (4) earlier), yield

$$W = \frac{2\pi^2 m e^2 E^2}{t^2 h^2} \qquad \omega = \frac{4\pi^2 m e^2 E^2}{t^3 h^3} \qquad 2a = \frac{t^2 h^2}{2\pi^2 m e E}$$

where m = mass of electron; e = charge of electron; E = charge of nucleus; t = a whole number. This shows that the energy of the system, the frequency of revolution, and the major axis of the electron's orbit all take on discrete values, as t varies from one whole number to another. That is, the stationary states are discrete and correspond to different energies, frequencies, and orbits of the electron.

How much of this does, or could, Bohr defend by appeal to Planck? Planck's oscillators, like Bohr's electrons, have states of oscillation in which no energy is emitted (although for Planck energy is continuously absorbed). However, there is a significant difference, as noted above. In the case of a Planck oscillator, unlike that of a Bohr electron, the frequency of oscillation of a single oscillator is independent of the energy in the system of oscillators. Planck does not assume $W = th\omega/2$. According to him, the frequency of the energy emitted by the system, *but not the frequency of oscillation of individual oscillators*, depends on the amount of energy in the system. For Planck what is quantized is the energy emitted by the system (and the frequencies of those radiations). The possible energies of the oscillator itself are continuous. To be sure, for Planck the *average* value of the kinetic energy of an oscillator is a whole multiple of $h\omega/2$. But this is an average, and Planck does not suppose that the kinetic energy of the oscillator takes on only discrete values.

Accordingly, the only assumptions Bohr does (or can) defend by appeal to Planck are these:

(a) Electrons in atoms have states of oscillation (stationary states) in which no energy is emitted or absorbed. (This is part of Bohr's first assumption, with the addition, from assumption 4, that electrons in the atom oscillate.)

(b) When energy is emitted from an atom the amount emitted is a whole multiple of $h\nu$, where ν = frequency of radiation; i.e., the energy emitted is quantized. (This is the central part of Bohr's third assumption.)

He does not (and cannot) defend the following assumption by appeal to Planck, since it is incompatible with a corresponding one Planck makes about oscillators:

(c) In an atom consisting of an electron rotating around a positive nucleus the stationary states (those in which no energy is emitted or absorbed) are such that the energy of the system is equal to a whole multiple of $\hbar\omega/2$, where ω = frequency of revolution. (This is equivalent to Bohr's fourth assumption.)

Although Planck makes a corresponding assumption for the average kinetic energy of his oscillator, he does not make it for all kinetic energies of the system. (c), together with other assumptions, commits Bohr to discrete energy levels of the system, which is incompatible with Planck's assumptions. Bohr admits that although Planck's claim involving $\hbar\omega/2$ is what led him to his own assumption (4), "when we consider how differently the equation is employed here and in Planck's theory it appears to me misleading to use this analogy as a foundation, and in the account I have given I have tried to free myself as much as possible from it" (p. 296). So Bohr does not defend (c) by appeal to Planck.

Even though Bohr does defend assumptions (a) and (b) by reference to Planck, it should be clear that he is employing analogical reasoning to do so. From Planck's ideas about thermal radiation emitted from black bodies Bohr is reasoning to certain ideas of his own regarding optical radiation emitted from atoms. Bohr's argument is perhaps this:

1. Heat radiation from black bodies is probably produced by electrical oscillators of some kind. (This was a generally accepted idea by 1913; and, as Bohr notes, Planck makes this assumption in deriving his experimentally correct law for black-body radiation.)

2. In Planck's theory of black-body radiation there are states of the oscillator in which no energy is emitted. When it is emitted the amount radiated is equal to a whole multiple of $h\nu$. (Bohr treats this as a reasonable hypothesis, noting that with it alone, making no other specific assumptions about what is oscillating, Planck's radiation law can be derived.)

3. Atoms produce optical radiation. (This is inferred from the existence of spectral lines.)

4. The optical radiation emerging from atoms is produced by the motions of electrons in the atom. (By 1913 the only other possibility, vibration of the atom itself, had been discredited.)

5. Rutherford's model of the atom, in certain respects, is probably correct. That is, it is probably the case that an atom consists of a small but massive nucleus surrounded by rotating electrons. (Bohr took this to be confirmed by alpha particle experiments.)

6. Since electrons in Rutherford's model are electrical oscillators, and since (by step 4) they produce the optical radiation, by analogy with Planck's electrical oscillators (in virtue of step 2), there is some reason to suppose that there are states of the electron in which no energy is emitted, and that when it is emitted the amount radiated is equal to a whole multiple of $h\nu$. (Bohr considers these characteristics of

Planck's oscillators—whatever else such oscillators are—to be the most probable, and the most extendable to other systems.)

If this interpretation is correct, then several points are worth stressing. First, it is an *argument* from analogy, and not simply an illustration or an analogy which led Bohr to these ideas. Bohr is defending some of his assumptions by appeal to Planck. Second, because there are disanalogies as well as analogies between Planck's oscillators and Bohr's electrons, the argument does not establish assumptions (a) or (b) or give them high probability. Nor does Bohr claim such virtues. But it does provide, or at least is intended to, some warrant for them. Third, such warrant does not derive from the fact that assumptions (a) and (b), if true, explain some phenomena. The analogical argument just constructed does not appeal to what these assumptions can explain. Rather it provides, as I shall put it, "independent warrant," i.e., warrant that is independent of the explanatory virtues of (a) and (b). Fourth, this is not a defense of all the assumptions Bohr makes. In particular, it is not a defense of assumption (c). Finally, by providing some support for assumptions (a) and (b), it contributes to showing that the theory is reasonable to pursue.

5. BOHR AND THE "LOGIC OF PURSUIT"

How does Bohr's defense of his theory square with the claims in section 1 about reasoning comprising the "logic of pursuit"? The first claim is that such reasoning is weaker than that which establishes a theory or shows it to be probable. Bohr's reasoning, I have suggested, was intended by him to be of this sort. He makes no claim that what he does in his trilogy shows his theory to be true or even probable. The second claim is that the reasoning occurs before the theory has been shown to be probable or improbable by experimental test. To be sure, Bohr alludes to experimental results which have a bearing on part of his theory. He notes that alpha particle scattering experiments render probable Rutherford's assumption, which he is also making, about the nucleus and surrounding electrons. And he considers Planck's radiation law to be experimentally established. But neither these results nor any others he invokes show that his four basic assumptions are probable. It is fair to say that Bohr's defense conforms to the second claim.

The third claim is that the reasoning is governed by a "logic," which turns out to be retroduction. One argues that a theory T is reasonable to pursue on the grounds that it explains certain phenomena. The fourth extends this by allowing the reasoning to include appeals not just to explanation but to other general methodological criteria such as simplicity, consilience, and coherence. The fifth introduces the practical costs of pursuit. (The sixth will be examined in section 7.) These ideas form the basis for the retroductive schema R_3 which makes reference not only to explanations but to other general methodological virtues and to practical costs:

R_3: T if true would correctly explain $O_1, \ldots, O_n$
T has other general methodological virtues
Practical costs of pursuing T are reasonable
    Hence
It is reasonable to pursue T

This schema only partially reflects Bohr's defense. Bohr does satisfy the first two premises. He provides explanations of phenomena, and these have some general methodological virtues, particularly consilience. Although there is no discussion of practical costs of pursuit, I assume Bohr thought these reasonable. The basic problem is that important parts of Bohr's defense go beyond what is covered in R_3.

For one thing, Bohr "motivates" his theory by saying in a general way what he is trying to do and why: he is trying to solve the instability problem in Rutherford's model, to derive formulas that will give atomic dimensions, and to apply this model to atoms and molecules generally. He is trying to do so because physicists are interested in developing models of the atom, central ideas of Rutherford's model are supported by experimental evidence, and that model has the problems Bohr notes. Stating his aims in a general way in the motivational part enables us to ask how well specific explanations and derivations in the retroductive part satisfy those aims. This, together with a defense of these aims in the motivation, helps to determine whether the theory is reasonable to pursue. We can ask why solving the instability problem is important and whether, or to what extent, Bohr's theory really does so.

Another part of Bohr's defense that goes beyond R_3 consists in non-retroductive arguments for individual assumptions. Bohr defends the assumption that electrons in atoms have states of oscillation in which no energy is radiated, as well as the assumption that when energy is radiated the amount emitted is a whole multiple of $h\nu$, by an argument from analogy with Planck's oscillators. (My reconstruction of this argument was given in section 4.) To be sure, Bohr goes on to show how these two assumptions of his theory, together with others, generate explanations of various phenomena, including the Balmer lines. But this is not the sole defense of these assumptions.

Although Bohr goes beyond what R_3 prescribes, in principle could he have shown his theory reasonable to pursue solely by such means? In effect, R_3 provides the following set of sufficient conditions:

(A) T is reasonable to pursue if
    (a) There is some set of observable phenomena that T if true would correctly explain
    (b) T satisfies some general methodological criteria (e.g., simplicity, consilience)
    (c) The practical costs of pursuing T are reasonable

Let me say more generally why I regard this (and hence R_3) as inadequate.

For any set of observable phenomena, some T satisfying (a), (b), and (c) above can always be constructed which is clearly not reasonable to pursue.[30]

Take any set of observable phenomena $O_1, \ldots, O_n$ you like, however unrelated they may seem.[31] Construct a theory T with two types of assumptions. The first is of the form "X exists," the second "If X exists, X causes it to be the case that $O_1, \ldots, O_n$." X is anything you like (sunspots, the Loch Ness monster, environmental pollution). Since T postulates the existence of a single cause, at least one central idea underlying simplicity is satisfied: "don't multiply entities beyond necessity." Not only is a single cause rather than a multiple set postulated for each phenomenon, but it is the same cause of all. If testability is required we can readily choose an X whose existence and causal efficacy with respect to the O's are testable. For consilience we can simply add new phenomena to the list of those caused by X which are different in type from ones that initially prompted this theory. Finally, to satisfy (c) we can choose an X the investigation of whose causal efficacy with respect to the O's will involve costs in time, money, and energy that are minimal.

Why aren't such theories reasonable to pursue? For one thing, virtually all of them are discredited or otherwise too crazy to pursue. In general there is no reason to suppose that some set of arbitrarily chosen phenomena, particularly those that seem completely unrelated, have any common cause, be it X or something else. Indeed, if such a set is arbitrarily chosen there will be strong reasons to suppose there is no common cause. Accordingly, scientists will have no justification whatever for, and hence little interest in, raising the question "What is the common cause of this set of phenomena?"

Suppose, then, we restrict the phenomena chosen to ones for which there is such a justification. (This, of course, goes beyond the conditions expressed in (A).) Even then, if X is arbitrarily chosen, and if there is strong reason to suppose X does not cause the O's, or even just no reason whatever to suppose it does, the theory constructed may be of insufficient interest to pursue, especially if there are other more promising theories to explain the phenomena.

Finally, consider a type of case in which neither the phenomena nor the cause X are arbitrarily chosen, and where there is some reason to suppose that the phenomena have a common cause and that this cause is X. This is still not sufficient to render the theory reasonable to pursue, even if it satisfies general methodological and practical cost criteria. To see this, let us return to Bohr and consider a set consisting of three "phenomena": (i) the lines in the visible spectrum of hydrogen described quantitatively by the Balmer formula; (ii) the lines in the ultraviolet region of the spectrum described by the Lyman formula; (iii) the lines in the infrared region described by the Paschen formula. Here is a theory to explain these phenomena:

> T: Hydrogen's being excited thermally or electrically causes it to emit radiation whose spectrum contains lines satisfying the Rydberg formula

$$\nu = R \left(\frac{1}{t_2^2} - \frac{1}{t_1^2} \right)$$

where R is a constant $= 109{,}677.581$ cm^{-1}, $t_2 = 2$, and $t_1 = 3, 4, 5, \ldots$ for the Balmer series; $t_2 = 1$ and $t_1 = 2, 3, 4, \ldots$ for the Lyman series; $t_2 = 3$ and $t_1 = 4, 5, 6, \ldots$ for the Paschen series.

Theory T correctly explains the spectral lines by citing a cause (excitation of hydrogen) and by producing a general equation from which the three sets of lines are derivable. The theory is simple (in that a single cause is postulated for all three phenomena); it is testable (the means of testing T was available to Bohr). Let us suppose that the costs of pursuit, in this case, the costs of devising and conducting experiments to test T, are not high. Finally, let us suppose there is some reason to assume that these three phenomena have the same cause, and that this is some type of excitation of hydrogen. Can we conclude that T is reasonable to pursue? Not necessarily. It depends on what is known, what is sought, and why.

By 1913 the Rydberg formula was widely accepted (Rydberg had developed earlier versions by 1890). It was also known that electrical and thermal excitation produces the hydrogen spectral lines. And it was widely accepted that the spectral lines are produced at the subatomic level by motions of electrons. The question that remained to be answered is how. Although in the initial version of his atomic theory in the Rutherford *Memorandum* Bohr does not address this question, he does so in the trilogy. Here one of the questions he is interested in pursuing is how to derive and explain the Balmer and other spectral formulas from assumptions about the structure of the hydrogen atom. And, for reasons given earlier, among the instructions Bohr attempts to follow is that this question be answered in terms of Rutherford's model of the atom. But if this model is to be used, then other questions need answering as well concerning the stability and dimensions of the atom—questions not addressed by theory T above. For these reasons pursuing T would not have been reasonable for Bohr to do. Not only was T already established, but it did not answer various questions that Bohr and others in 1913 sought to answer; and even if it answered others (e.g., what causes the spectral lines?), it did not do so in an appropriate way, one that satisfied instructions calling for answers at the subatomic level.

Accordingly, what is missing from schema (A) are factors of a sort provided in the "motivation": information about questions to be raised, instructions to be satisfied, and what justification there is for doing so. Also missing from (A) is mention of arguments for individual assumptions that do not consist simply in an appeal to explanations of phenomena or to other general methodological criteria—what I have called "independent warrant." How might these be taken into account so as to provide a set of conditions more adequate than (A)?

6. A NEW SCHEMA FOR PURSUIT

Included among instructions are ones invoking

(i) general methodological criteria

(ii) specific empirical conditions (e.g., that the answer conform to Rutherford's model)

So far category (i) has been taken to include criteria such as simplicity, testability, and coherence. But "explains some set of phenomena" and "has independent warrant" are also general methodological considerations. These will now be included in category (i) as well. Moreover, practical considerations such as time, money, and energy may be incorporated into instructions, so that the latter will also include

(iii) practical constraints on answers to questions. For example, the instructions may call for answers the testing of which is economical given the resources available.

Now, as noted, the "motivation" for a theory T offered by a scientist will include information about what questions that scientist seeks to answer by reference to T, what instructions he seeks to impose, and what justification he has for doing so. And, we have seen, this information is relevant in determining whether it is reasonable for that scientist to pursue T. Accordingly, we might replace schema (A) with

(B) Theory T is reasonable for scientist S to pursue if
    (a) There is a set of questions that S seeks to answer and a set of instructions that S seeks to impose with respect to these answers.
    (b) S is justified in raising these questions and imposing these instructions, and in believing that T provides answers to these questions in a way that satisfies these instructions.

Conditions (a) and (b) are highly contextual. What questions a scientist seeks to answer, what instructions he seeks to impose, and whether he is justified in doing so, depends on his knowledge, interests, and resources. As far as instructions are concerned, which specific empirical conditions (category (ii)) and practical considerations (category (iii)) one is justified in imposing on answers to questions depends on the information and resources available.[32] Even the extent to which general methodological criteria (category (i)) should be satisfied can vary contextually. It may be reasonable for a scientist to pursue a theory despite a lack of independent warrant for any of its assumptions, or despite the fact that means for testing it are unavailable.[33] A balancing of factors is required. If the scientist regards it as very important to answer certain questions so that specified empirical conditions are satisfied, and if T is the only theory known to do so, it may be reasonable for that scientist to pursue T despite the fact that it does not yet have independent warrant or a known means of testing.

In view of the fact that interests, resources, and available information vary contextually, there is a prudent course of action on the part of someone defending a theory in the absence of experimental tests that provide substantial support. This involves showing what questions the theory will attempt to answer, subject to what empirical conditions, what justification there is for raising these questions and imposing these conditions, and what answers the theory provides; to what extent it is consilient, simple, etc.; and what, if any, independent warrant it has. Let others decide whether these benefits are of sufficient interest or importance to them to warrant their own costs of pursuit. This was essentially Bohr's strategy.

Schema (B), I suggest, is more adequate than schema (A), since by introducing questions and instructions it takes into account factors cited in the motivation as well as independent warrant (if any). Will it permit the unwanted cases, such as the Rydberg theory, noted in the previous section that are allowed by schema (A)?

Consider the question

Q: How is the Balmer formula derivable?

which in 1913 Bohr wished to answer. From the set of instructions Bohr sought to impose on the answer delete those calling for answers appealing to the subatomic structure of hydrogen (e.g., answers in terms of Rutherford's model of the atom). The remaining set will contain, among other things, general methodological constraints (e.g., explanation, consilience) and some empirical conditions (e.g., conservation of energy). If Bohr was justified in imposing his original set of instructions on answers to T, he was justified in imposing this subset. And let us suppose, as we have been doing, that Bohr was justified in raising Q. Now the Rydberg theory answers Q in a way that satisfies this truncated set of instructions. And in 1913 Bohr would have been justified in believing this. Accordingly, schema (B) would allow us to conclude that in 1913 the Rydberg theory was reasonable for Bohr to pursue. Obviously, we need to consider the *complete* set of instructions Bohr seeks to impose on answers to Q, not a partial subset. (a) in schema (B) should be so understood. With this understanding the Rydberg theory (and the other unwanted cases of section 5) should be disallowed. Whether other changes in schema (B) are required to render it sufficient I shall not pursue further.[34]

I will conclude this section by briefly noting a theory of the atom that Bohr himself did not regard as reasonable to pursue. It was developed by the astrophysicist J. W. Nicholson in a series of papers in 1911 and 1912. Independently of Rutherford, but like him, Nicholson proposed that the positive electricity of atoms exists "in small spherical volume distributions of uniform density, whose radius is small in comparison even with that of the electron, and gives rise to nearly the whole mass of an atom. . . ."[35] Like Bohr, Nicholson introduced Planck's quantum into his theory, but did so by assuming that the ratio between the energy of the system and the frequency of rotation of the electrons is a whole multiple of Planck's constant. (He assumes $W = th\omega$, while Bohr assumed $W = th\omega/2$.) Nicholson also assumed that spectral lines were produced by electrons vibrating in a direction perpendicular to the plane of their orbital motion. (By contrast, in Bohr's theory, the lines are produced when electrons pass from one stable orbit to another.) With his assumptions

Nicholson derived and explained lines in the spectra of certain stellar nebulae as well as lines in the spectrum of the solar corona.

Bohr notes all these facts in his trilogy, but then raises various objections. The most important is the stability problem of the ring of electrons (p. 183). Bohr admits that although perpendicular displacements will not produce instability if the number of electrons is not great, Nicholson's theory provides no stability for displacements of the electrons in the plane of the ring. So his theory does not adequately solve the stability problem. Secondly, Bohr notes that Nicholson's theory, unlike his, does not yield the Balmer and Rydberg formulas for line spectra.[36] In short, Bohr believed that Nicholson's theory was unable to provide answers to two important questions that Bohr raises about "nuclear" theories of the atom of the sort proposed by Rutherford. So far as he is concerned, schema (B) is not satisfied.

7. TESTS OF BOHR'S THEORY

I turn finally to the sixth claim concerning the "logic of pursuit," viz. that the reasoning used to show that a theory is reasonable to pursue is different in kind from that used to establish it or show that it is probable. In the months and years following his trilogy, as a result of new experiments, Bohr employed a number of arguments which together he believed showed the theory, or at least central parts of it, to be probable. I shall briefly note two of the most important ones.

From his assumptions Bohr derived the following general formula for the frequency of radiation ν emitted by electron transitions:

$$(1) \qquad \nu = N^2 \frac{2\pi e^4 M m}{h^3 (M + m)} \left(\frac{1}{n_1^2} - \frac{1}{n_2^2} \right)$$

where N = the number of electrons, e = the charge of each electron, M = mass of nucleus, and m = mass of electron.[37] If $N = 1$, formula (1) yields the spectrum of hydrogen. Now, Bohr notes, if $N = 2$ we should obtain a helium spectrum (since there are two electrons in a helium atom). And he writes:

> The formula is found very closely to represent some series of lines observed by Fowler and Evans. These series correspond to $n_1 = 3$ and $n_1 = 4$.[38]

These lines had originally been observed by Pickering in 1896 in some star spectra and by Fowler in 1912 in a vacuum tube containing a mixture of hydrogen and helium. Initially they were taken to be spectral lines of hydrogen. In 1913, however, Bohr attributed them to helium, and in a letter to Rutherford on March 6 he suggested the need for an experiment to help settle the issue. Later that year in such an experiment Evans obtained the lines from pure helium, and no trace of hydrogen lines appeared.

In his 1915 paper quoted above Bohr also notes that experiments of Rau in 1914 on minimum voltages necessary for producing spectral lines showed that about thirteen volts were sufficient to excite lines of the Balmer series of hydrogen, while about 80 volts were necessary to excite the new lines under question. And he points out that these values are in close conformity to the values calculated from his theory to remove the electron from the hydrogen atom (13.6 volts) and to remove both electrons from the helium atom (81.3 volts).

A second important argument Bohr employs derives from experiments conducted by James Franck and Gustav Hertz in 1913 and 1914. These experiments involved bombarding mercury atoms with electrons. When an electron collides with a mercury atom it does so elastically, i.e., loses no energy, if its own energy is less than 4.9 volts. When the bombarding electron is accelerated to 4.9 volts it loses all its energy to the mercury atom, which then emits radiation consisting of an ultraviolet mercury line of wavelength 2536. Bohr notes that in accordance with the Planck quantum condition $E = h\nu$ assumed in his own theory, if the frequency associated with this line is multiplied by Planck's constant h, we obtain $E = 4.9$ volts, which represents the energy of the radiation emitted. This accords with the experimental value of 4.9 volts which is the total energy lost by the electron which causes radiation to be emitted.[39] Bohr considered the experiments of Franck and Hertz to provide significant support for central ideas in his first and third assumptions, viz. that an atom does not absorb radiation continuously, but only when there is a transition from one stationary state to another; and during this transition the energy emitted is quantized and satisfies $E = h\nu$.

What sorts of arguments are these, and how, if at all, do they differ from earlier ones that Bohr used in showing his theory to be reasonable to pursue?

The "helium" argument involves a derivation of a theoretical formula for frequency of radiation emitted by a helium atom. This formula is obtained from the more general theoretical formula (1) together with the assumption that for the helium atom N (the number of electrons) = 2. Bohr then argues that the helium formula is satisfied by a series of known lines in star spectra, and even more convincingly, by the spectral lines obtained by Evans in experiments on pure helium. In effect, then, Bohr's argument appeals to an explanatory virtue of his theory. Not only does the theory offer an explanatory derivation of spectral lines of hydrogen, but it does so for helium as well. Moreover, the argument satisfies consilience. The spectral lines of helium (as well as of hydrogen) represent phenomena of a different sort from the phenomena cited in the "motivation" that prompted his theory in the first place. But these considerations are no different in kind from those given in the trilogy (e.g., the derivation of the Balmer lines in hydrogen) to show the reasonableness of pursuing the theory. Appeals to what the theory can explain and how, and to consilience, are among those cited in the retroductive account of the "logic of pursuit."

As for the experiments of Franck and Hertz, Bohr does not spell out the argument in detail. Here is my reconstruction:

1. The experiments of Franck and Hertz show that when mercury atoms are bombarded with electrons the latter lose no energy if their energy is less than 4.9 volts; also at such electron energies the mercury atoms emit no radiation.

2. From 1, and the assumption of conservation of energy, since below 4.9 volts electrons lose no energy, mercury atoms absorb no energy from electrons below 4.9 volts.

3. The experiments of Franck and Hertz also show that when the bombarding electrons are accelerated to an energy of 4.9 volts they lose their energy and the mercury atoms emit radiation.

4. From 3, and the assumption of conservation of energy, when bombarding electrons receive an energy of 4.9 volts the mercury atoms absorb energy.

5. From 2, 3, and 4, mercury atoms absorb energy from electrons, and then emit energy in the form of radiation, discontinuously.

6. Generalizing this to all atoms and to other bombarding agents: Therefore, all atoms absorb and emit energy discontinuously.

If this is Bohr's reasoning, then the argument is not a retroductive one. It does not appeal to the fact that the fundamental assumptions of Bohr's theory explain the experimental results. Rather it involves an application of the principle of conservation of energy to these experimental results to arrive at the conclusion in step 5 that mercury atoms absorb and emit energy discontinuously; there is then an inductive generalization of this to all atoms. So reconstructed the argument provides "independent warrant" for central parts of Bohr's assumptions 1 and 3. Moreover, this argument is stronger than the one in section 4 which also provides such warrant. The latter involves an analogical inference from Planck's mechanism for the emission of thermal radiation from black bodies to Bohr's mechanism for the emission of optical radiation from atoms. The present argument is much more direct, since it involves inferences from experimental information about radiation emitted and absorbed by atoms themselves. Moreover, Planck's oscillators were still at least somewhat speculative, although Bohr thought certain assumptions about them probable. However, the Franck-Hertz argument is based on more firmly established empirical results.

This argument and the previous "helium" one represent two standard ways that experimental evidence can be shown to contribute to the probability of an hypothesis. They involve, respectively, providing independent warrant for the hypothesis and a demonstration that some phenomenon can be explained by derivation from the hypothesis. The situation can be represented in more general probabilistic terms as follows.[40] Suppose that some observed phenomenon O_1 that provides independent warrant for an hypothesis h renders that hypothesis probable to degree k (i.e., $p(h/O_1) = k$). Suppose further that some phenomenon O_2 can be explained by derivation from h. Then, according to probability theory, the probability of h given both O_1 and O_2 is at least

$k(p(h/O_1 \ \& \ O_2) \geq k)$. Moreover, suppose O_2 is a prediction from h whose truth is not known with certainty, i.e., $p(O_2) < 1$. Then it follows that the probability of h given both O_1 and O_2 is greater than k. In short, if certain phenomena that provide independent warrant for an hypothesis confer upon it some probability, then explaining other phenomena by derivation from that hypothesis guarantees that this probability is at least sustained. Accordingly, both independent warrant and explanatory derivations contribute to the probability of an hypothesis. And if the derivations include some predictions this probability increases. The situation is the same whatever probability is conferred. Appeals to independent warrant and explanatory derivations are made both in arguing that an hypothesis is reasonable to pursue and in arguing that empirical tests confer high probability. The essential difference lies in the strength of the arguments.

Despite this, the sixth claim—that the reasoning used to show that a theory is reasonable to pursue is different from that used to establish it or show it probable—is not thereby refuted. Indeed, I suggest, claim 6 is true. This can be seen from our discussion of such reasoning in previous sections. Whether a theory is reasonable to pursue depends in part on what questions scientists seek to answer, on what instructions they want satisfied, and on the practical costs of pursuit. But these factors are irrelevant for determining whether a theory is established or made probable by the evidence. By 1913 the Rydberg theory, noted in section 5, was not a reasonable one for Bohr to pursue, since it did not address the questions he and others wanted answered. Yet the theory was established by the evidence, something unaffected by its failure to answer desired questions and by the practical costs of pursuing the theory.

To show that a theory is probable one may invoke independent warrant and explanatory power—both of which can also be cited in showing that it is reasonable to pursue. But in the latter case there are additional pragmatic considerations that are irrelevant to the former.

8. CONCLUSIONS

1. Bohr's attempt to defend his first quantum theory of the atom in 1913 and 1914 involves arguments that purport to show that the theory is reasonable to pursue. His reasoning satisfies the first two claims of defenders of a "logic of pursuit": it is weaker than that which establishes a theory or shows it probable; and it occurred before experimental tests showed the theory to be probable.

2. Claims 3, 4, and 5 assert that "pursuit" reasoning is governed by a "logic," which involves an appeal to what the theory can explain, to (other) general methodological virtues it displays, and to practical costs of pursuing it. Schema R_3 represents the general form of this reasoning. But it only partially reflects Bohr's reasoning. Although he does defend his theory by providing explanations of various phenomena and he does appeal to methodological virtues of his theory, Bohr goes beyond this in two important respects. He "motivates" his theory by indicating what general type of theory he seeks and

why such a theory is of interest. And he gives non-retroductive arguments that provide independent warrant for individual assumptions (e.g., an argument from analogy with Planck's oscillators); these do not appeal to what the theory can explain, to methodological virtues such as consilience, or to practical costs of pursuit.

3. Not only does retroductive schema R_3 fail to fully represent Bohr's reasoning, but it falls short more generally, as the Rydberg example shows. In its place I suggest a more contextual schema—(B) in section 6—that takes into account "motivation" and "independent warrant."

4. The 6th claim of defenders of a "logic of pursuit" is that reasoning comprising such a logic is different in kind from that employed to establish a theory or show it probable. The arguments from experiments that Bohr later used which, he believed, showed his theory probable involved appeals to phenomena explainable by the theory, other methodological virtues such as consilience, and independent warrant. But appeals of these general sorts were made earlier in arguing that his theory was reasonable to pursue. The difference is simply one of strength (e.g., more analogical reasoning in the pursuit argument, more direct, non-analogical reasoning later). Nevertheless, claim 6 is true. Reasoning purporting to show that a theory is worth pursuing is essentially different from that used to establish it. Although they may share certain elements, including appeals to explanation and independent warrant, "logic of pursuit" reasoning is contextual in a way that "logic of establishment" is not.[41]

NOTES

1. Here and in what follows when "probability" is mentioned, I mean "high probability," not "increase in probability." The authors in question might agree that the reasoning they have in mind increases the probability of its conclusion. What it does not do is render it high. In *The Nature of Explanation* (New York, 1983), I criticize a theory which says that *e* is *evidence* for an hypothesis *h* if and only if *e* increases *h*'s probability. Evidence, I argue, requires high probability, not increase in probability. However, in what follows we do not need to reject the increase-in-probability account of evidence. What defenders of a "logic of pursuit" are saying is that even if the reasoning they have in mind provides "evidence" for an hypothesis it is very weak evidence. It does not make it reasonable to believe an hypothesis.

2. Charles Sanders Peirce, *Collected Papers*, edited by Charles Hartshorne and Paul Weiss (Cambridge, Mass., 1960), 5.171; italics his.

3. N. R. Hanson, *Patterns of Discovery* (Cambridge, 1958), 85.

4. John Stuart Mill, *A System of Logic* (London, 1959), 325.

5. Ibid., 327; Mill's italics.

6. Peirce, *Collected Papers*, 5. 181. In text citations on Peirce are to this work.

7. Martin V. Curd, "The Logic of Discovery: An Analysis of Three Approaches," in T. Nickles, *Scientific Discovery, Logic, and Rationality* (Dordrecht, 1980), 203.

8. Peirce writes: "Long before I first classed abduction as an inference it was recognized by logicians that the operation of adopting an explanatory hypothesis—which is just what abduction is—was subject to certain conditions. Namely, the hypothesis cannot be admitted, even as a hypothesis, unless it be supposed that it would account for the facts or some of them" (5.189).

9. N. R. Hanson, "The Logic of Discovery," reprinted in P. Achinstein, *The Concept of Evidence* (Oxford, 1983), 60. Among the "phenomena" explained Hanson includes mathematically expressed laws and other formulas. And among his explanations are those which involve derivations of such laws or formulas from the hypotheses "retroduced." Whether retroductive reasoning includes the derivations themselves, or just the claim that $p1, p2, \ldots$ can be derived from H, is not clear in Hanson (or Peirce). In what follows both sorts of cases will be counted.

10. On some occasions Hanson gives schemas with Peircean conclusions; e.g., *Patterns of Discovery*, p. 86.

11. Curd, "Logic of Discovery," 203.

12. For example, he writes that in retroduction "... one infers that a certain state of things *may be* true and that indications of its being so are sufficient to warrant further examination" (Peirce, *New Elements of Mathematics*, edited by Carolyn Eisle [The Hague, 1976], vol. 3, p. 203). See also the quotations following point 4 below, where Peirce speaks of taking up hypotheses for examination.

13. William Whewell, *Philosophy of the Inductive Sciences* (New York, 1967), vol. 2, 65.

14. Ibid.

15. This is not to deny that Whewell accepted retroductive reasoning. My point is only that unlike Peirce he took such reasoning to be capable of establishing an hypothesis with certainty. Peirce was, however, influenced by Whewell. As Menachem Fisch notes:

> In his early Harvard lectures ... Peirce made much of Whewell's philosophy and praised him continually for stressing, contrary to Mill, the abductive nature of scientific induction—although, of course, Peirce would coin the term only much later. But once Peirce's own account of the scientific method was formed all reference to Whewell was dropped. (Menachem Fisch, *William Whewell Philosopher of Science* [Oxford, 1991], 110)

16. Nicholas Rescher, in *Peirce's Philosophy of Science* (Notre Dame, Ind. 1978), interprets Peirce as providing a "cost-benefit" account.

17. Curd reformulates Peirce's schema R_1 so that while the premises are the same the conclusion is that there are "*prima facie* grounds for pursuing A." He then identifies "other factors in the logic of pursuit" invoked by Peirce.

18. Hanson, *Patterns of Discovery*, 85–86.

19. *Philosophical Magazine* 26 (1913): 1–25, 476–502, 857–875; reprinted in *Niels Bohr Collected Works*, edited by L. Rosenfeld (Amsterdam, 1981), vol. 2, 161–233. Page references to Bohr will be from this volume.

20. Bohr speaks of energy-quanta. These are not to be thought of as localized "particles," but simply as discrete amounts of radiated energy.

21. *Collected Works*, 530–31.

22. Numbers of other physicists reached the same conclusion. For example, James Jeans claimed:

> The only justification at present put forward for these assumptions is the very weighty one of success. ... The series of results obtained in this way are, I think, far too striking to be dismissed merely as accidental. At the same time, it would be futile to deny that there are difficulties, still unsurmounted, which appear to be enormous. (*Niels Bohr Collected Works*, 124.)

Bohr's brother Harald reported to Bohr in the autumn of 1913 that

> People here [in Gottingen] are still exceedingly interested in your papers, but I have the impression that most of them ... do not dare to believe that they can be

objectively right; they find the assumptions too "bold" and "fantastic." If the question of the hydrogen-helium spectrum could be definitively settled, it would have quite an overwhelming effect. . . . (127). [In section 7 below the latter question will be discussed.]

23. What to count as questions and what as instructions is to some extent arbitrary. We might formulate one of Bohr's questions as "How can the line spectrum of hydrogen be derived?" and say that Bohr's instructions call for answering it on the basis of Rutherford's model. Alternatively, Bohr's question might be formulated as "How can the line spectrum of hydrogen be derived from Rutherford's model?" where the previous instructions are built in. What is important is that conditions for answering the question are imposed, whether or not they are incorporated into the question itself. For more on "instructions" and how they operate, see my *Nature of Explanation*, 53–56.

24. *Collected Works*, p. 301. This is from an address before the Physical Society in Copenhagen, December 20, 1913.

25. See John L. Heilbron, *Historical Studies in the Theory of Atomic Structure* (New York, 1981), 43. These collected essays of Heilbron, particularly essay 1 ("Lectures on the History of Atomic Physics") and essay 3 ("The Genesis of the Bohr Atom," written with T. S. Kuhn), have been very helpful to me.

26. See Thomas S. Kuhn, *Black-Body Theory and the Quantum Discontinuity, 1894–1912* (Oxford, 1978), 5.

27. Max Planck, *The Theory of Heat Radiation* (New York, 1959), 153. This is Planck's so-called "second theory." Unlike the first one it assumes that only emission, but not absorption, is discontinuous.

28. "On the Spectrum of Hydrogen," *Collected Works*, vol. 2, p. 296.

29. There he writes: "Now the essential point in Planck's theory of radiation is that the energy radiation from an atomic system does not take place in the continuous way assumed in ordinary electrodynamics, but that it, on the contrary, takes place in distinctly separated emissions, the amount of energy radiated out from an atomic vibrator of frequency ν in a single emission being equal to $th\nu$, where t is an entire number, and h is a universal constant." (p. 164).

30. See my *Particles and Waves* (New York, 1991), 237, note 7.

31. Peirce and Hanson make no requirement that the phenomena explained seem related, only that they are "surprising."

32. See my *The Nature of Explanation*, chapter 4, for a general discussion of the contextual nature of evaluating explanations.

33. For cases of these sorts, see my *Particles and Waves*, Essay 8.

34. Clearly (B) does not provide a set of necessary conditions, since it only covers cases in which S is aware of the theory T and has in mind a set of questions and instructions. But this is not necessary. In 1913 it might have been reasonable for a physicist who did not yet know Bohr's theory to pursue that theory. For such cases, we might consider what S's questions and instructions would be if S came to be aware of T.

35. J. W. Nicholson, "The Spectrum of Nebulium," *Monthly Notes of the Royal Astronomical Society* (1912): 49.

36. P. 117. This claim is also made by Bohr in an unpublished draft of a letter to *Nature* in 1913, p. 270.

37. See "On the Quantum Theory of Radiation and the Structure of the Atom," *Philosophical Magazine* 30 (1915): 400; in *Collected Works* vol. 2, p. 398.

38. *Collected Works*, vol. 2, p. 399.

39. Franck and Hertz assumed that 4.9 volts represented the energy necessary to ionize the mercury atom, i.e., to remove an electron completely from an atom. But on Bohr's theory the latter figure should be 10.5 volts. Bohr interprets 4.9 volts not as the ionization

potential of mercury but simply as the energy required to produce a transition from the normal state to some other stationary state. He observes in a footnote (p. 409) that his theoretical calculation of 10.5 volts for the ionization potential of mercury is the same order of magnitude as the value obtained by McLennan and Henderson to be the minimum enorgy necessary to produce the usual mercury spectrum.

40. See *Particles and Waves*, Essay 4.

41. I am indebted to Laura Snyder and Robert Rynasiewicz for very helpful discussions.

Empiricism, Objectivity, and Explanation

ELISABETH A. LLOYD AND CARL G. ANDERSON

Wesley Salmon, in his influential and detailed book, *Four Decades of Scientific Explanation*, argues that the pragmatic approach to scientific explanation, "construed as the claim that scientific explanation can be explicated entirely in pragmatic terms" (1989, 185) is inadequate. The specific inadequacy ascribed to a pragmatic account is that objective relevance relations cannot be incorporated into such an account. Salmon relies on the arguments given in Kitcher and Salmon (1987) to ground this objection. He also suggests that Peter Railton's concepts of the ideal explanatory text and explanatory information (Railton 1981) can provide what the pragmatic approach lacks. This suggestion is not a conclusion of course; we read it as the promotion of part of a research program aimed at forging a greater consensus on scientific explanation—an admirable goal. However, we do not see the pragmatic approach as inadequate. We will show that a synthetic account inspired by Salmon's adaptation of Railton would be equivalent to van Fraassen's pragmatic account in three respects: accepting or rejecting requests for explanation; the practice of giving scientific explanations; and the evaluation of the goodness of explanations. We include all three under the general rubric of explanatory "practice." Admittedly these are not the only three features by which an account of explanation might be evaluated. Roughly, we mean to show that a synthetic account cannot do a better job of accounting for the scientific practices which are of importance to the constructive empiricist, and therefore no argument can be presented to the constructive empiricist to convince her that by her own standards the synthetic account is superior.

Nevertheless, consensus is still possible, for we believe that an account developed along the lines suggested by Salmon can give an equally adequate account of the practice of scientific explanation as well. What distinguishes such an account from a purely pragmatic one is a commitment to a certain conception of objectivity. Though this conception of objectivity is manifested in the metaphysics of the account, it does not affect the account of explanatory practice. Therefore, advocates of both realism and anti-realism can see that each account of explanation is an adequate description of explanatory practice, and that one's prior commitments alone determine which account will be found more sympathetic. That is also to say that if one wishes to debate the merits of realism and anti-realism, one ought not to look to an account of scientific explanation for knock-down arguments, for one can only find there what one brings to the account to begin with.[1]

1. CRITICISMS OF A PRAGMATIC APPROACH TO EXPLANATION (OR: THE INCOMPLETENESS PROOF OF THE PRAGMATIC APPROACH)

Kitcher and Salmon (1987) claim to show that there are serious shortcomings of van Fraassen's pragmatic account of explanation. They focus on the solution that van Fraassen offers for the problem of the "asymmetries of explanation." Briefly, the problem runs as follows. There are cases in which very closely related arguments, equally well supported and equally satisfactory as far as a given account of explanation goes, are perceived to have completely different explanatory value. In van Fraassen's example, the challenge is to distinguish between: (1) the explanation that derives the length of the shadow from the height of the tower; and (2) the explanation that derives the height of the tower from the length of the shadow (1980, 104, 132). The mathematical calculations involved are the same, but, under the usual interpretation, one seems to be an explanation, while the other does not. In addressing the problem of asymmetry in explanation, one is required to explain, in a principled manner, the basis for this distinction.

Van Fraassen's solution is to remind us that explanations are relative to contexts, and what counts as an explanation in one context will not in another. His story of the Tower and the Shadow is designed to demonstrate that, contrary to our usual intuitions, in *that* context, the height of the tower is explained by the length of the shadow (1980, 132). Kitcher and Salmon seem persuaded that this example does describe a context in which a strange explanation can be accepted, but they deny that this approach will solve the "*traditional* problem of asymmetries of explanation" (1987, 316). They believe that the fundamental flaw in van Fraassen's account of explanation is that he has no adequate way to restrict relevance relations. Let us examine how they reach this conclusion.

In van Fraassen's formal framework, a request for explanation is a question Q, characterized as an ordered triple $\langle P_k, X, R \rangle$, where P_k is the topic of the question, X is the contrast class of propositions (including P_k) and R is a relevance relation between propositions and the ordered pair $\langle P_k, X \rangle$. The question can be phrased in quasi-natural language as: Why is it the case that P_k, in contrast to the other members of X, given R?

First, Kitcher and Salmon note that whether an answer A is relevant "depends solely on the relevance relation R. If A bears R to $\langle P_k, X \rangle$ then A is, by definition, the core of a relevant answer to Q" (1987, 318). They then claim that van Fraassen "incorporates no relevance requirement on R in the formal characterization" (1987, 318). Finally, Kitcher and Salmon claim to show that "the lack of any constraints on 'relevance' relations allows just about anything to count as the answer to just about any question" (1987, 319). Such a conclusion involves a serious misreading of van Fraassen, as has been pointed out by Alan Richardson (1992). We will review briefly just where this misreading occurs, and what consequences it has for the evaluation of a pragmatic account of explanation.

Kitcher and Salmon argue persuasively for the importance of placing restrictions on relevance relations. Without some restrictions we would be committed to the conclusion that "for any pair of true propositions, there is a context in which the first is the (core of the) only explanation of the second" (1987, 319). Hence, some restrictions on relevance relations are necessary. Kitcher and Salmon then investigate what restrictions are offered on van Fraassen's account, and conclude that, since on van Fraassen's account only the contrast class of the questioner restricts the relevance relation, "his theory is committed to the result that almost anything can explain almost anything" (1987, 322).

They offer an example intended to support their conclusion. Consider why JFK died on Nov. 22, 1963, where P_k = JFK died 11/22/63, $X = \{$JFK died 1/2/63,. . ., JFK died 12/31/63$\}$, and R is a relation of astral influence. Kitcher and Salmon conclude that an answer may consist in a *true* description of the positions of the stars and planets, and the prediction of JFK's death may follow from astrological theory.

Kitcher and Salmon claim that van Fraassen must accept such an explanation because he has only subjective limitations on relevance relations. Now, what is wrong with the JFK explanation? Kitcher and Salmon's diagnosis of the problem is that "in the context of twentieth-century science, the appropriate response to the question is rejection. According to our present lights, astral influence is not a relevance relation" (1987, 322). They claim this to be an example of the crucial deficiency of van Fraassen's pragmatic account of explanation, and they chide him that he "ought to be equally serious about showing that relevance is not completely determined by subjective factors" (1987, 324). They recommend that "the set of genuine relevance relations may itself be a function of the branch of science and of the stage of its development" (1987, 325).

In summary, Kitcher and Salmon conclude that, since a pragmatic account cannot delineate what ought to count as objective relevance relations, it cannot provide an adequate account of explanation. Salmon relies on this very same argument in his long discussion of the virtues and vices of various theories of explanation (1989).

The problem for Kitcher and Salmon, however, is that, as Alan Richardson has shown, van Fraassen provides much more than purely subjective means for delimiting relevance relations. In fact, van Fraassen has explicitly recognized two restrictions on relevance relations:

Explanatory factors are to be chosen from a range of factors which are (or which the scientific theory lists as) objectively relevant in certain special ways—but. . . the choice is then determined by other factors that vary with the context of the explanation request. To sum up: no factor is explanatorily relevant unless it is scientifically relevant; and among the scientifically relevant factors, context determines explanatorily relevant ones (van Fraassen, 1980, 126).

In other words, there are two levels of delimitation. The background of accepted scientific theories at the time serves to limit what counts as "scientifically relevant"; and, *within those limits*, salience to the interests of the questioner further constrains the R relation. Going back to the Kitcher and Salmon example, then, van Fraassen's view of explanation would *also* reject the JFK question, on the grounds that astrology is not an accepted scientific theory, just as Kitcher and Salmon would (Richardson [1992] gives a careful and full discussion of this point).

Kitcher and Salmon actually quote the crucial part of the above passage in their discussion of this very problem (1987, 324). However, instead of recognizing this as a constraint on requests for explanations, they consider it only on the level of the evaluation of answers. In their article it is clear that Kitcher and Salmon mistakenly believe that van Fraassen cannot reject the JFK question. They further confuse the situation by investigating how answers to the question would be evaluated on van Fraassen's account. We must stress that it is a basic logical error to criticize van Fraassen's account of the evaluation of answers to a question, when everyone agrees that that question ought to be rejected, and thus not answered at all. In that misguided criticism, Kitcher and Salmon attribute to van Fraassen the position that the astrological answer to the JFK case can only be rejected if A violates the background knowledge K (1987, 324). Kitcher and Salmon therefore take themselves to deliver a decisive blow against van Fraassen by pointing out that A and K are consistent, thereby apparently depriving van Fraassen of any means to criticize the astrological explanations of JFK's death.

Kitcher and Salmon have missed the role scientific theories play in the rejection of requests for scientific explanation. Their argument involves a clear violation of the logic of van Fraassen's explicit limitations on relevance relations, quoted above. The JFK-astrology question itself would have been rejected by van Fraassen as being outside the boundaries of objective scientific knowledge.

It may be that Kitcher and Salmon believe that the only reason not to answer a request for scientific explanation is that the question does not *arise* in the given context. At one point in their article they state that "If the question arises in the given context, it is normally appropriate to provide a direct answer" (1987, 318). According to van Fraassen, a why-question Q *arises* in a context K if (1) the background knowledge K implies both the topic of the question P_k and the falsity of all other members of the contrast class X; and (2) K does not imply that all propositions bearing the relevance relation to $\langle P_k, X \rangle$ are false (see 1980, 144–46). To establish by condition (2) that Q does not arise requires a relatively strong position. There may be cases where science has not established this position, and thus a question arises although there is nothing in science to provide an answer to it. Because such an intermediate position is possible, there may be contexts in which a question does arise, though when interpreted as a request for scientific explanation it is inappropriate to provide a direct answer.

Stepping back from the details of the criticism, we can see that Kitcher and Salmon's criticism of a pragmatic account of explanation expresses typical realist concerns. They believe that only *objective* relevance relations can pick out those answers and theories that are *true* (1987, 329). They seem to assume that van Fraassen, because of his empiricism, cannot possibly have or use "objective" relevance relations in his pragmatic account of explanation. The plausibility of that assumption depends on the demand for a strong metaphysical view of objectivity, which will be discussed below.

2. RAILTON AND THE IDEAL EXPLANATORY TEXT

In *Four Decades of Scientific Explanation*, Salmon uses the arguments discussed in the previous section to support his conclusion that a pragmatic account of explanation must be augmented by some account which provides objective relevance relations, such as Peter Railton's (Salmon 1989, 161, 185). Railton does appear to have a metaphysically richer account of scientific explanation. Railton explicitly appeals to the concept of the ideal explanatory text in order to distinguish between correct explanations and merely pragmatically acceptable ones (1981, 243). This distinction will be the focus of our considerations, for it is the basis of Salmon's hope that Railton can supply the means to provide the objectivity which he thinks the pragmatic account lacks. To fix our interpretation of Railton's account, let us examine the role of the ideal explanatory text and of explanatory information in Railton's "Probability, Explanation, and Information" (1981). We will not rehearse the finer details of Railton's presentation, nor of Salmon's summary in his 1989 book.

Railton sees the ideal explanatory text as defining one end of a spectrum of explanatoriness. Railton acknowledges, as everyone must, that the sentences used in the concrete scientific explanations given in practice are terse, abbreviated, and often ambiguous. These features of concrete explanations contrast sharply with the ideals of rigor and exhaustiveness embodied in the deductive-nomological model that Railton and many others view as a paradigmatic formalization of the traditional conception of explanation. Railton's idea is that this contrast arises from the distance between these two types of explanation on a continuum of explanatoriness. To his credit, Railton rejects the simplistic partition of this continuum into classes of "explanations" and "non-explanations":

> The answer lies in not drawing lines, at least at this point, and instead recognizing a continuum of explanatoriness. The extreme ends of this

continuum may be characterized as follows. At one end we find what I will call an *ideal D-N-P text*. . . . At the other end we find statements completely devoid of what I will call 'explanatory information' (1981, 240).

To give substance to the intuition that explanatoriness is a matter of degree, Railton introduces the notion of explanatory information.[2] A candidate explanation is explanatory to the extent that it conveys explanatory information:

Consider an ideal D-N-P text for the explanation of fact p. Now consider any statement S that, were we ignorant of this text but conversant with the language and concepts employed in it and in S, would enable us to answer questions about this text in such a way as to eliminate some degree of uncertainty about what is contained in it. To the extent that S enables us to give accurate answers to such questions . . . we may say that *S provides explanatory information concerning why p* (1981, 240).

We may question the metaphysical commitments of such an account. Of course, there need be no single answer. One may take a fictionalist line on the ideal explanatory text, or one may take a Platonist line (as considered by Salmon, 1989, 194, n. 25). Our suggestion is that, given the work Salmon sees it as doing, the ideal explanatory text be considered as that theoretical entity which dictates the content of the objective statements demanded by our practices of scientific explanation.[3] If this is the role of the ideal explanatory text in an account of explanation, we can easily see that the conception of objectivity brought to an account will affect its metaphysical status.

Let us consider how the ideal explanatory text might be seen as the provider of objectivity. It now appears that there is wide agreement that explanations—understood concretely—are to be viewed as tripartite relations involving theories, facts, and question contexts, or what Salmon more carefully calls "salience" (1989, 161). What aspects, then, are abstracted away in the idealization that leads from concrete explanations to ideal explanatory texts?

The obvious answer is—the salient factors. Intuitively, we recognize salient factors as the category of the most particular and specific interest-laden features of an explanatory context that influence the content of concrete explanations. We contrast these with the more general features such as the theories held by conversants and the facts of the explanandum situation. Regarding these two, we can say with confidence that the facts, by their objectivity, are clearly not to be abstracted from. The case of theories requires a more involved investigation.

In one formulation, Railton presents a schema for an ideal text that is tied to a specific theory; this schema includes a theoretical derivation of a probabilistic law (1981, 236). It is natural to presume that if any such text must be in a theoretical language, the idealization to an ideal explanatory text does not abstract from theory. Railton's account is not committed to the strongest version of that conclusion, which holds that an ideal explanatory text is couched in terms of the theories held by the conversants in an explanation.

One could pursue the line that the theories of the ideal explanatory text are themselves idealized entities.[4] Railton suggests this line when he writes of the "standard context" in which the concepts and terms of an ideal explanatory text are understood (1981, 246), and of different scientific disciplines filling out different portions of the same ideal explanatory text (1981, 247). Such a position includes a stronger conception of the ideal explanatory text, and it opens the possibility that the ideal explanatory text could fit into the framework of the ontic view of explanation which Salmon advocates. Briefly, the ontic view identifies explanations with the states of affairs referred to in concrete explanations, and views the language used in concrete explanations as a necessary, though variable, concession to our linguistic forms of communication (see Salmon 1989, 86).

The alternatives sketched above correspond to various conceptions of objectivity. The advocate of intersubjectivity will be satisfied that the ideal explanatory text is sufficiently objective if it is couched in terms of (or is a part of) scientific theories accepted by the scientific community of which the conversants in an explanation are a part. Those who favor stronger versions of objectivity, such as Salmon, may write their own standards into the characterization of the ideal explanatory text. Even a Platonist could embrace a version of the ideal explanatory text.

We conclude that the ideal explanatory text cannot abstract from theories altogether, though according to advocates of stronger versions of objectivity it need not be couched in terms of the theories actually held by conversants. With the understanding that we do not think that the metaphysics of an ideal explanatory text account of scientific explanation are fixed, we can now proceed to an examination of what the implications for scientific practice are of such an account.

3. THE PRAGMATICS OF EXPLANATION

Salmon embraces the concept of the ideal explanatory text precisely because it is not to be interpreted as 'subjective'. "The ideal explanatory text contains all of the objective aspects of the explanation; it is not affected by pragmatic considerations. It contains all *relevant* considerations" (1989, 161). Salmon allows that salience is determined by pragmatic factors, but maintains that relevance is an objective relation, and that pragmatics has no role in determining objective relations. Van Fraassen responds that a pragmatic account can include as part of the context of concrete explanations those relations the conversants justifiably accept as objective relations. If the ideal explanatory text is interpreted as included in the theory, or bundle of theories, held by the scientific community of the conversants in an explanation, then a pragmatic account can include that text in the characterization of the general background of theories and knowledge. Railton acknowledges that we must include such an element in a pragmatic characterization of concrete explanations. "Whether in a given context we *regard* a proffered explanation as embodying explanatory information, in light of the interpretation we impose on it and our epistemic condition generally, is a matter for the pragmatics of explanation" (1981, 243). We want to argue for the stronger position that any influence the ideal explanatory text has on explanatory practice can be captured by the pragmatics of explanation.

Our scientific theories list certain relations as objective, and in our practices of considering requests for scientific explanation, generating scientific explanations, and evaluating scientific explanations, we are required to take those theoretical dictates as compelling. The main point of this section can be stated briefly: not only must we take what our theories say seriously, we have no other means at our disposal to determine what counts as an objective relation.

The advocates of a strong conception of objectivity will still protest that a pragmatic account cannot make the crucial distinction we recognize between scientifically correct explanation and (merely) pragmatically acceptable explanation. We respond that the pragmatic account covers our own practice of explanation as well as that of hypothetical agents. Why do Kitcher and Salmon believe that a request for an astrological explanation of JFK's death ought to be rejected? Because "in the context of twentieth-century science" astral influence does not count as an objective relevance relation (1987, 322). That is just to say that they, two agents who are part of a scientific community holding certain theories and having specific background knowledge, believe that the best scientific theories exclude such relations. Van Fraassen's pragmatic account does not demand that all agents at all times must reject such a request. Indeed, if an agent has no scientific grounds for judging the relation of astral influence to be excluded from the class of objective relations, then by the account that agent ought not to reject the request for explanation.

A pragmatic account does succeed in making sense of a phenomenon all parties recognize, namely that in different contexts the appropriate response to a request for scientific explanation varies with the accepted scientific background theories and factual information of the conversants. Kitcher and Salmon note that there is an assumption in van Fraassen's discussion to the effect that the questioner and the answerer are operating in a common context with a common body of background knowledge K "determined roughly by the state of science at the time" (Kitcher and Salmon 1987, 319). Notice that by the terms of van Fraassen's account, a debate over whether a request for explanation contains an objective relevance relation can occur only if there is some disagreement in the scientific theories accepted by the debaters. Thus what different scientific communities will accept as legitimate requests for scientific explanation may vary as the theories held by those communities vary.

For example, suppose Dick, Jane, and Bill are examining the results of a study on smoking histories and lung cancer. Bill notices that in a certain group of k subjects only one subject, S_k, developed lung cancer. He asks Dick and Jane for an explanation, where his request question is characterized by the following.

P_k = Subject S_k developed lung cancer

X = {Subject S_1 developed lung cancer,. . ., Subject S_k developed lung cancer}

R = smoking history

Jane, a member in a scientific community that holds a biomedical theory which marks smoking history as an objectively relevant factor in the development of lung cancer, accepts the request and gives as an answer "because subject S_k smoked two packs a day for twenty years."

Now imagine that Dick declares, " 'I firmly believe'[5] that smoking history is irrelevant to the development of lung cancer." Can he thus reject the request for explanation? We do not yet know. Dick cannot justify his rejection on the grounds that his tarot card reader told him that smoking history is irrelevant to lung cancer. If he has no scientific basis for ruling out R, then the question in this example does *arise*. Dick cannot reject the question because it does not arise, nor because the relevance relation is wildly non-scientific.[6] At most, he may decline to answer the question on the grounds that his scientific theories give him no answer. That is, he may admit that he does not know the answer. If, on the other hand, he justifiably holds a scientific theory which disagrees with Jane's and marks smoking history as not relevant, then he may justifiably reject the request.

A Salmon-Railton synthetic account cannot explicate the practice of scientific explanation better than van Fraassen's. Even if we view ideal explanatory texts as realist entities we have no a priori access to these texts. We have only the current development of empirical science to give us the best explanatory text. This means that the practices of considering requests for scientific explanation, giving scientific explanations, and evaluating the goodness of scientific explanations can be fully explicated by a pragmatic account. If we believe a certain relation, such as astral influence, merits the rejection of a request for explanation, our only justification can be the acceptance of scientific theories which mark those relations as non-existent. Otherwise, the appropriate answer to a request for an astrological explanation is "I don't know." Similarly, when we give scientific explanations, we must look to the theories we accept for guidance on issues of relevance.

Consider finally the evaluation of explanations. On any interpretation of ideal explanatory texts the evaluator of an explanation must look to current scientific theories to determine whether or not something is explanatory information. How do we know when an explanation *correctly* answers questions about the relevant ideal explanatory text? Railton admits that "Not knowing fully what the relevant ideal text looks like, we evidently will be unable in many cases to say how much explanatory information a given proffered explanation carries. But again, it is not the job of an analysis of explanation to settle questions beyond the reach of existing empirical science" (1981, 243). To sum up: how an agent ought to act in these explanatory practices can be fully determined by the information available in the context. Railton, and anyone else who chooses to embrace the ideal explanatory text, must admit that a pragmatic account can adequately explicate these practices.

4. CONCLUSION

We conclude with a diagnosis of the current situation. There is significant disagreement over what should be demanded of an account of scientific explanation, but all parties agree that it must save the phenomena of explanatory practice. Non-empiricists may of course demand more, but they should not be surprised when empiricists do not demonstrate sympathy for their demands. The debates thereby joined will sound familiar, though to call such debates familiar is not to suggest that they are uninteresting or unimportant. For example, a realist may point out that for a given scientific explanation, facts about the relevant ideal explanatory text give an answer to the question: Why is this explanation scientifically correct? But note that in the eyes of the constructive empiricist this is a suspect why-question. It merits the same treatment that the constructive empiricist gives the demand for an explanation of the regularities observable in nature. The empiricist response to such questions is rejection (see van Fraassen 1980, 203). Moreover, a conception of the ideal explanatory text as something not contained in current accepted theory (whatever the metaphysical status of that something) strongly suggests that the ideal explanatory text is to be discovered. A central aim of constructive empiricism is to emphasize that scientific theories are constructed, and to replace the language of discovery with that of construction wherever possible. The constructive empiricist must therefore reject such a conception of the ideal explanatory text, since it suggests the language of discovery and by her own standards it is useless metaphysical baggage.

Critics of a pragmatic account of scientific explanation have brought an antecedent commitment to some form of metaphysical realism to the analysis of explanation. Given those commitments, they demand a strong notion of objectivity be incorporated in that analysis. Those commitments may also influence what is counted as an adequate analysis. We have demonstrated that due to the constraints on science as a human activity, any synthetic account of explanation designed to import a stronger notion of objectivity into a pragmatic account will be equivalent to van Fraassen's on the practical issues of rejection of explanation requests, giving explanations, and the evaluation of explanations. Realists may still conclude that the empiricist account of explanation is faulty because it lacks the realist conception of objectivity—see for example Philip Kitcher (1989). Similarly, empiricists may conclude that the realist account of explanation is faulty because it involves unnecessary metaphysical commitments, and because it includes notions which make no contribution to saving the phenomena of scientific practice, and which encourage the entrenchment of the language of discovery in an account of theory development.

We find the opposition between the two views to include significant common ground. The pragmatic account insists that objectivity is to be determined by empirical science. Since theories are held by agents in communities, the actions of those agents ought to be explicated in terms of what the theories they hold say about objective states of affairs. So far there is no quarrel. But the empiricist now declares that since we are able to fully explicate the practice of explanation without appeal to any fuller conception of objective relevance than "something marked by our scientific theories," there is no need for further metaphysical commitments. We suggest that the empiricist is on solid ground so far as scientific explanation is concerned. If the realism–anti-realism controversy is to be resolved, the focus cannot be on the practice of explanation. The different conceptions of objectivity that have been brought to the analyses of explanation are responsible for the difference in the development of accounts of explanation. Better we spend our time examining the deep and long-standing debates over those conceptions directly.

NOTES

1. Peter Railton has offered a similar suggestion in his essay "Explanation and Metaphysical Commitment" (1989, 224).

2. One might question whether or not explanation is a matter of degree, or how the maximum of explanatoriness is to be defined—we believe that the terse statements of a natural language explanation may be counted as fully explanatory in the contexts in which they are requested. Railton is of course free to demand that explanatory information, as a term of art, be defined as quantitative. Railton himself admits that there is no satisfactory account of this notion of information available (1981, 240), and we shall not pursue the prospects for the development of such an account.

3. In Salmon's words, "The ideal explanatory text contains all of the objective aspects of the explanation" (1989, 161).

4. Such as those theories waiting for us at "the end of enquiry."

5. These are Salmon's words used in his justification of the rejection of astral influence as a relevance relation. See (1989, 162).

6. One might argue that in the JFK example the relevance relation of astral influence is simply not a candidate for use in a scientific explanation. We take it that no such argument applies here.

REFERENCES

Kitcher, Philip. 1989. "Explanatory Unification and the Causal Structure of the World." In *Scientific Explanation*, edited by P. Kitcher and W. Salmon, 410–505. Minneapolis.

Kitcher, Philip, and Wesley Salmon. 1987. "Van Fraassen on Explanation." *Journal of Philosophy* 84: 315–30.

Kitcher, Philip, and Wesley Salmon. 1989. *Scientific Explanation*. Minneapolis.

Railton, Peter. 1981. "Probability, Explanation, and Information." *Synthese* 48: 233–56.

Railton, Peter. 1989. "Explanation and Metaphysical Controversy." In *Scientific Explanation*, edited by P. Kitcher and W. Salmon, 220–252. Minneapolis.

Richardson, Alan. 1992. "Explanation: Pragmatics and Asymmetry." Delivered to the Pacific Division Meetings, American Philosophical Association, Portland, Oregon.

Salmon, Wesley. 1989. *Four Decades of Scientific Explanation*. Minneapolis. Also published in 1989 as "Four Decades of Scientific Explanation." In *Scientific Explanation*, edited by P. Kitcher and W. Salmon, 3–219. Minneapolis.

van Fraassen, Bas C. 1980. *The Scientific Image*. Oxford.

MIDWEST STUDIES IN PHILOSOPHY, XVIII (1993)

Theoretical Explanation

R. I. G. HUGHES

By a *theoretical explanation* I mean an explanation of a regularity in the world, or in our account of the world, in terms of a scientific theory. An account of theoretical explanation should satisfy various desiderata. For example:

(1) The account should be recognizable as a part of a broader picture of explanation.

(2) It should also be linked to a general account of the structure of scientific theory.

(3) It should be descriptively accurate; it should accommodate anything we would normally consider a theoretical explanation.

(4) It should allow for the evaluation of explanations; it should show where, and on what grounds, such evaluations are made.

(5) It should be sensitive to the achievements of earlier accounts of scientific explanation, to how they succeeded, and where they fell short.[1]

These desiderata have guided the account of theoretical explanation I offer here. In capsule form, it runs as follows.

We explain some feature X of the world by displaying a model M of part of the world and demonstrating that there is a feature Y of the model that corresponds to X, and is not explicit in the definition of M.

1. EXPLANATION

All manner of things require explanation. We can ask someone to explain to us what made the Beatles preeminent, what distinguishes analytic cubism from synthetic, how to make an omelette, how the West was won, why giraffes have long necks, why the seven stars are no more than seven, and so on and so forth.

We make such requests in the hope of increasing our understanding. The explanation may also help us to acquire a skill, like the ability to make an omelette, but, in asking for an explanation rather than a demonstration, we are seeking an increased cognitive awareness of the process involved. Acquiring the skill may involve breaking a few eggs.

To provide an explanation is often to answer a question, as my examples show. Contrary to standard views, however, this is not always the case. Nonnative speakers of soccer sometimes ask me to explain the offside rule to them. I may be self-indulgent on these occasions, but there is no individual question to which my response is the answer. Likewise, if I were asked to explain the periodic table of the elements or the special theory of relativity, the request would seem perfectly legitimate, but not reducible to some conjunction of what, how, or why-questions.

Of course, some explanations of general topics provide answers to questions. Newton's explanation of the rainbow answers the question, "How is the rainbow formed?" But Newton himself describes his project very generally as, "By the discovered Properties of Light to explain the Colours of the Rainbow" (Newton [1730] 1952, 168). His explanation answers, not just a single question, but a large and not clearly specifiable family of questions. Consider Cotton Mather's encomium on Newton's achievement, written in 1712:

> This rare Person, in his incomparable Opticks, has yet further explained the Phenomena of the Rainbow; and has not only shown how the Bow is made, but how the Colours (whereof Antiquity made but three) are formed; how the Rays do strike our Sense with the Colours in the Order which is required by their Degree of Refrangibility, in their Progress from the Inside of the Bow to the Outside: the Violet, the Indigo, the Blue, the Green, the Yellow, the Orange, the Red.[2]

Mather distinguishes the question, "how the Bow is made," from various others to which Newton provided the answer; he could have cited still others, all embraced by Newton's explanation of the rainbow. My own use of the phrase "the explanation of the rainbow" is not idiosyncratic. Carl Boyer (1987), in his book on the rainbow from which the quotation is taken, talks of Kepler "believing that the explanation of the rainbow would be expedited . . ." (183), of Newton "beginning his explanation of the rainbow" (252), and of Airy, Miller, and Potter producing "the definitive explanation of the rainbow" (294). Nor is this list exhaustive.

I return to the rainbow, if not the offside rule, in section 3. Setting aside for now explanations of general topics like these, we may agree that many, perhaps most, explanations provide answers to questions. These questions may be of various kinds; the examples listed earlier all began with the interrogatives "what," "how," or "why," but other openings are possible.[3] Among philosophers of science, however, there has for forty years been a consensus verging on unanimity, not only that all scientific explanations provide answers to questions, but that the only questions they answer are why-questions.

Why should we believe this? Look again at Cotton Mather's tribute to Newton. In the list of questions to which Newton provided answers, there is not a why-question to be found. All are how-questions. Were Newton's explanations of these phenomena not properly scientific?

One might argue that to every how-question that is a request for a scientific explanation there corresponds a synonymous why-question. This would be a bad response, for two reasons. The first is that it is not obviously true. Maybe the question, "Why do rainbows occur?" is the same question as Mather's, "How is the Bow made?" but there is no why-question that naturally corresponds to, "How do the kidneys clean the blood?" Certainly it is not, "Why do the kidneys clean the blood?"

The second reason is that, even if we find a means of translating how-questions into why-questions, the result may mislead us, since the logic of why-questions is not the logic of how-questions. If we consider why-questions on their own, then it may seem obvious that every why-question presupposes a contrast class, and that implicit in, "Why this?" is the phrase, "rather than that." We may then go on to build this fact into our theory of explanation.[4] But "How this, rather than that?" is a very peculiar construction indeed; there is no plausible contrast class that we can associate with, "How do the kidneys clean the blood?" The price paid for apparently minor artificialities of translation is a skewed account of scientific explanation.[5]

As my title suggests, this essay deals only with a limited class of explanations. This class, however, contains answers to how-questions as well as why-questions. Even so, there are many scientific explanations that fall outside it. In the first place I confine myself to what it is for a *theory* to explain some feature X of the world. Secondly, I consider only cases where X is a *general regularity*, rather than the occurrence of a specific event. Each of these restrictions requires some comment.

On the analysis proposed by Peter Achinstein (1983), *explaining* is a speech act. More precisely, it is an illocutionary act. Achinstein proposes (C), below, as the fundamental condition necessary for a speaker S to have perfomed the act of explaining q by uttering u. Perhaps surprisingly, the condition makes no reference to the intended audience.

(C) S utters u with the intention that his utterance of u render q understandable (Achinstein 1983, 16).[6]

On Achinstein's account the act of explaining is logically prior to the concept of an explanation: a given proposition is only an explanation if it is expressed in an illocutionary act of explaining. Achinstein (1983, 83–88) calls this the "ordered pair" view; an explanation is an ordered pair of the form ⟨proposition, act of explaining⟩. His account emphasizes the pragmatic nature of explaining and explanation.[7] With certain reservations, it is a view I accept. It does, however, raise an immediate difficulty for my project, in that I talk of a theory explaining X, but would never claim that a theory performed a speech act.

The difficulty is more apparent than real. Even if we adopt Achinstein's approach, we may still inquire what kinds of utterances can figure in acts of explaining, or equivalently, what kinds of propositions can be the first member of an explanation. Thus we can regard "Theory T explains X" as an ellipsis for, "A speaker could use the resources of T in explaining X." My project is then to unpack the phrase "using the resources of T." For reasons that will appear I prefer that mode of expression to one that speaks in terms of the propositions, or the propositional content of T; that apart, the project knits quite happily with an analysis like Achinstein's.

If there is a problem it is whether, given the differences in acumen and background knowledge from one listener to another, one can say anything useful about making a topic understandable. Hempel, for one, thought that there was no place in the logic of explanation for pragmatic notions like comprehensibility, which varied from one individual to the next.[8]

In a sense, Achinstein has an answer to this. For him, explaining involves only the intention to make something understandable; as I have noted, his fundamental condition makes no reference to the audience. But should we accept his claim that explaining is unequivocally an illocutionary, rather than a perlocutionary act?[9] Just as we make unsuccessful attempts to persuade, we may make unsuccessful attempts to explain, and our failure may be due, not to ineptitude on our part, but to the obduracy of the listener in the first instance, and his obtuseness in the second. Or so we may think.[10] Note that even on Achinstein's account the audience cannot be utterly ignored, if only because we cannot intend what we know to be impossible.

To return to the original problem, it is true that an explanation of, say, the Aharanov-Bohm effect that appears in *Scientific American* is not the same as one that appears in Sakurai's *Advanced Quantum Mechanics*, simply because the two publications have different readerships. But, first of all, the notion of a readership itself implies some shared background knowledge and a commonality of interests. Although within a given readership individuals will differ, a general account of explanation can accommodate those differences by noting that "understandable" can be adverbially modified. The explanation appearing in *Scientific American* will make the effect completely understandable to some readers, barely understandable to others, and something in between to the majority.

Secondly, it will prove a virtue of my account of theoretical explanation that it allows for the possibility of alternative explanations of the same effect, each pitched at a level appropriate to a particular audience.

I turn now to the second restriction on my account. My reasons for dealing with general regularities rather than particular events are the same as Michael Friedman's. In his essay, "Explanation and Scientific Understanding" (1974, 188–89), he points out that explanations of generalities are much more typical of the scientific literature than explanations of particular events, yet discussion of the latter has dominated the philosophical literature on explanation. Standard accounts, from Hempel and Oppenheim onwards, contain merely a "promissory note" (Salmon 1989, 10) concerning the former. Like Friedman, I would like to make good that promise, although in doing so I will depart considerably from what was once the received view.

2. MODELS

David Pines (1987, 65) describes a plasma as "a gas containing a very high density of electrons and ions." He goes on to write,

> In any approach to understanding the behaviour of complex systems, the theorist must begin by choosing a simple, yet realistic model for the behaviour of the system in which he is interested. Two models are commonly taken to represent the behaviour of plasmas. In the first, the plasma is assumed to be a fully ionized gas; in other words, as being made up of electrons and positive ions of a single atomic species. The model is realistic for experimental situations in which the neutral atoms and impurity ions, present in all laboratory plasmas, play a negligible role. The second model is still simpler; in it the discrete nature of the positive ions is neglected altogether. The plasma is thus regarded as a collection of ions moving in a background of uniform positive charge. Such a model can obviously only teach us about electronic behaviour in plasmas. It may be expected to account for experiments conducted under circumstances such that the electrons do not distinguish between the model, in which they interact with the uniform charge, and the actual plasma, in which they interact with positive ions. We adopt it in what follows as a model for the electronic behaviour of both classical plasmas and the quantum plasma formed by electrons in solids. (Pines 1987, 67)

This is a very revealing passage. Four things in particular are worth noting. The first is that the theorist seeks to understand a system's behavior. The second is that the achievement of this understanding involves the choice of a model to represent this behavior. The third is that more than one model can be used, and that these models may be at odds with one another; in this example, one model represents positive charge as discrete and associated with particles of matter, while the other represents it as uniformly distributed through the plasma. The fourth is that the model Pines adopts involves, so our best theory tells us, a much higher degree of misrepresentation than does the other.

On one issue scientists and philosophers of science are now in accord. Both agree that to present a theory is to display a model, or a class of models.[11] Philosophers of science refer to this view as the *semantic view* of theories. To a *theoretical definition* (Giere's term), which specifies a certain model or class of models, is added a *theoretical hypothesis* to the effect that a certain part of the world, or of the world as we describe it, may be regarded as a system of that kind.[12] The phrases "part of the world" and "may be regarded" are deliberately vague. I will discuss each in turn.

The part of the world that the theory deals with I will call the *subject* of the model. It may be very narrowly specified, as with Bohr's model of the hydrogen atom, or more generally, as in the case of Pines's model of a plasma, or very generally indeed; the intended subjects of Newtonian mechanics included all systems of material bodies.

Theories, and the models they define, are of various kinds and are related in various ways. Consider again the model of a plasma that Pines adopts. This is a *constitutive model*.[13] It gives us a picture of how matter, in one particular state, is made up. We also need to know how the model behaves. Here again, Pines gives us two choices; we can regard it either as a classical or as a quantum system.

I will use the classical case for illustration. In this case the behavior of the plasma is regarded as governed by Newtonian mechanics, augmented by Coulomb's law for the forces between electrical charges. I will call the resulting theory "N^+." Although a practising physicist, in applying a familiar theory like this, does not need to think how it can best be articulated, on the semantic view of theories the most perspicuous way to do so is again in terms of models. The particular variant of the semantic view that I shall use here is known as the "state space" version of the semantic approach.[14] Though this presentation departs from everyday scientific practice, it is not one that a physicist would find peculiar.[15]

On this account we specify first the elements of an abstract system and the way they are to be described, and secondly the way the state of that system varies with time. An N^+-system consists of a set of points (each to be thought of as a particle) moving through space. To each point we associate two numbers which do not change with time (the mass and charge of the particle).[16] The instantaneous state of the system is defined by the position and velocity of each particle at that instant. The theory then tells us how the state will evolve through time. The evolution is governed by the forces that the environment exerts on the particles and the forces they exert on each other—equivalently, by the Hamiltonian of the system.

This specification of the state of a system and its evolution comprises the theoretical definition of the models of the theory. When Pines, making the second of his two choices, investigates a classical plasma, he is adopting the theoretical hypothesis that the kind of system he is looking at can be regarded as an N^+-system. This example, incidentally, shows the need for the phrase "the world as we describe it" that appears in the gloss on a theoretical hypothesis. Classical physics is here being applied to the *model* of a plasma that Pines has adopted.

The class of N^+-systems is a subset of the very general class of Newtonian systems, systems governed by the theory of Newtonian mechanics. Models of this degree of generality of application I call *foundational models*.[17] At a deeper level yet are models of more abstract subjects. Newtonian mechanics presupposes a particular account of space-time. This too is described in terms of a class of models. It is represented as a four-dimensional structure involving a three-dimensional Euclidean space, and an independent dimension of time isomorphic to the real number line. Though the distinction between this kind of model and a foundational model is not a sharp one, I call these models *structural models*.

I have traced the relation between a specific model of a plasma and the models posited by one of the most fundamental of all our theories. Equally important are the models of the theoretical and instrumental processes which connect the kinds of theories I have been discussing to the phenomenal world. To discuss these, however, would take us too far afield. For present purposes I need only draw attention to what all these diverse models have in common.

When we say that the subject of the model "can be regarded" as a system of the specified kind, we mean that the model *represents* the subject, in at least two senses of the word. In the first place it represents it in the way that a portrait represents the sitter. This representation, however, is always representation-as. The kinetic theory of gases represents a gas *as* a Newtonian system in the same way that Joshua Reynolds represented Mrs. Siddons *as* the Muse of Tragedy.[18]

The model can also represent the subject in another sense, a sense similar to that in which someone can represent me at an event I cannot myself attend. When we are satisfied that our theory is a good one, the model can stand in for its subject, not, it is true, in circumstances identical to those the subject might encounter, but in their analogues. That is to say, models may be used not merely to reproduce known patterns of behavior, but to simulate behavior under unfamiliar conditions—hence their usefulness in prediction. As with an understudy, the respects and the degree to which we expect the behavior of the model to capture that of its subject will vary. We can rely on Newton's model of the solar system to conform to the kinematic behavior of the actual solar system very precisely indeed, barring small discrepancies involving the planet Mercury. In contrast, Pines adopts a model of a plasma which he expects to be adequate only under restricted conditions, that is, "under circumstances such that the electrons do not distinguish between the model, in which they interact with the uniform charge, and the actual plasma, in which they interact with negative ions" (Pines, 1987, 68).[19]

My understudy and the model differ, however, in that in the latter case various ontological attitudes towards its components are possible. We may take them to correspond to items in the world, or we may not. My emphasis on representation might suggest a firmly anti-realist attitude. In fact, I merely think it unlikely that there is an argument to be made for the reality of these constituents on the basis of the empirical adequacy, or the explanatory power of the model, however remarkable these may be.[20] Arguments for realism must seek support from elsewhere. In particular, I hold the view that the questions of realism and of the adequacy of explanation are independent of one another, though I shall not argue for it here.

3. THE RAINBOW

As a case study in explanation I will use Newton's explanation of the rainbow. I choose this example because it is simple enough to be presented in full, yet rich enough to illustrate a large number of issues. Though formulated in terms of an out-dated theory of light, it is not just of historical interest. Versions of it still appear in highly reputable school textbooks, and I will compare Newton's presentation with the one offered by Noakes in a textbook "written for the use of Sixth Form students [British students in their last two years of high school] who are specializing in science" (Noakes 1953, v), and with another appearing in *Physics*, a text devised in the 1960s by the Physical Sciences Study Committee (PSSC) on behalf of American students of the same age but with fewer mathematical skills.[21] A composite version runs as follows. I apologize if it reminds the reader of days best forgotten.

Rainbows are formed when the sun shines on falling rain. To observe them, the observer must face the rain, with his back to the sun. They appear because some of the sun's rays enter the raindrops, are reflected within them, and then emerge towards the observer.

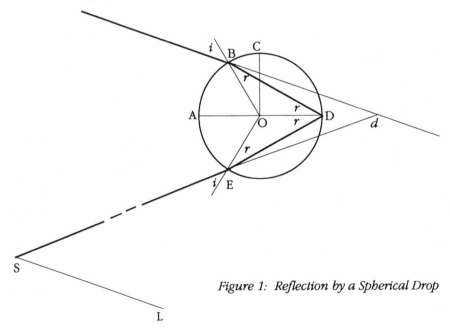

Figure 1: Reflection by a Spherical Drop

(A) Consider a ray striking a spherical drop at the point B with angle of incidence i, and continuing inside the drop with angle of refraction r (see figure 1). From Snell's law, $\sin i = \mu \sin r$, where μ is the refractive index (for Newton, the "degree of refrangibility") from air to water for the light. The ray is internally reflected at D. The law of reflection, and the geometry of the drop, together guarantee that, when the ray leaves the drop at E, the angle of incidence on the water-air surface is r. Since the trigonometrical relationship expressed by Snell's law is independent of the direction of travel of the light, the angle of emergence is i. The ray suffers three deviations; it is refracted on entering the drop, reflected at the rear surface of the drop, and refracted again on leaving it. By simple geometry, the total deviation d is equal to $180° + 2i - 4r$.

(B) Since r is obtainable from i by Snell's law, it follows that d is a function of i. It turns out that, as the point of incidence B moves up the sphere from A to C, the angle d decreases from 180° to a minimum value of about 140°, and then increases again to about 165°. Elementary calculus shows that, at minimum deviation, $\cos i = \sqrt{\{(\mu^2 - 1)/3\}}$. For a wide range of angles of incidence, the deviation will be at or near this minimum value. Hence we would expect the drop to reflect a lot of light in this particular direction. Conversely, we would expect a spectator looking at a rain shower to receive most light from those drops situated at places from which light reflected with minimum deviation would reach him.

We are now in a position to answer individual why-questions: Why do these drops form a bow? And why does the light appear colored?

(C) To answer the first question, imagine a spectator receiving a ray of light reflected at minimum deviation from a drop. We may represent the spectator's eye by a point S on the emergent ray in figure 1. The radius of the drop is very small compared with its distance from the spectator, so that SO is effectively collinear with the emergent ray. It follows that any drop whose center O' lies on SO will also reflect light at minimum deviation towards S. If we now draw a line SL through S parallel to the incident ray, i.e., parallel with the sun's rays, and then rotate the plane of the paper about the axis SL, then (a) the position of S will remain unchanged, (b) the line representing an incident ray will remain parallel to SL, and so can continue to represent a ray of light from the sun, and (c) the points O, O', etc., will move in circles about the axis SL. Just those drops whose centers are represented by points on these circles will reflect light with minimum deviation towards the spectator. Since these circles all lie on the surface of a cone with apex S, the spectator will see these drops as a bright bow in the sky.

(D) The colors of the bow are now easily explained. The refractive index μ for light going from air to water varies with the color of the light. For red light it is about 1.333, for violet light about 1.346. For this reason, the value of d at minimum deviation differs according to the color of the light; the minimum deviation suffered by red light is about 2° more than that suffered by violet. Hence "the position of the drop determines not only whether it will contribute light to the rainbow but also the color of its contribution" (PSSC 160, 223). The result is a bow whose inner edge is violet, and whose outer edge is red.

For simplicity I have omitted from this account any discussion of bows of higher order. In addition to the primary bow discussed above, there are bows of the second, third, and higher orders, produced by rays that have been reflected within the drop twice, three times, and so on. One can extend the minimum deviation explanation without difficulty to cover these bows, and, in fact, Newton, Noakes, and the PSSC all do so.[22]

These authors differ, however, in the detail in which they present the minimum deviation explanation. The most obvious differences appear in their presentations of the material of paragraph (B). Noakes, for example, works carefully through the proof that, at minimum deviation, i is given by $\cos i = \sqrt{\{(\mu^2 - 1)/3\}}$, and then adds, rather laconically, "The value of dd/di will be small in the neighbourhood of the minimum value of d, and thus for each colour the direction of minimum deviation will be that of greatest intensity" (Noakes 1953, 118).

In contrast, Newton ([1730] 1952, 171) simply quotes the expression for i and assures his readers that, "The truth of all this Mathematicians will easily examine." His explanation of why minimum deviation is associated with greatest intensity is, however, considerably more ample than Noakes's. He writes,

> Now it is to be observed, that as when the Sun comes to his Tropicks [i.e., at the solstices], Days increase and decrease but a very little for a great while together; so when by increasing [the angle i] the Angles [of deviation] come to their Limits, they vary their quantity but very little for some time together and therefore a far greater number of the Rays which fall upon all the points B in the Quadrant AC, shall emerge in the Limits of their Angles than in any other Inclination. (Newton [1730], 1952, 171)

Newton talks here of "Angles" in the plural, because his analysis is not confined to the primary bow.

The PSSC makes no gesture towards a mathematical derivation of the minimum deviation result. Instead they include a photograph of an experiment to show that the result is to be expected. A pencil of rays is passed through a hole in a screen, and strikes a small refracting sphere. The light reflected back to the screen illuminates a circular area, with most of the light concentrated at the outer edge of the circle. One may ask to what extent this demonstration is explanatory—especially when compared with the explanations offered by Newton and Noakes—and I shall return to this question in the next section.

The differences between the three presentations reflect differences in the authors' assessments of their intended audiences. Nonetheless, all three versions share a common core. In each case the rainbow is explained in terms of a model in which a parallel beam of light is reflected by a set of refracting spheres. The behavior of the light is then modeled in terms of the ray theory.

Some of this behavior is summarized by the laws of refraction and reflection of light, which are brought into play in paragraphs (A) and (D). For this reason it is tempting to give a Hempelian analysis of the explanation, and to say that the explanandum was deduced from a set of laws and initial conditions, none of which were superfluous to the derivation. But this temptation should be resisted. In the first place, as Cartwright has pointed out (1983, 46–47), the simple form of Snell's law used here can only be applied to refraction within an optically isotropic medium; the universality we require in a law is absent. If, however, we think of Snell's law simply as part of the specification of a model that represents the behavior of light in a restricted set of circumstances, no such requirement exists. This brings me to my second point.

The so-called "laws" employed here are all expressed in terms of the particular model we are using. The angles of incidence, refraction, and reflection that enter our calculations are angles between normals to the drop's surface and *rays* of light. Since the terms in the laws refer to elements of a theoretical model, there is no interpretation of the laws which is independent of theory.[23] A similar point is made by Nagel; from our perspective, the passage where he does so ends with a splendidly ironic twist:

> Experiments on beams of light passing from a given medium to a denser one show that the index of refraction varies with the source of the beam. Thus, a beam issuing from the red end of the solar spectrum has a different index of refraction than has a beam issuing from the violet end. However, the experimental law based on such experiments is not formulated in unquestionably observational terms (e.g. in terms of the visible colors of light beams) but in terms of the relation between the index of refraction of a light ray and its wave frequency. (Nagel 1961, 82–83)

There is also a further assumption about light concealed in paragraph (B) of the explanation. Intensities of light are assumed to be monotonically additive: the more rays emerging in a given direction, the greater the intensity of the light emitted in that direction. On a Hempelian analysis this assumption would have to figure in the explanation as a law of nature. Yet the fact that light intensities do not, in general, behave in this way is now so well established that no one would grant this assumption any nomological force. In contrast, within Newton's model, the assumption is perfectly acceptable. He writes, "By the Rays of Light I understand its least Parts, and these as well Successive in the same Lines as Contemporary in several Lines" (Newton [1730] 1952, 2). That light has such "least parts" is, on our account, just part of the theoretical definition of the model Newton is working with. A natural inference from this "least parts" assumption is that the intensity of light on a given surface is proportional to the number of rays striking it per unit time.

There are, of course, models with greater empirical adequacy than Newton's, notably the wave model developed in the nineteenth century by Young, Fresnel, and their successors. An explanation of the rainbow in terms of the wave theory, generally known as "Airy's theory," was available from about 1840. This theory explained more and was in closer agreement with observation than Newton's.[24] Yet, as Boyer records (1987, 309), ten, even twenty years after it appeared, workers in the field expressed their frustration that textbooks still explained the rainbow in Newtonian terms. But in their complaints we should distinguish two demands, one legitimate, the other not. The legitimate demand was for a textbook that presented Airy's theory. The illegitimate demand was that the Newtonian explanation be "banished from the teaching of science."[25]

The demand is illegitimate because, as Noakes and the PSSC recognized a century later, Newton gave us an understanding of the rainbow. In doing so he did not explain all the phenomena associated with rainbows; he said nothing, for example, about supernumerary bows, colored bands sometimes observed within the inner edge of the primary bow and beyond the outer edge of the secondary.[26] Without inconsistency, Noakes can acknowledge the omission, even as he proffers the minimum deviation account as his explanation of the rainbow. His section on the rainbow ends with this paragraph:

> A complete theory of the rainbow embracing both the true and the supernumerary bows has been worked out, and according to this the angular radius of the primary bow should be a little less than the value given by the minimum deviation explanation and that of the secondary bow a little greater. (Noakes 1953, 120)

Noakes is fully aware that the "complete theory of the rainbow" has little in common with the minimum deviation explanation. Indeed, a large part of his book is concerned with the wave model of light. Nonetheless, in writing this paragraph, he is not suggesting that students should simply forget what they have just read. He regards the minimum deviation explanation, as I do, as a genuine explanation, albeit one couched in terms of a theory of limited usefulness. How good an explanation it is, and in what respects, remains to be seen.

4. UNDERSTANDING

In 1963 Mary Hesse put forward "a view of theoretical explanation as metaphorical redescription."[27] This proposal has never received the attention it deserves.[28]

Hesse draws on the account of metaphor given by Max Black, on which a metaphor involves a non-literal description of a primary system in terms of a secondary system. Thus when Prospero says of his brother Antonio,

He was
The ivy which had hid my princely trunk
And suck'd my verdure out on't

a primary system, Antonio and Prospero, is seen in terms of a secondary system, ivy and tree-trunk. On this account a metaphor does not merely compare the two systems; it is not replaceable by a statement of the points of similarity between the two. Metaphors are enlightening because they present us with a new framework of ideas within which to locate the primary system. Reynolds's portrait of Mrs. Siddons as the Tragic Muse aims at the same effect. It invites us to achieve a greater awareness of the sitter by seeing her in terms of a new set of associations.

Thus metaphor has a cognitive function. A similar function is served, Hesse argues, by the use of models in science. We may take a plasma, or the transmission of light, as our primary system (the *subject* of the model, in my terminology), and represent the one as a collection of electrons in a sea of positive charge, the other as a wave motion in a luminiferous ether. By redescribing them in this way we make their behavior understandable; that is, we explain it.

The major problem for this account of explanation, and a reason, I suspect, why it has found few advocates, is that metaphors differ from scientific models in important respects.[29] Metaphors, as Hesse points out, are intended to be understood. If a metaphor is to be successful, the secondary system must be already familiar to us, so that it arrives with a rich store of associations. No doubt some scientific models, like the fluid theories of heat and electricity in the early nineteenth century, are like this; they explain the unfamiliar in terms of the familiar. But very often the reverse is true. To repeat a point that many others have made before me, scientific explanations typically explain a familiar occurrence in terms of a markedly less familiar model. We are familiar with the fact that a lump of iron usually stays in one piece instead of crumbling out againe to his Atomies. Yet only quantum mechanics can explain why it does so. And the problem is very pronounced for someone like myself, who countenances the use in explanation of highly abstract models. The models described in section 2, for example, of a Newtonian system, or of space-time, would not normally be said to carry with them a rich store of associations; their very abstractness serves to defamiliarize them.[30]

These problems raise the question: in the absence of the web of associations evoked by a familiar secondary system, whence does a scientific model acquire the cognitive function that Hesse and I ascribe to it?

To answer this we need to look more carefully at the similarities and differences between models and metaphors. They are alike in one key respect: each of them is a representation-as. The difference between them, crudely put, is that the secondary system of a metaphor is a pre-existing entity, whereas a scientific model is a free creation.

This characterization of the difference, however, requires immediate qualification. On the one hand, although a successful metaphor requires a familiar secondary system, not only are the ideas associated with it transferred to the primary system, but the complex of these associations is itself affected by the use of the metaphor. Joshua Reynolds's painting changes the way we think of the Tragic Muse as well as the way we see Mrs. Siddons. This is why Black calls his view of metaphor an "interaction view." Similarly, when the wave model of light was adopted, this affected not only the way scientists thought about light, but the way they thought about waves.

On the other hand, to describe our choice of a model as "free" is to overestimate human inventiveness. "Even when painters try to create sirens and satyrs with the most extraordinary bodies," writes Descartes, "they cannot give them natures which are new in all respects; they simply jumble up the limbs of different animals."[31] Again, Huygens's wave model of the transmission of light was suggested to him by what he knew about sound.

Notice that both these qualifications bring metaphors and modeling closer together, and to that extent undermine the original objection. For the sake of argument, however, I will take the extreme case of a mathematical model of a phenomenon, abstract enough for any associations that may have once suggested it to have withered away. What, cognitively, takes their place?

I will make two suggestions. Both involve something I have not so far emphasized, the use of mathematics in physical theory.

My first suggestion is that, when a scientific model is first presented to us, its resources are not immediately visible; we have to put it to use to discover them. We do so by matching the behavior of the model to that of its subject. Imagine that we set out to account for a particular phenomenon within the domain of a theory in terms of a model that the theory supplies. We apply to the model a set of constraints that correspond to the circumstances surrounding the phenomenon. Given these constraints, the behavior of the model can be deduced mathematically from its theoretical definition. If this behavior matches that of its subject, we are encouraged to repeat the process with another phenomenon, and then another. As an example of the process, consider how the wave theory of light is applied successively to the phenomena of interference, diffraction, and polarization.

When we learn physics, some of these applications are worked through on our behalf, others are set as examples. In either case the hoped-for result is the same. We become aware of the resources of the model. Whereas, in most cases, the significance of a metaphor takes only a brief time to sink in, a model's resources are made available to us gradually as we come to see how much can be deduced from its theoretical definition. As a succession of phenomena are shown to be successfully modeled by the theory, the models become progressively more familiar, and we become able to apply the theory ourselves; increasingly we see the subject in terms of the model, and vice versa.

The individual episodes that increase our understanding are, typically, occasions when acts of explaining take place. We learn from a textbook, a teacher, or fellow student how the theory can be applied to a specific topic. I take it to be a virtue of this account of explanation that it has the following corollary: the more we already know about Newton's theory of light, the more we will understand about the rainbow after reading his explanation of it. Another of its virtues is that it endorses our intuition that theoretical unification brings about an increase in understanding.[32]

My second suggestion is that, paradoxically, an abstract mathematical model can carry with it a greater volume of associations than any concrete secondary system. This is because the same mathematical model can have an indefinite number of realizations. In one of his lectures on physics, Feynman shows that the same differential equations can be used to represent damped mechanical oscillations and damped electrical oscillations.[33] Given this isomorphism, any intuitions we have about the former will apply to the latter, and vice versa. This strategy, of establishing isomorphisms between topics in areas of physics apparently remote from one another, is widespread in physics. And the more realizations that a mathematical model permits, the greater the range of the associations we can bring to any one of them.

With this in mind, let us return to the different explanations given by Newton, Noakes, and the PSSC for the fact that the angle of minimum deviation is also the angle of maximum intensity of the reflected light.

The PSSC presents us with a photograph of an experiment to demonstrate this and so gives us good reason to believe it to be the case. They do nothing, however, to show its connection with other phenomena and so do nothing to explain it. Newton explains it with a simile rather than a metaphor. The situation, he says, is like one involving the number of hours of daylight in each day; the length of the day does not vary much around the solstices, so that a large proportion of days of the year are approximately the same length as either the longest or the shortest day. Similarly, we should expect a large number of rays to be deviated at or around the angle of minimum deviation.

Implicit in Newton's argument is the general argument that Noakes supplies. I earlier described Noakes's analytic treatment of the topic as "laconic." So it is, but it also carries more associations than either of the others. It says that, whenever y varies continuously with x, then around an extremal point, when dy/dx is at or close to zero, the value of y does not change significantly with small changes in x. It thereby invites the reader to associate the deviation of a light ray by a spherical drop, not only with the length of the day near the solstices, but also with the height above ground of a projectile, the distribution of height in a human population, and the temperature of a bucket of water into which someone has dropped a red hot horseshoe. For some readers, of course, the mathematical analysis will itself be enough.

5. EVALUATION

In the introductory section of this essay I summarized my account of theoretical explanation as follows.

We explain some feature X of the world by displaying a model M of part of the world, and demonstrating that there is a feature Y of the model that corresponds to X, and is not explicit in the definition of M.

This bald statement can now be amplified. The "features of the world" I am concerned with are general regularities. I have ignored problems that arise in connection with explanations of particular events, particularly the problem of explaining the occurrence of an event of low probability. These are the problems that statistical relevance models and causal accounts of explanation were designed to deal with. Insofar as my discussion deals with probabilities at all, it does so by showing how statistical laws might be explained.[34]

In my discussion I distinguished constitutive, foundational, and structural models. These examples were chosen to show not only that there are different kinds of models, and that the theories associated with them can differ widely in domain, but, more importantly, that the distinction sometimes drawn between scientific models and foundational theory is better regarded as a distinction between different kinds of models; common to all of them is the fact that they should be thought of as representations. However, the boundaries between the categories I specified are not definite, nor is the list of categories exhaustive. Models of processes, of the formation of a rainbow for instance, do not fit readily into any of them—though in the case of the rainbow it is clear that Newton's emission theory, or a wave theory of light, will function as a foundational theory.

The term "demonstration" is used in the summary in its seventeenth-century sense of mathematical proof. Now the limiting case of a mathematical proof is repetition of one of the premises. Hence, to avoid explanation by stipulation, we disallow appeals to a feature of the model explicit in its theoretical definition. We do not explain the rectilinear propagation of light merely by adopting a ray theory.

By building a particular feature into the theoretical definition of the model, however, we may do more than simply rule out its use as an explanation of the corresponding feature of the world. We may actively reject a demand for such an explanation.[35] Compare Newton's attitude to his law of gravity with that of physicists a century later: whereas Newton thought that a complete physics would explain the law, by the nineteenth century the demand for an explanation was regarded as illegitimate.

My summary account of theoretical explanation spoke to its form. It said nothing about what made an explanation good or bad, nor did it raise the question of whether, and in what circumstances, a proferred explanation might conform to the specification but nevertheless be rejected.

Two sets of considerations, one pragmatic, the other theoretical, govern our evaluation of an explanation. The pragmatic are the simplest to deal with. They are summed up in the obvious precept that an explanation should be pitched at a level appropriate to its intended audience. What to Murray Gell-Mann is a splendid explanation may be no explanation at all to me. It follows that the model appealed to in an explanation should be one with which the listener is familiar, or with which he could be reasonably expected to become familiar as the explanation proceeds. Our best theory may not, in every case, provide the best explanation.

But by what standards are we to judge theories? Two criteria often canvassed are empirical adequacy and explanatory power. But the latter is not an entirely straightforward concept. For if we believe that our best explanations may not be provided by our best theories, then explanatory power cannot simply be equated with a capacity for providing good explanations. We can cash it out in terms of the number and diversity of the phenomena representable within the models that the theory supplies, but, while I shall argue that we are right to do so, two questions need to be addressed along the way. First, why does this give a measure of explanatory power rather than of the domain over which the theory is empirically adequate? Secondly, what justifies our disregard of the pragmatic dimension of explanation?

With regard to the first, we may note that scientists and historians of science do in fact distinguish the two criteria. Bohr's model of the hydrogen atom, for instance, was (and is) held to have displayed considerable explanatory power, despite the fact that the only strikingly accurate predictions it made were the wavelengths of the lines of the hydrogen spectrum. It provided explanations for a dozen other things, but for the most part got them right only within an order of magnitude (Hughes, 1990, 77–79).

It also seems plausible that, under certain circumstances, a theoretical innovation that accounted for something we already knew would be valued even if it produced no new empirical knowledge. Imagine that a new representation of fundamental particles was produced, whose only novel contribution was to tell us why massless particles never carry electric charge. This fact we know already, and so the theory would not add to our empirical knowledge. Even so, would we not value it for the understanding it provided? My surmise is that, if the model the theory provided seemed to us *ad hoc*, then we would not. If, on the other hand, the theory seemed a natural extension of one we already had on the books, or if, by postulating some new symmetry, let us say, it used a strategy with which we were already familiar, then we would. In enabling us to understand a significant fact, the theory would give an explanation of it.

Nonetheless, the proposal that explanatory power should be considered an aim of science, on a par with, but not identifiable with empirical adequacy, has sometimes met with resistance.[36] This resistance derives in part from a suspicion that an insistence on explanation is a covert, and sometimes not so covert, demand that our representations of natural processes should be governed by a set of a priori, even (perish the thought) metaphysical commitments. But the proposal I canvassed in the previous section, that theoretical explanation be viewed as metaphorical redescription, contains no metaphysical agenda. It suggests that theoretical explanations allow us to understand the world, not by showing its conformity to principles external to the theory, but by representing it in terms of the models the theory itself supplies. As we become aware of the resources of these representations, so we come to understand the phenomena they represent. Hence, as I argued earlier, the greater the variety of the contexts in which we see the theory applied, the deeper this understanding becomes.

A parallel can be drawn between linguistic and scientific understanding and will help to sever the latter from disreputable metaphysical associations. Linguistic understanding is manifested in terms of competence: our ability to recognize whether a word is correctly or incorrectly applied, to paraphrase sentences in which the word appears, and to apply it ourselves in appropriate, but not wholly familiar situations. Similarly, our understanding of a theory is displayed by the ease with which we follow explanations presented in terms of the theory, and—most significantly—by our capacity to paraphrase these explanations and to find analogues for the behavior of the models to which they appeal. We may also learn to apply the theory ourselves, albeit at the humble level of one who points out that, on the minimum deviation account of the rainbow, only the outer edge of the primary bow will be a pure color.

We may accept the notion of explanatory power and agree to measure it by the range and diversity of phenomena representable within the theory. We are still left, however, with the second question. How are we to justify our disregard of the pragmatic dimensions of explanation? Explanatory for whom? we may ask. And if the answer is that, in talking of explanatory power, we make an implicit reference to an ideal audience, the first problem merely makes way for another. If explanatory power is assessed in terms of this ideal audience, why is it relevant to the evaluation of explanations offered to a real audience? For an ideal audience the wave theory of light has greater explanatory power than Newton's emission theory; however, given Noakes's intended audience, he was quite right to present his explanation of the rainbow in terms of the latter.

Essentially the same question can be posed in a different way. How is the emphasis on theoretical unification in my account of understanding to be squared with a willingness to allow explanations of the same phenomenon to be made in terms of different theories for different audiences?

The answer is this. Among the theories that deal with a certain set of phenomena, we distinguish, first of all, a *best* theory, and secondly a class of *acceptable* theories. Explanatory power (with respect to an ideal audience) is one of the criteria by which we choose our best theory, the other being empirical adequacy. (I do not mean to suggest that the choice is always straightforward.) An acceptable theory is one that bears a relation, yet to be specified, to our best theory of this set of phenomena. When a number of competing explanations of a phenomenon are available, we should consider only those provided by acceptable theories. Among these we should choose, on any specific occasion, the one we judge to be pitched at the level appropriate to the intended audience. Should there be more than one such explanation we should choose the one provided by the better theory, again assessed in terms of empirical adequacy and explanatory power. Explanations in terms of unacceptable theories are to be rejected.[37]

To complete this account, I need to indicate what it is for a theory to be acceptable, but before doing so I will give a general idea of what the notion involves. Models, like paintings, are neither true nor false; instead they offer more or less adequate representations of the world. To say that a theory is acceptable is to say that, within a restricted ambit, the representation it offers is adequate. The perspective from which we make this judgment is the one offered by our best theory. From this perspective we may find the ontology of another theory extremely dubious, but nevertheless regard the models provided by that theory as perfectly adequate for certain purposes. The advent of general relativity, for example, may make us harbor doubts about the assumptions underlying Newton's theory of gravity. Yet it would be lunatic to deny the adequacy of Newton's theory for virtually all the purposes that concern NASA. The fate of Newtonian mechanics should also remind us that even our best theory can only be regarded as acceptable, despite the fact that it is, for now, the theory by which others are judged.

Unacceptable theories include the theory of continuous creation in cosmology and the Bohr-Kramers-Slater theory of radiation. They are unacceptable not just because they are flawed; we have good inductive evidence that even our best theories are flawed. They are unacceptable because we cannot reproduce within our best theories approximations to their fundamental postulates, that throughout the universe matter is continuously being created and that individual interactions violate the principle of conservation of energy.

In contrast, Newton's emission theory of light, though flawed, is in many contexts acceptable. It is acceptable from the perspective of the wave theory of light because a wave model can mimic the behavior of a model in which light travels in straight lines that bend towards the normal on entering an optically denser medium. From our present viewpoint, in turn, the wave theory is acceptable because quantum theories of the electromagnetic field can mimic the behavior of a model in which transverse waves are propagated through an elastic medium.

Acceptability, relative to a theory, does not imply reducibility, as that term was once used, and so avoids the difficulties that beset that concept. There is certainly no requirement that the vocabulary of one theory be translatable into the vocabulary of the other. Notice that, in each of the examples given, what is being mimicked is behavior that is typically appealed to in explanations provided by the theory in question. Thus acceptability is always qualified; to be used in an explanation, a theory must be acceptable in salient respects. The acceptability of one theory with respect to another is a reflexive relation; the question my examples suggest, whether, in general, it is also transitive, is left as an exercise for the reader.[38]

NOTES

1. For a historical account of work on the topic of scientific explanation, see Salmon (1989).

2. Cotton Mather's *Thoughts for the Day of Rain* is quoted by Boyer (1987, 262).

3. Bromberger (1962, 74) suggests "whence" and "whither."

4. As does van Fraassen (1980, 141–42).

5. On this issue I was awakened from dogmatic slumbers by a paper by Maria Trumpler (1987); her examples came from recent work in physiology on how the nerves transmit information.

6. As Achinstein emphasizes (1983, chap. 2), this condition is not sufficient.

7. In contrast, on the Hempel-Oppenheim view explanation was either a syntactic or syntactic-cum-semantic affair, depending on one's view of entailment.

8. See Hempel (1965, 413). For another response to this problem see Friedman (1974, 188). Page references to the latter are to Pitt (1988).

9. The mark of a perlocutionary act, like persuading or convincing, is that it brings about an effect on the listener. See Austin (1965, 101–102).

10. Similar considerations show that explaining is not unequivocally an illocutionary act either.

11. Compare, e.g., van Fraassen (1980, 64) with Hawking (1988, 9). Some philosophers, like Mary Hesse, have been in step with scientists for longer than others.

12. The terms are due to Giere (1985, 80). This vocabulary was proposed (and accepted) as a friendly amendment to van Fraassen's account of theories (see n. 15, below). For Giere,

the theoretical hypothesis is that, "The designated real system is *similar* to the proposed model in specified *respects* and to a specified *degree.*" I am unhappy with this formulation; like Goodman (1968, 3–10), I reject the view that a representation is similar to, or resembles its subject.

13. I take this term from the title of Bohr's trilogy of papers, "On the Constitution of Atoms and Molecules." Bohr's model of the hydrogen atom is a paradigm case of a constitutive model; see Hughes (1990).

14. This version of the semantic view was first proposed by Ewart Beth (1949), and later developed by Bas van Fraassen (1980, chap. 2).

15. In fact it conforms pretty much to the Hamilton-Jacobi presentation of classical mechanics. I took the state-space approach in my presentation of quantum theory (Hughes 1989a, chap. 2). It has also been successfully applied to evolutionary theory in books by Elisabeth Lloyd and Paul Thompson; for brief but instructive comments on the approach see Lloyd (1988) and Thompson (1988).

16. The Pines model also needs to specify the charge density per unit volume of space, since positive charge is represented as uniformly distributed.

17. The taxonomy is not cut and dried, and there is no standard nomenclature for these classes; see section 5.

18. Representation-as and the ambiguity of "This picture represents Winston Churchill as an infant" are explored by Goodman (1968, 27–31).

19. In fact, all this passage says is that we expect the model to be adequate except where it is inadequate. Nonetheless, my point still stands.

20. See Nancy Cartwright's essay, "When Explanation Leads to Inference" (1983, essay 5). As must be obvious, I have also learned a great deal from essay 8: "The Simulacrum Account of Explanation."

21. See, respectively, Newton ([1730] 1952, 168–85); Noakes (1953, 116–20); PSSC (1960, 222–23). To make notation uniform, I quote these authors with minor changes in, e.g., the labeling of points in the diagram.

22. None of them, it should be added, mention that to see the third and fourth orders the spectator would have to look towards the source of light. This was first pointed out by Newton's contemporary Halley. See Boyer (1987, 249–50).

23. I take this to be a general characteristic of "laws of nature."

24. See Boyer (1987, 301).

25. F. Raillard in the *Comptes Rendus* of the Academie des Sciences (44 (1857), 1142–44), quoted by Boyer (1987, 308).

26. These had been remarked on by Witelo as early as the thirteenth century, and a careful observation of them was made by Langwith, Rector of Petworth, in 1722; see Boyer (1987, 277–78). Langwith makes the astute comment, "I begin now to surmise . . . that the suppos'd exact Agreement between the Colours of the Rainbow and those of the Prism is the reason why [this phenomenon] has been so little observed."

27. (Hesse 1965, 111). Page references are to (Hesse 1980).

28. It is not mentioned, for example, in Salmon's 1989 overview of accounts of explanation.

29. Another, no doubt, is that any self-respecting philosopher of science finds metaphor a distastefully squishy subject.

30. They are *like* metaphors, however, in that by their use the primary system is defamiliarized—in the sense in which that word is used by Shklovsky and by Brecht. But I will not pursue that line of thought here.

31. Descartes, *First Meditation* (1988, 77).

32. Friedman (1974, 195) takes the "unifying effect of scientific theories" to be the "crucial property" that gives rise to understanding.

33. See Feynman, Leighton, and Sands (1969, 1: 25.8).

34. In fact, part of the stimulus for this account was the problem of explaining the statistical regularities that quantum mechanics deals with. See Hughes (1989a, chap. 8.9; 1989b).

35. See van Fraassen (1980, 111–12).

36. In particular, I am thinking of Duhem ([1906] 1962) and of van Fraassen (1980).

37. Louise Anthony first made me realize what now seems to me obvious, that not all explanations are acceptable.

38. I presented many of the ideas in this essay in the spring of 1992 at a faculty–graduate student seminar at the University of South Carolina, and at a Colloquium at North Carolina State University. I am grateful to the participants at those events for helpful suggestions, and in particular I would like to thank Davis Baird, who generated the diagram for me on his computer.

REFERENCES

Achinstein, Peter. 1983. *The Nature of Explanation*. New York.

Austin, J. L. 1965. *How to Do Things with Words*. Oxford.

Beth, Ewart. 1949. "Towards an Up-to-Date Philosophy of the Natural Sciences." *Methodos* 1: 178–85.

Boyer, Carl B. 1987. *The Rainbow*, 2nd ed. Princeton.

Bromberger, Sylvain. 1962. "An Approach to Explanation." In Butler (1962), 72–105.

Butler, R. S. Ed. 1962. *Analytic Philosophy, Second Series*. Oxford.

Cartwright, Nancy. 1983. *How the Laws of Physics Lie*. Oxford.

Churchland, Paul M., and Clifford A. Hooker. Eds. 1985. *Images of Science*. Chicago.

Cushing, James T., and Ernan McMullin. Eds. 1989. *Philosophical Consequences of Quantum Theory*. Notre Dame, Ind.

Descartes, René. 1988. *Selected Philosophical Writings*, translated by J. Cottingham, R. Stoothoff, and D. Murdoch. Cambridge.

Duhem, Pierre. 1906. *The Aim and Structure of Physical Theory*, translated by P. P. Wiener, New York, 1962.

Feynman, Richard P., Robert B. Leighton, and Matthew Sands. 1969. *The Feynman Lectures on Physics*, 3 vols. Reading, Mass.

Fine, Arthur, and Jarrett Leplin. Eds. 1989. *PSA 1988*, 2. East Lansing, Mich.

Friedman, Michael. 1974. "Explanation and Scientific Understanding." *Journal of Philosophy* 71: 5–19, reprinted in Pitt (1988), 188–98.

Giere, Ronald M. 1985. "Constructive Realism." In Churchland and Hooker (1985), 75–98.

Goodman, Nelson. 1968. *Languages of Art*. Indianapolis.

Hawking, Stephen W. 1988. *A Brief History of Time*. New York.

Hempel, Carl G. 1965. *Aspects of Scientific Explanation*. New York.

Hesse, Mary. 1965. "The Explanatory Function of Metaphor." In Hesse (1980), 111–22.

Hesse, Mary. 1980. *Revolutions and Reconstructions in the Philosophy of Science*. Bloomington, Ind.

Hiley, B. J., and F. David Peat. Eds. 1987. *Quantum Implications, Essays in Honour of David Bohm*. London.

Hughes, R. I. G. 1989a. *The Structure and Interpretation of Quantum Mechanics*. Cambridge, Mass.

Hughes, R. I. G. 1989b. "Bell's Theorem, Ideology, and Structural Explanation." In Cushing and McMullin (1989), 195–207.

Hughes, R. I. G. 1990. "The Bohr Atom, Models, and Realism." *Philosophical Topics* 18: 71–83.

Kitcher, Philip, and Wesley C. Salmon. Eds. 1989. *Minnesota Studies in the Philosophy of Science*, 13. Minneapolis.

Lloyd, Elisabeth A. 1988. "The Semantic Approach and its Application to Evolutionary Theory." In Fine and Leplin (1989), 276–85.

Nagel, Ernest. 1961. *The Structure of Science*. New York.

Newton, Isaac. 1730. *Opticks*, 4th ed. New York, 1952.

Noakes, G. R. 1953. *A Textbook of Light*, rev. ed. London.

Physical Sciences Study Committee. 1960. *Physics*. New York.

Pines, David. 1987. "The Collective Description of Particle Interactions." In Hiley and Peat (1987), 65–84.

Pitt, Joseph C. Ed. 1988. *Theories of Explanation*. New York.

Sakurai, J. J. 1967. *Advanced Quantum Mechanics*. Redwood City, Calif.

Salmon, Wesley C. 1989. "Four Decades of Scientific Explanation." In Kitcher and Salmon (1989), 3–219.

Thompson, Paul. 1988. "Explanation in the Semantic Concept of Theory Structure." In Fine and Leplin (1989), 286–96.

Trumpler, Maria. 1987. "Explanation in Physiology." Unpublished, Yale University.

van Fraassen, Bas C. 1980. *The Scientific Image*. Oxford.

Selective Scientific Realism, Constructive Empiricism, and the Unification of Theories

STEVEN SAVITT

Michael Friedman's book, *Foundations of Space-Time Theories*,[1] has been greeted warmly. It won the Machette prize and shared the Lakatos prize. Reviews have been enthusiastic. For instance Larry Sklar (1988, 158) wrote:

> Michael Friedman's imaginative, thoughtful, and carefully worked out study of the fundamental philosophical issues at the foundations of contemporary space-time theories in physics constitutes a very major contribution to the literature in this field.

Sklar then makes an important point:

> Since Friedman correctly understands that the resolution of the most important of these philosophical questions can only be done in a broader context of a foundational philosophical investigation into the nature of theories in general, the book should be given the careful attention of not only those interested in the philosophy of physics or of space and time, but of those concerned with more general issues of realism and conventionalism in our naturalistic account of the world.

As one interested in one kind of realism, I propose to follow Sklar's advice and give Friedman's book some of the attention it merits. More specifically, Friedman presents an argument for *scientific* realism that can be seen as a response to Bas van Fraassen's constructive empiricism.[2] I propose to disentangle as far as I am able this line of argument, which is of general interest, from the formal context of space-time theories in which it is presented.[3]

I shall begin by sketching van Fraassen's well-known views, and I shall emphasize the important philosophical presuppositions shared by Friedman and him, thus making Friedman's response all the more cogent. In particular these shared presuppositions allow van Fraassen and Friedman to agree on the precise nature of the issue between them; the realist views observable or phenomenological structures as literally embedded in higher order theoretical structures whereas the anti-realist views these higher order theories as mere representational devices.

Having defined the issue, I will then present rather abstractly Friedman's argument that embedding, in the context of the unification of theories, provides a significantly more highly confirmed world-picture than representing. I will try then to give this argument some intuitive content by applying it (as Friedman does) to the kinetic theory, before I try to use an event in the development of that theory to undermine the principle that supports his account of confirmation. Finally, I will suggest that reconstructed history of science may not be able to settle the realist-empiricist dispute decisively.

I. CONSTRUCTIVE EMPIRICISM

According to van Fraassen, certain things or kinds of things in the world (paradigmatically, persons and middle-sized objects in space and time) are observable. We find out what things are observable through ordinary experience and through systematic examination and reflection, through science. Like any scientific hypotheses our hypotheses concerning the nature and reliability of our procedures of observation may change through time, but mutable though the line between what we can and cannot observe may be, that some boundary exists is bedrock.

Scientists often speak of entities on the far side of that line. They construct theories of hidden realms, of unobservables (paradigmatically atoms or sub-atomic particles), in order that we may better understand or predict or control the behavior of those things that we can observe. Of course, we may employ these theoretical languages to speak about the observables. It may even be that all language is somehow anchored in these theoretical flights, but that fact about language—if it is a fact—should not obscure the fundamental distinction between observable and non-observable things, shifting though that distinction may be.

The theoretical languages employed by scientists are best thought of as specifications of classes of models. Each model will have a part, the empirical substructure, intended to represent the relations between or the behavior of the observables, the appearances. A theory is true if the world (or some portion of the world) is one of its models. A theory is *empirically adequate* if its empirical substructure accurately represents that which we can measure. The aim of science is to find theories that are empirically adequate, and the acceptance of a theory by a scientist involves (or should involve?) only the belief in its empirical adequacy, not belief in its truth *tout court*.

Why not truth? First, van Fraassen argues that all the virtues we could reasonably seek from theories can be provided just as well by empirically adequate theories as by true theories. Second, he uses an enticing principle of epistemic conservatism to argue that one should get by with empirically adequate theories if one can. There is already epistemic risk in limning the appearances; why embrace some additional (theoretical) description of a hidden realm, given that empirical adequacy will enable one to deal with all that one will experience? As van Fraassen put it, "[I]t is not an epistemological principle that one might as well hang for a sheep as for a lamb"(*SI*, 72).

Friedman also endorses the model-theoretic approach to theories, the theory-ladenness of language, and the relativity of the observable–non-observable distinction. Although he shares this common ground with van Fraassen, Friedman ingeniously undermines van Fraassen's application of his principle of epistemic conservatism by arguing that on a realist view of theories (which will be described in more detail below, but which in brief is encapsulated in the formula that acceptance of a theory is belief in its truth) the description of the unobservables will, in successful theories and under certain circumstances, be more highly confirmed than the descriptions of the appearances would be if theories are construed merely as ways of representing (or saving) the appearances. We shall now see how this could possibly be.

II. REDUCTION AND REPRESENTATION

Friedman construes the traditional realist-relationalist debate over the existence of absolute space as a debate over whether the quantifiers of physical theory may range over all points, occupied and unoccupied, in a (Newtonian) container space or whether they are to be restricted, as Leibniz would have it, to the set of physically occupied spatial points. In more modern and general form, if $\mathcal{M}$ is a four-dimensional spacetime manifold and $\mathcal{P}$ is the set of physically occupied spacetime points, the question is whether $\mathcal{P}$ is literally a subset of $\mathcal{M}$ (a 'reduction' in Friedman's terminology) or whether $\mathcal{M}$ is merely useful, as van Fraassen would have it, as a mathematical representation of $\mathcal{P}$ in virtue of some mapping of $\mathcal{P}$ into $\mathcal{M}$ in which the properties and relations in $\mathcal{P}$ are reproduced. At first blush the relationalist seems committed to a more cautious ontology—$\mathcal{P}$ rather than all of $\mathcal{M}$—with no loss of predictive or explanatory power (if $\mathcal{P}$ is indeed isomorphically embeddable into $\mathcal{M}$). Friedman tries to show, however, that this initial advantage is more than offset by the fact that (realist) embeddings or reductions interact with one another over time in such a way they are more highly confirmed than their corresponding representations.

First, a logical point. If we let $\mathcal{A} = \langle A, R'_1, \ldots, R'_n \rangle$ be a theoretical structure and $\mathcal{B} = \langle B, R_1, \ldots, R_m \rangle$ a phenomenological structure and Δ_i a class of models, then the following (realist) inference is valid:

$$\langle B, R_1 \rangle \subseteq \mathcal{A} \text{ and } \mathcal{A} \in \Delta_1$$
$$\langle B, R_2 \rangle \subseteq \mathcal{A} \text{ and } \mathcal{A} \in \Delta_2$$
$$\overline{\langle B, R_1, R_2 \rangle \subseteq \mathcal{A} \text{ and } \mathcal{A} \in \Delta_1 \cap \Delta_2.}$$

The corresponding non-realist inference is not valid:

$$\exists\, \mathcal{A} \,\exists\, \phi : \langle B, R_1 \rangle \rightarrow \mathcal{A} \text{ and } \mathcal{A} \in \Delta_1$$
$$\exists\, \mathcal{A}' \,\exists\, \psi : \langle B, R_2 \rangle \rightarrow \mathcal{A}' \text{ and } \mathcal{A}' \in \Delta_2$$

$$\exists\, \mathcal{A}'' \,\exists\, \chi : \langle B, R_1, R_2 \rangle \rightarrow \mathcal{A}'' \text{ and } \mathcal{A}'' \in \Delta_1 \cap \Delta_2.$$

The existential quantifiers in the premises assure the invalidity of the second inference (246).

It follows that when two phenomenological theories are modeled in the same domain, their conjunction must also have a model in this same domain. But if two phenomenological theories are represented, even in the same domain, the claim that their conjunction is represented in that domain is a new claim, one that is not a consequence of the two separate representations. This difference between the logical relations of conjoined reductions and conjoined representations to their components, it is argued, profoundly effects their relative degrees of confirmation.

We must see—schematically at first, but a detailed example will follow— how this difference in degree of confirmation arises. Suppose that we have two reductions, A and B, and that each has received some "boost" in confirmation—A at time t_1 say, and B at t_2. Later we conjoin the two and find a new, successful observational prediction at t_3. Each of the conjuncts A and B has now received repeated "boosts" in confirmation—A, for instance, at both t_1 and t_3. Friedman illustrates this situation with the diagram in figure 1.

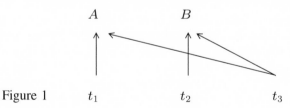

Figure 1 t_1 t_2 t_3

The story for representations, according to Friedman, is quite different. Suppose that the representations A and B receive some confirmation at t_1 and t_2 respectively and that the joint representation is formulated at t_3. Diagrammatically we now have:

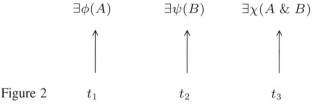

Figure 2 t_1 t_2 t_3

Since the joint representation $\exists\chi(A \ \& \ B)$ has precisely the same observational consequences as the joint reduction $\lceil A \& B \rceil$, a constructive empiricist might wonder why the situation should not be depicted like this:

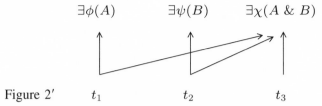

Figure 2' t_1 t_2 t_3

But Friedman's point is that even though the joint representation does have the same observational consequences as the joint reduction, it is less well confirmed at t_3 than the joint reduction because either (a) it is confirmed at t_3 *only* by the *novel* prediction(s) that it makes at t_3 and not by the successful predictions which earlier confirmed A and B or (b) the earlier successful predictions of A and B count so little towards the confirmation of the essentially new hypothesis $\exists\chi(A \ \& \ B)$ that one can always be sure that $\lceil A\&B\rceil$ will be more highly confirmed. This is the natural way to understand Friedman's remark that "the only way in which the representation $\exists\chi(A \ \& \ B)$ could be as well confirmed as the reduction $\lceil A\&B\rceil$ is if some genius had already formulated $\exists\chi(A \ \& \ B)$ before t_1, that is, only if temporal evolution is completely removed from our picture"(247–48).

In the absence of such a genius we are left with an anti-realist picture of confirmation's being divided between the representations $\exists\phi(A), \exists\psi(B)$, and $\exists\chi(A \ \& \ B)$: whereas the confirmation of the components A and B aggregates in the joint reduction $\lceil A\&B\rceil$. We can therefore see how on Friedman's realist picture the unified theory $\lceil A\&B\rceil$ may come to accumulate more confirmation (from its conjuncts separately and from the new test that they face in conjunction) than the phenomenological theories derivable from the various representations severally possess if higher-order theories are mere representational devices. *If Friedman's argument is correct, then a realist can show that epistemic conservatism need not invariably dictate a preference for phenomenological theories over theories postulating genuine observables.*

As Friedman emphasizes throughout his discussion, this abstract does not merely indicate a logical possibility but also illuminates actual (or, more realistically, reconstructed) historical reasoning. "These are the considerations, it seems to me," writes Friedman, "that explain our preference for a literal interpretation of kinetic theory over a non-literal, purely representative interpretation"(248). Let us then next see how Friedman's abstract argument applies to his familiar example.

III. THE KINETIC THEORY OF MATTER AND HEAT

In the mid-nineteenth century James Joule was able to derive certain empirical laws concerning the behavior of gases from a number of postulates concerning their micro-structure.[4] Joule postulated a realm of molecules of negligible size but very large in number and in moving randomly. These molecules were supposed to obey Newtonian laws (in particular, collisions between them or between molecules and any containing vessel were supposed to be perfectly elastic), but forces between molecules not in contact were supposed to be negligible.

If (as Friedman would put it) we embed empirical gases in this molecular realm, then one can identify the pressure P on the wall of any vessel containing a gas with the force per unit area exerted by the molecules in their collisions with the container, and then one can readily prove that

$$P = \frac{N m_0 \overline{v^2}}{3V}, \tag{1}$$

where N is the number of molecules in the container, m_0 is the mass of each molecule, $\overline{v^2}$ is average of the squares of the molecular speeds (or the mean-square speed) and V is the volume of the container. If we assume that $\overline{v^2}$ remains constant when the temperature of the gas remains constant, then (1) rewritten as

$$PV = \frac{N m_0 \overline{v^2}}{3}, \tag{2}$$

expresses Boyle's law, for all the quantities on the right-hand side of the equation are constant at constant temperature.

In addition to Boyle's law, the kinetic theory was shown to imply the equation of state of a gas, Dalton's law of partial pressures, Graham's law of diffusion, and much more. Of course, these successes do not depend upon taking molecules literally; a representational approach can (and did) yield the same results. It is a later episode, according to Friedman, that distinguishes these two approaches. But before we turn to it, let me derive one relation that will prove useful later.

If we note that $\overline{E}_k$, the average translational kinetic energy of a molecule, is equal to $(m_0 \overline{v^2})/2$ and if we use the empirically established relation

$$PV = rT, \tag{3}$$

where T is the temperature of the gas and r is a constant which depends upon the kind and the mass of the gas under consideration, we can show that

$$\overline{E}_k = \frac{3rT}{2N}. \tag{4}$$

It is possible to show that the ratio r/N is constant, which allows us to rewrite (4) in the useful form:

$$\overline{E}_k = \frac{3kT}{2}, \tag{5}$$

where k is Boltzmann's constant.

Now back to Friedman's story. Early in the twentieth century, Ernest Rutherford and Niels Bohr developed a model of the atom with the mass and positive charge concentrated in a nucleus and with negatively charged electrons orbiting the nucleus in shell-like structures. We need not dwell on the details of this model or its many successes. For our purposes, what we must note is that if we identify molecules of chemical elements with (structures of) Rutherford-Bohr atoms—that is, if we provide a finer structure for the molecules of the kinetic theory discussed above—we are able to explain of valence, the regularities of the periodic table and the phenomenological laws of chemical combination of the various elements. We are, in Friedman's view, adding these atomic assumptions to those of the classical kinetic theory, and the increased appeal (i.e., confirmation) of the combined theories can be accounted for only by invoking the conjunctive pattern that presupposes the literal embedding of empirical in theoretical domains. Since this is precisely what I intend to question, I shall quote Friedman's summation of the point of this example:

On a literal interpretation the temporal evolution of our theory displays a characteristic conjunctive pattern:

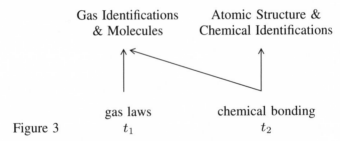

Figure 3

On a nonliteral interpretation there is no conjunctive pattern:

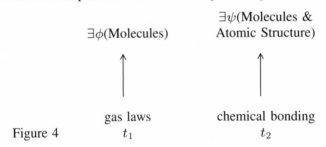

Figure 4

In particular, the representation $\exists\psi$(Molecules & Atomic Structure) is not implied by the conjunction $\lceil\exists\phi$(Molecules) & $\exists\chi$(Atomic Structure)$\rceil$. As a consequence, the hypotheses involved in the non-literal version of molecular theory receive fewer boosts in confirmation. Moreover, as time goes on, as our molecular hypotheses play a role in more and more derivations, this disparity in confirmation becomes greater and greater. The difference in logical form between reductions and representations . . . makes all the difference in the world. (248)

IV. CONJUNCTION OR CORRECTION?

My aim in this section is to cast doubt upon the crucial argument supporting the idea that the confirmation of representations over time is more properly depicted in figure 2 rather than in figure 2' above. I shall do this by considering the history of the kinetic theory in greater detail. If, for convenience, we refer to the classical kinetic theory represented in the left column of figure 3 as 'A' and the atomic theory represented in the right column as 'B', then I wish to consider a modification of A that I will call A^* that appeared between t_1 (1846, Joule's derivations) and t_2 (say 1916, when G. N. Lewis introduced the explanation of valence in terms of electron sharing).

Although the classical kinetic theory A, as indicated above, made many successful predictions, it encountered some problems as well. In particular, it entailed an exact but not generally correct value for the specific heat capacity of any gas.

The specific heat capacity c_v of any substance is the quantity of heat absorbed or given up per unit mass of a material when its temperature is raised or lowered one degree. For a gas, as long as volume is kept constant, we would then have

$$H = \Delta T \times c_v \times m, \qquad (6)$$

where H is the quantity of heat added, T is the change in temperature (in degrees Kelvin), and m is the mass. Since H, T, and m can all be measured, c_v can be experimentally determined.

But c_v can also be predicted theoretically. Equation (5) tells us that the mean kinetic energy of any molecule of gas is equal to $3kT/2$. If the temperature of the gas is raised by one degree Kelvin, the average increase in kinetic energy per molecule is $3k/2$ or $2.07 \times 10^{-23}j$. This is the value of the heat capacity per molecule of *any* gas, according to the kinetic theory. But since each mole of a gas contains the same number of molecules (Avogadro's number, N_0), the heat capacity per mole of any gas should be equal to $N_0 3k/2$ or $12.5j/\text{mole } K^o$.

This prediction of the kinetic theory only partially agreed with measured values. For certain gases (helium, argon, and mercury vapor—all monatomic gases) the predicted value agreed quite well with the measured values. But for other gases (hydrogen, oxygen, nitrogen—all diatomic gases) the measured values were all roughly 5/3 the predicted value. Let me quote Holton and Roller here.

> Here we have a breakdown of the model, a flaw in the theory as developed so far. Joule, who brought the theory up to this point, was stumped by the discrepancy in the heat capacities of polyatomic gases, the more so because in his day no monatomic gases had yet been generally identified. In fact his first thought was that the experimental values of c_v then available might be in error, but when new determinations only widened

the gap between the predicted and experimental values, he seems to have been so discouraged that any further work on the kinetic theory appeared to him to be futile. His model had not made provision for a molecule to store energy in any other form than the kinetic energy of translational motion, and at that time no reason was apparent for revising this simple idea. (1958, 454)

Rudolf Clausius in 1857 provided the necessary emendation of the original theory. He supposed that diatomic molecules had a structure that would permit them two rotational degrees of freedom in addition to the three translational degrees of freedom possessed by monatomic molecules. (By 'degree of freedom' is meant an independent way in which a particular molecule can absorb or lose energy.) If one further assumed that available energy distributes itself in equal shares to each active degree of freedom (equipartition), then it follows that the specific heat per mole of diatomic gases must be 5/3 that predicted for monatomic gases, in good agreement now with the measured outcomes.

Let us refer to Clausius's theory as A^*, and let us consider the transition from A to A^*. It resembles the transition from A to $(A + B)$, to put it neutrally, in that a deeper level of molecular structure is being exploited in order to produce new predictions, but it is easy to see that the transition from A to A^* cannot be represented by the conjunctive pattern of figure 3. If, after all, A^* contained A as a subcomponent (that is, if $A^* = (A \& T)$, where T is any other theoretical (sub)component), A^* would have to be inconsistent since A by itself entails that the heat capacity per mole of (e.g.) hydrogen is $N_0 3k/2$ but A^* entails that it is $N_0 5k/2$.

Although it is impossible to construe the move from A to A^* as conjunction, it is straightforward for van Fraassen to construe it as correction, as a theory being made more empirically adequate. But if we take that tack, we must confront Friedman's point about confirmation.

Is A^* confirmed only by its novel predictions? The answer seems to be "no," however 'novel' is understood. There is a strong sense of 'novel' in which a prediction is novel if it was suggested to members of the relevant scientific community by the theory that entailed it and had not occurred to them before. But A^*, to the best of my knowledge, makes no novel predictions in this sense.

In a more controversial sense of 'novel' a prediction is *use-novel* if it is not used in the construction of the theory from which it is a prediction. Since accounting for the heat capacities of (what turned out to be) diatomic gases was the motivation for Clausius's work, the new predictions clearly are not novel in this sense.[5]

There is a final sense of 'novel' in which a theory generates a novel prediction if it is historically the first theory to entail that prediction. This seems to be the sense of 'novel' that Friedman has in mind. And in this sense A^* does indeed generate novel predictions concerning the heat capacity per mole of diatomic gases. In this sense, however, Friedman is committed to the view that all the novel predictions of A that are also entailed by A^* provide little or no confirmation for A^*, since A, as it were, beat it to the punch. Significant confirmation for A^* could come only from *its* novel predictions concerning the heat capacities of polyatomic gases. Although Friedman only sketches a theory of confirmation in his discussion of relationalism, it is hard to see how his picture could be completed in such a way as to avoid the unacceptable conclusion that A was more highly confirmed than A^* in 1857.

I believe this example, in which A^* corrects A, shows that Friedman cannot maintain in general that only the novel predictions of a theory confirm it. But if he cannot maintain this general principle, then he no longer has any argument to show that the reduction $\lceil A \& B \rceil$ must be more highly confirmed than the joint representation $\exists \chi (A \ \& \ B)$, i.e., that figure 2 rather than figure 2' depicts the confirmation of representations through time, and this principle (as my discussion above should have made clear) is the pivot on which his argument for scientific realism swings.

V. CONJUNCTION AND CORRECTION?

Even if the argument of the previous two sections should prove incorrect, there is a general consideration suggested by the example of Clausius's work which shows that Friedman's argument against constructive empiricism cannot be decisive. The question that Friedman puts to himself is this:

> [P]hysicists themselves distinguish between aspects of mathematical structure that are meant to be taken literally—that really correspond to pieces of the physical world—and aspects of mathematical structure that are purely representative. What is the rationale for this distinction?(238)

The rationale, as I understand Friedman's argument, is found in methodological principles that are justified by being in reflective equilibrium with reconstructed history. In particular, from his consideration of space-time theories Friedman concludes that

> . . . the principle of parsimony functions as an *upper limit* on the process of theoretical postulation, a process that also has a *lower limit* generated by the need for theoretical unification.(338)[6]

These two principles and his reconstructed history make a powerful case for his scientific methodology, with much the same authority that a solution to a double crostic has when a quotation emerges from the fit of answers to clues on the one hand and guesses at words or phrases in the patterned spaces on the other.

While it is reasonable to suppose that the solution to a double crostic is unique, there is less support for uniqueness claims depending upon reconstructed history of science. Referring to work of Nancy Maull (1976) and Richard Burian (1977), Ernan McMullin wrote,

> Are there independent sources of warrant [for philosophy of science], then, one of these being history of science? This is too simple. For the

writing of history of science is shaped by the preconceptions that are brought to it. It is no more theory-neutral than is observation in science itself. (1979, 65)

Can empiricist preconceptions lead to a (reconstructed) history of kinetic theory emphasizing the search for empirical adequacy while muting the role of unification? Clausius's contribution, the postulation of structure to gain empirical adequacy, suggests that it can. Further support comes from the work of Mary Jo Nye who remarks (1976, 266) that in the revitalization of the atomic hypothesis after 1895, "The key was in the revelation of the long-called-for sub-units of a now divisible atom, and in visible evidence inferentially tied to real molecules, real atoms." She considers new experimental tools and heuristic power relevant as well.[7]

It was not merely a gestalt switch or an intuitive leap which suddenly made the atomic hypothesis more than an indifferent hypothesis. It was experiment, some of it due to new instruments like the vacuum tube and the ultramicroscope, which revitalized the atomic programme. Atomism became extraordinarily fruitful in not only better explaining well-known phenomena, but in suggesting entirely new kinds of information and forms of research which in turn tested, refined, and further tested the hypotheses. (1976, 268)

This suggestion that there may be a reconstruction of the history of the atomic hypothesis that is more congenial to the constructive empiricist, of course, amounts only to a promissory note rather than currency. Its point is only that while the elegance and coherence of Friedman's reconstruction provides a stiff challenge to the constructive empiricist, it does not consign her to the ranks of the unemployed.

VI. CONCLUSIONS

My discussion in section V of the transition from A to A^* at least suggests, though it does not show, that the transition from A to A^* to $(A^* + B)$ might be reconstructed as correction at least as well as conjunction. Since Friedman cites the development of the kinetic theory of matter as one of the paradigmatic exemplifications of the conjunctive pattern of inference on which he bases his case for (selective) scientific realism, the possible availability of a non-realist mode of reconstruction does lend some support to a remark of van Fraassen's. "[T]here seems to me no doubt," he writes, "that the aim of empirical adequacy already requires the successive unification of 'mini-theories' into larger ones, and that the process of unification is mainly one of correction and not of conjunction" (SI, 87).

Second, I believe that my discussion in sections III and IV of the transition from A to A^* does show that even if the transition from (say) A^* to $(A^* + B)$ is conjunctive rather than corrective, Friedman has offered no compelling argument to show that the joint reduction will always be more highly confirmed than the joint representation.

Finally, however, let me remind you that if my argument is successful, then one argument in favor of scientific realism has been undermined. This does not show that scientific realism is false or otherwise untenable.[8]

NOTES

1. Further references to this book will be in parentheses in the text.

2. See van Fraassen (1980), for example, which will be cited as '*SI*' in the text.

3. Friedman's argument is discussed in its intended context in Weingard and Smith (1986), as Carl Hoefer pointed out to me.

4. My account of the kinetic theory follows Holton and Roller (1958).

5. If its ability to account for the heat capacities of diatomic gases counts positively toward the confirmation of A^*, then use-novelty is not necessary for confirmation. This historical example seems to support the line of argument in section 4 of Mayo (1991).

6. Hence my designation *selective* scientific realism for his view. My argument, of course, is concerned only with his lower limit.

7. Gardner (1979) may be viewed as a stab at a pluralist reconstruction as well.

8. I am grateful to Cliff Hooker for providing careful and helpful criticism of a previous draft of this paper.

REFERENCES

Burian, R. M. 1977. "More Than a Marriage of Convenience: On the Inextricability of History and Philosophy of Science." *Philosophy of Science* 44: 1–42.

Friedman, M. 1983. *Foundations of Space-Time Theories*. Princeton.

Gardner, M. R. 1979. "Realism and Instrumentalism in 19th-Century Atomism." *Philosophy of Science* 46: 1–34.

Holton, G., and D. H. D. Roller. 1958. *Foundations of Modern Physical Science*. Reading, Mass.

McMullin, E. 1979. "The Ambiguity of 'Historicism'." In *Current Research in Philosophy of Science: Proceedings of the P. S. A. Critical Research Problems Conference*, edited by P. D. Asquith and H. E. Kyburg, 55–83. Philosophy of Science Association. East Lansing.

Maull, N. 1976. "Reconstructed Science as Philosophical Evidence." In *PSA 1976*, volume one, edited by F. Suppe and P. D. Asquith, 119–130. Philosophy of Science Association. East Lansing.

Mayo, D. G. 1991. "Novel Evidence and Severe Tests." *Philosophy of Science* 58: 523–52.

Nye, M. J. 1976. "The Nineteenth Century Atomic Debates and the Dilemma of an 'Indifferent Hypothesis'." *Studies in History and Philosophy of Science* 7: 245–68.

Sklar, L. 1988. "Review of *Foundations of Space-Time Theories*." *Journal of Philosophy* 85: 158–64.

van Fraassen, B. 1980. *The Scientific Image*. Oxford.

Weingard, Robert, and Gerrit Smith. 1986, "Critical Notice: Michael Friedman's *Foundations of Space-Time Theories*." *Philosophy of Science* 53: 286–99.

Essentially General Predicates

PETER RAILTON

I

In various areas of the philosophy of science, it has been thought that significant difficulties could be overcome if we could give a credible account of the distinction between genuinely and nominally general predicates. Two over-familiar examples:

Confirmability. Some have thought that the so-called "new riddle of induction" posed by Nelson Goodman might be solved, at least for a significant class of cases, if we could give a workable distinction between a predicate like 'green', which appears to be essentially general, and 'grue', which appears to involve in some essential way reference to a particular time. After all, 'grue' is introduced by Goodman with the following definition (for some specific time, t):

(1) x is grue iff (x is examined before t and x is green) or (x is not examined before t and x is blue).[1]

Goodman is quick to point out, however, that it is no easy matter saying whether a given predicate involves essential reference to a particular time, since, given:

(2) x is bleen iff (x is examined before t and x is blue) or (x is not examined before t and x is green),

the following equivalence holds:

(3) x is green iff (x is examined before t and x is grue) or (x is not examined before t and x is bleen).

This is hardly the end of the matter, of course, and various authors have attempted to show that 'grue' and 'bleen' involve essential temporal reference in some deep or intrinsic way that 'green' does not, despite (3).

Nomologicality. The problem of distinguishing laws from accidental generalizations, though dismissed as a pseudo-problem by some, continues to attract attention in part because of the seemingly different status of laws when it comes to confirmation, explanation, the support of counterfactuals, and so on. One might hope to rule out a large class of accidental generalizations by claiming that the predicates involved in genuine laws are essentially general, making no reference to any particular time or place. Thus even though both

(4) All grass is green

and

(5) All the coins now in my pocket are silver

would typically be represented in extensional logic as universal generalizations of the form $\lceil (x)(Fx \supset Gx) \rceil$, only (4) is taken to be a candidate for lawhood, since (5) makes reference to a particular place and time. Moreover (5), unlike (4), does not support appropriate counterfactuals, e.g., 'This coin,' said of a penny, 'would be silver if it were in my pocket' is false (contrast 'This garland,' said of a daisy chain, 'would be green if it were grass'). One might think it easy to rule out (5) as a possible law because it contains express terminology of time and place. Suppose, however, the following annoying predicate is introduced:

(6) x is coinowket iff x is a coin now in my pocket.

Using (6), one can formulate the generalization

(7) All coinowkets are silver,

which contains no express spatio-temporal terminology.

Such a crude evasion could be blocked if we rejected not only generalizations that contained explicit reference to a particular time or place, but also those that are "merely equivalent" to a generalization which does involve such reference. The difficulty now is that (4) is "merely equivalent" to:

(8) All grass in London and elsewhere is green.[2]

In both the case of confirmability and the case of nomologicality, one senses something contrived about examples such as (3) and (8). One might, for example, be inclined to blame the difficulties in formulating a satisfactory constraint for ruling out reference to particulars on the limitations of extensionalist language. Alternatively, one might be tempted to think that the unwanted examples could be ruled out even within an extensionalist language with just a bit more ingenuity.

I propose to look at the matter somewhat differently, by questioning in a preliminary way the very project of seeking to exclude from confirmation or law-formulation predicates that can be seen upon analysis to make some reference to particular places or times. For according to some rather plausible analyses of some very familiar predicates (indeed, some of the very predicates used by philosophers as paradigms for making a contrast with contrivances like 'grue' and 'coinowket'), these predicates involve an essential reference to a particular place or time. I do not intend here to defend these analyses, or to state a criterion of "essential reference," or even to explore in any depth the appropriate response to the sorts of examples I will be considering. The point I wish to make is very simple, and cautionary. It will be enough if the analyses mentioned are strong contenders. What I say will, I hope, be obvious—if not always noticed.

II

Consider the predicate 'x is green'. A number of philosophers have thought that something like the following is *a priori* true:

(9) x is green iff x is disposed to produce, in normal human observers under standard conditions, sensations of green-ness.

Owing to the appearance of 'green' on both sides of the biconditional, (9) is certainly not a reductive analysis of 'x is green'. But it may nonetheless be true, and may even contribute informatively to an analytic understanding of 'green'. For (9) exhibits an equivalence between being green and eliciting color experiences of a certain kind in humans under ordinary circumstances. Thus, (9) might figure as part of an explanation of the difference between a "secondary quality" like color, and a "primary quality" like shape, since

(10) x is square iff x is disposed to produce, in normal human observers under standard conditions, sensations of squareness

seems false, unless we understand 'standard conditions' in a very special way— a way that trivially guarantees (10), e.g., by specifying whatever it takes to ensure veridical perception.[3] By contrast, 'standard conditions' in (9) is understood in a substantive, nontrivial way—ordinary daylight illumination, ordinary perceptual apparatus, etc. We do not, that is, pretheoretically think of shape as constitutively linked to perceptual experience after the manner of color.[4]

But exactly how do we understand this constitutive connection in the case of color? If we imagine a world just like ours in terms of the physical constitution of emeralds and light, but different in respect of typical sensory mechanisms, is that a world in which emeralds *are red* (say), or a world in which green things *look red* to the inhabitants? Perhaps intuitions differ here, but there is a very strong case for saying that the color of objects is rigidly fixed by *actual* normal humans and standard conditions, as various philosophers have remarked.[5] That is, we should really have written (9) as:

(11) x is green iff x is disposed to produce, in normal human observers and under standard conditions *as they actually are*, the sensation of green-ness.

If this is right, then for reasons quite unconnected to 'grue'-some examples (and, parenthetically, reasons that have nothing to do with the limitations of an extensional language), 'green' seems on a non-perverse analysis to involve an essential temporal particularity or "positionality."

One philosophical response to Goodman's original riddle has been to attempt to come to the rescue of 'green', and to repudiate 'grue', by giving an account of projectibility according to which a predicate whose analysans essentially involves a positional element cannot legitimately be projected. Present reflections suggest that this response may be misguided, since this way of repudiating 'grue' would appear to drive out 'green' as well. And should we have a theory of induction that rules, *a priori*, against projecting predicates like 'green'?

III

Now consider the predicate '*x* is grass'. It might seem possible to give a reductive account of any given grass-predicate, say '*x* is timothy (*Phleum pratense*)' in terms of possession of certain biochemical/structural properties. Such a reduction purportedly would establish the predicate's "essentially general" or "purely qualitative" character. But amino acids, and arrangements thereof, can come into existence in many ways. Would an appropriately shaped bit of matter with the same biochemical composition as *Phleum pratense*, but brought into existence in interstellar space by a random—and admittedly very surprising—fluctuation, be a blade of timothy grass? Well, would an appropriately shaped bit of matter with the same biochemical composition as *Homo sapiens*, but brought into existence in interstellar space by a random fluctuation, be a human being? Again, intuitions vary, but there is a strong case for saying that ordinary species concepts are partly evolutionary/historical in character. Other planets, with no shared evolutionary history, could contain biologically and structurally similar species to those on Earth, but not co-members of the same species.

This issue arises not only for folk biological taxonomies, but for biological theory as well. Contemporary biologists have debated the appropriateness of competing definitions of 'species'—phenetic, cladistic, or evolutionary. There is no need for us to decide among the claims of these competing definitions here. Indeed, decision may in any event be unnecessary since each of these possible ways of characterizing species boundaries appears to have its particular usefulness, so that one might admit all three, while insisting upon care about which one is to be employed in which setting and for which purposes.[6] The narrow point relevant to the present discussion is simply that each of these three ways of defining 'species' seems to be perfectly *admissible*—if not identically useful—for purposes of biological theorizing, even though two of them (the cladistic and evolutionary) involve reference to a particular evolutionary history, and therefore to particular places and times.

Moreover, it would seem as if the three predicates are on par with respect to projectibility and possible involvement in laws. Nothing about the reference to a spatio-temporal particular (the actual historical sequence of evolution on Earth) in cladistic or evolutionary definitions seems to have affected the

willingness of biologists to project hypotheses involving such a species designation to new cases. Earthlings can, it seems, as readily assemble evidence for
projecting the hypothesis that

(12) All *Phleum pratense* is green

as for the hypothesis that

(13) Everything with biochemical/structural composition C[= the composition of *Phleum pratense*] is green.

Similarly in the case of lawfulness. Nothing in biological practice suggests
any special problem about claiming nomological status for (12), at least when
it is suitably fitted out with non-trivial standard conditions. Of course, another
generalization is also true, namely (13). And it might be objected that

(14) Everything grown in Cleveland possessing biochemical/structural
composition C is green

is plainly not nomological, yet it differs from (13) in the same way, formally,
as (12). However, this objection ignores the question of the potentially special
significance of geneology for *biological* description and explanation. Presumably, philosophers should not attempt to legislate in advance whether it will
turn out that biological theory is best developed using a phenetic, cladistic, or
evolutionary taxonomy. (Similarly, philosophers should not attempt to legislate
in advance whether the semantics of natural language should make appeal to
actual causal histories in establishing relations of sameness and difference in
meaning among terms.) The question would seem to be one best dealt with
cautiously, with an eye to actual, emerging empirical theory.

IV

Still, one might think that species predicates are sufficiently peculiar to make
the biological example of limited interest. Aren't genuine natural kind terms
essentially general, and do they not form the core of theorizing in the physical
sciences? Perhaps they do form the core, but the question remains whether they
are in all relevant senses "essentially general."

Consider the predicate 'x is water', directing attention not to its reference, but to its meaning. Of course, little can be said about the meaning of
'water' without some degree of controversy. Rather than attempt to formulate
an uncontroversial thesis about the meaning of 'water', let us ask whether,
among the various views about the proper approach to the semantics of 'water'
currently receiving serious attention, there are some in which the analysans
involves essential reference to particulars.

Think of 'water' as it occurred in ordinary English prior to the discovery
of water's chemical composition. A dictionary or well-schooled native informant might have told us that water is a clear, colorless, tasteless, and odorless
substance, liquid at room temperature, essential for plant and animal life, widely
used as a solvent, and abundant on or near the surface of the earth—being found

in wells, springs, rain, rivers, and oceans. Call this the 'folk-theoretic conception of water'. Perhaps this folk theory had other components as well—let us say, some survival of the Greek view that water is an *elementary* substance.

Interestingly, this folk-theoretic conception contains several spatio-temporally specific terms: 'abundant on Earth', 'widely used [by us, presumably] as a solvent', and (in the case of evolutionary/historical taxonomies) 'essential for plant and animal life'. Are all of these particularistic elements inessential to the meaning of 'water'?

On one significant approach to the meaning of 'water', Kripke-Putnam natural kind semantics, some element of particularity is indeed essential. A term like 'water' functions like a proper name, and reference is fixed via historically particular introducing episodes involving dubbings, causal attributions, or paradigm cases, and by actual causal/intentional chains of continuity in use. According to this approach, the folk-theoretic conception of water might be wrong in many respects, but this need not disrupt reference, since the term should not be understood as functioning descriptively, as an abbreviation for the folk conception of water or for any subsequent scientific conception.[7]

An alternative, the Ramsey-Lewis approach, does treat terms like 'water' as abbreviated definite descriptions, and this might be thought to remove a certain kind of particularity.[8] Yet what would the appropriate definite description be in the case of the natural language term 'water'? At least prior to the nineteenth century, it might simply be given by the folk conception: "Water is the unique clear, colorless, tasteless, and odorless elementary substance, liquid at room temperature, essential for plant and animal life, widely used as a solvent, and abundant on or near the surface of the earth—being found in wells, springs, rain, rivers, and oceans." For our purposes, it is important to note that this sort of descriptivist account of the term 'water' preserves the essential particularity of the folk conception, since the description contains clauses that function indexically.

Now it might be thought that this particularity is not shared by those theoretical terms coined by science, as opposed to those, like 'water', originating in natural language. And perhaps it is true that many introduced terms in scientific theory lack particularity. But if we consider some central theoretical terms in science, it is quite plausible that the description by means of which their Ramsey-Lewis meaning is given would include various essentially particular elements. For example, the appropriate description for 'electron' would include not only a general theoretical role, but also some particular causal attributions: electrons are, *inter alia*, those entities *actually* responsible for certain well-known effects.

None of this is utterly uncontroversial, but it is widely believed that some sort of rigidity is part of a plausible semantics for natural kind terms like 'water' and 'electron'. Such terms do seem to have, among their semantic functions, the job of helping us to fix reference in such a way as to make room for descriptive error or incompleteness. Indeed, in part for reasons of this kind, a plausible version of the Ramsey-Lewis approach would hold that anything really deserving the name of a folk-theoretic conception of water would be less definite than the account sketched thus far suggests.[9] For the folk-theoretic term was not abandoned when scientists discovered that water is a compound rather than elementary substance, or that water is rarely found in nature in a pure form, even in "pure water" springs, or that water surfaces and volumes have some chromatic properties, or that water forms a mixture or emulsion—rather than a solution—with certain substances we familiarly speak of as "dissolving" in water. And so on. So our folk-theoretic conception of water might better be given along the following lines: water is the substance found hereabouts (if there be a fairly unified one) which comes closest—and comes close enough—to being a clear, colorless, tasteless, odorless liquid (at room temperature), widely used as a solvent, essential for life, abundant on Earth, etc.

This sort of descriptive flexibility gives natural kind terms a special use-fulness in science, since they enable scientific language to maintain continuity of meaning across significant changes or controversies in theory. Notice that this sort of utility depends significantly upon the indexical component of the meaning of such terms, and therefore upon their spatio-temporal specificity: they are able to pick out "*that* stuff" via reference-fixing means that are not entirely general.

Presumably we do not wish to exclude such natural kind terms from scientific discourse, confirmation, or explanation. If that is right, then once again we have found reason for caution against any general program of attempting to identify and exclude predicates whose non-perverse analysans contains "intrinsically particular" elements.

V

Now it is fair enough to say, in response, that not all particularity is of a kind: some predicates whose analysans involves spatio-temporal particularity seem fishy as candidates for projection or nomologicality (e.g., 'grue' and 'coinowket'), while others (e.g., 'green', 'grass', and 'water') do not. So the philosophical task would become one of separating the sheep from the goats: discovering the ways in which particularity in the analysans disqualifies. This project would replace the project of developing an all-purpose characterization of spatio-temporal particularity for the purpose of blanket exclusion.

Fair enough, I say. This is an important project for developing an un-derstanding of scientific discourse and practice. But is this a recipe for an autonomous philosophical program? What reason have we to think that philos-ophy possesses a method that would enable it to separate the sheep from the goats *a priori*? Predicates involving a historical component may, for example, prove not to be useful in some kinds of scientific theorizing, but indispensable in others. Moreover, not all of the *prima facie* admissible predicates involving specificity are cut from the same cloth. A species predicate like '*Phleum pratense*', when construed as containing an essential evolutionary/historical component, is particularistic in a way quite different from the particularity of 'water', and both are quite different in their own ways from the particularity of a secondary quality predicate like 'green'. 'Water' applies indifferently to anything with a certain chemical composition, H_2O, wherever and whenever it is found and whatever its causal history; '*Phleum pratense*' applies to things with the biochemical composition of *Phleum pratense* only if they share an evolutionary history on earth; and neither 'water' nor '*Phleum pratense*' depend constitutively upon a temporally and species-specific *response* on our part, as 'green' appears to. No doubt philosophy should attempt to develop an understanding of the various ways predicates differ in their particularity and how such differences are associated with differences in subject matter, explanatory aim, and so on. But can philosophers expect to be able to say much in advance about which kinds of particularity will find niches in which areas of scientific theorizing? This is not a recommendation of philosophical quietism, but of a philosophical activism more directly engaged with the task of understanding how different elements of scientific theorizing function, and why.

VI

A somewhat different sort of objection would hold that the exclusion of terms whose analysans involve spatio-temporal particulars is to be motivated by the idea that such terms—whether 'green' or 'grue', '*Phleum pratense*' or 'coinowket'—are "ultimately dispensable" since they will not figure in our "ultimate, explanatory, comprehensive scientific theory of the world." This ultimate theory, one is to imagine, will mention only fundamental laws, token-indistinguishable elementary particles, and the four basic forces.

What intrigues me initially in this objection is not so much the faith it displays that we will find a comprehensive theory, but rather the conception it presupposes of what such a theory would be like. A *comprehensive* explanatory theory would presumably have living organisms within its scope. But the history of natural selection, told at the subatomic level, would not only be bewildering in complexity in ways that would be a practical obstacle to understanding, it would, more importantly, leave out some significant structural and modal features of the world. For example, self-replicating organisms operating competitively within environmental constraints will, as a class, display some confirmable, counterfactual-supporting regularities, e.g., principles of ecology and population biology. These commonalities extend across species boundaries in ways that show how the behavior of physically rather different systems can be non-accidentally similar *if* those physically different systems realize interacting self-replicating organisms. Such principles (like principles in statistical mechanics) will be robust with respect to certain variations in the microphysics, and so support a distinctive "level" of understanding of systems. An allegedly comprehensive theory that omitted the entire vocabulary of organisms, reproductive strategies, gene frequencies, selection pressures, etc., would also omit all regularities formulated in this vocabulary, and would be explanatorily impoverished. This can be so whether or not the "ultimate" theory affords some sort of "reduction" of the higher-level theory.

In any event, it is unclear why a predicate's participation in an ultimate theory should bear on its aptness for participating at all in laws and explanations, or, especially, in projection and confirmation. Take projection and confirmation first. Even if we suppose that an ultimate theory will eventually be within our grasp, nonetheless, reaching it patently will require many years of unfinished theories and theorizing. If nothing can genuinely—or differentially—be confirmed which would not figure in the finished theory, it is quite unclear how we are ever to get there.

Questions of explanatoriness and nomologicality are perhaps more difficult. Explanatoriness, I have argued elsewhere, is a matter of communicating explanatory information, and many sorts of statements with no prospect of participating in an ultimate theory can accomplish this.[10] Suppose we wish to explain why grass snakes have the skin pigmentation they do. We say: this pigmentation makes them appear green, and grass, their preferred habitat, is also (though, notably, owing to a different biochemistry) green; green things look alike not only to us but apparently also to those animals with a more active and long-standing interest in the gustatory possibilities of grass snakes; and this simple but effective bit of camouflage has contributed to the species' reproductive success relative to competitors. Such remarks are laden with information about the etiology of grass-snake pigmentation, even though they explicitly cite no law and even though some of the predicates involved have analysans not entirely general.

Turning to nomologicality, we find it to be a common desideratum of philosophical accounts of lawlikeness that they capture to a significant degree patterns of usage in actual science with respect to which generalizations are, or are not, called laws. Many paradigm laws concern magnitudes that would not figure in an "ultimate" physics of elementary particles and basic forces. The ideal gas law, for example, concerns "nonultimate" variables of pressure and temperature; Ohm's law makes reference to electric current in a circuit; Snell's law employs a term for the refractive index of various media; Bernoulli's law involves terms for pressure difference and rate of flow in a liquid; and so on. One could, of course, stipulate that the term 'law' belongs only to the most basic generalizations of elementary particle physics, but then one would have to invent a new vocabulary to distinguish those generalizations within the class of "non-laws" which scientists regularly call 'laws', cite in explanations, use to support counterfactuals, confirm by their instances, and so on. It would be enough for my purposes here that generalizations containing predicates whose analysans involve particulars could be in this distinguished "non-law" group.

VII

If predicates like 'x is grue' and generalizations like 'All coinowkets are silver' are not to be excluded from proper scientific practice by an *a priori* philosophical theory segregating "essentially general" from "essentially particular" predicates, or by an *a priori* philosophical theory segregating good from bad "essentially particular" predicates, how, then, are they to be excluded?

One answer, which perhaps affords little comfort, is, "By no general *a priori* method at all—merely, if at all, by inductive preconception and inductive practice." We approach inquiry with certain prior expectations about what sorts of regularities will continue to hold. We then assess the import of new experience in light of these expectations. We do not—indeed, we cannot—pay attention to anything like the full range of *a priori* possible hypotheses when revising our beliefs in the face of experience—even the community of scientists, viewed collectively, does not have anything like this scope. Rather, rational, limited believers, lay and scientific, begin with what they now believe, and then go on to seek and assimilate further experience. In this way, it is possible for prior expectations both to shape inquiry, and in turn, as experience grows, to be shaped by the results of that inquiry. Nothing can be done to ensure that all predicates will get a fair chance, or to guarantee that, in any actual (rather than idealized) short or long run, all hypotheses with *prima facie* anomalous predicates will be winnowed out. Indeed, some *prima facie* anomalous predicates may turn out to be just what we need to describe and explain our world adequately.

These meager observations are not much of a substitute for a powerful exclusionary theory. But this may be a more tenable state of affairs than one in which philosophers propose formal principles of exclusion that threaten the place in empirical learning and explanation of such familiar and useful predicates as 'green', 'grass', and 'water'.

NOTES

1. See Nelson Goodman, *Fact, Fiction, and Forecast* (Cambridge, Mass., 1955), 74. This way of defining 'grue' makes objects be grue once and for all. An alternate definition would introduce a time variable. That complication is not necessary for current purposes, and time variables will be suppressed throughout.

2. Again, see Goodman, *Fact, Fiction, and Forecast*, 78. See also the discussion of A. J. Ayer, "What is a Law of Nature?" reprinted in *Readings in the Philosophy of Science*, edited by Baruch A. Brody (Englewood Cliffs, 1970).

3. For discussion of the "whatever it takes" reading of 'standard conditions', and of the primary/secondary quality contrast in general, see Crispin Wright, "Realism, Antirealism, Irrealism, Quasi-realism," in *Midwest Studies in Philosophy*, vol. 12, edited by Peter French, Theodore Uehling, and Howard Wettstein (Minneapolis. 1988), 25–49.

4. Of course, it might be a surprising result of philosophical investigation that we should challenge this commonsense understanding of the shape-color contrast.

5. The first discussion of this point of which I am aware is Martin Davies and Lloyd Humberstone, "Two Notions of Necessity," *Philosophical Studies* 38 (1980): 1–30.

6. See, for example, Ernst Mayr, "Biological Classification: Toward a Synthesis of Opposing Methodologies," *Science* (1981): 510–16.

7. For discussion, see Saul Kripke, "Naming and Necessity," in *Semantics of Natural Language*, edited by Donald Davidson and Gilbert Harman (Dordrecht, 1972) and Hilary Putnam, "The Meaning of 'Meaning'" and "Explanation and Reference," in his *Mind, Language, and Reality: Philosophical Papers*, vol. 2 (Cambridge, 1975).

8. See David Lewis, "How to Define Theoretical Terms," *Journal of Philosophy* 67 (1970): 427–46.

9. I am indebted here to conversations with David Lewis.

10. See P. Railton, "Probability, Explanation, and Information," *Synthese* 48 (1981): 233–56. See also David Lewis, "Causal Explanation," *Collected Papers*, vol. 2 (New York, 1986).

On Philosophy and Natural Philosophy in the Seventeenth Century

HOWARD STEIN

That seventeenth-century usage did not distinguish science from philosophy is, I suppose, widely known; but perhaps it is less widely regarded as a matter of consequence. I think the usual view is that the practice of restricting the sense of the word "philosophy" to some set of more general or more fundamental kinds of inquiry makes a useful distinction, and one that is in effect to be found in practice throughout what we regard as the "modern" period of philosophy.

I do not wish to claim that this is entirely wrong. A difference that has some kinship with our own departmentalization of disciplines can certainly be discerned among works of seventeenth-century authors; if that were not so, the view that I have called the usual one could hardly have been maintained at all. But I think the view is misleading—which is a more insidious thing than to be conspicuously wrong. I want to present two broad theses, which I shall try to make plausible by a number of concrete examples: first, that the historical understanding both of what *we* call "philosophy" and of what *we* call "science" in the early modern period is significantly better served if we pay more attention than has usually been done to the relation they had for their practitioners; second (and, for me, more important), that a certain shift of perspective *upon philosophical issues*—issues that we still trace back to the seventeenth and eighteenth centuries—may result; a shift that I myself regard as a philosophic gain and that at any rate I hope will, less controversially, be seen as worth reflecting upon.

Let me begin with a rather blatant example—one that I have discussed before (as long as twenty-five years ago). It is taken from a work no longer much regarded, but quite influential in its day; a work to whose stimulation I owe something, for at the time I was a student it stood as a pioneering attempt to bring the history of philosophy into contact with that of science. I speak of the late E. A. Burtt's *The Metaphysical Foundations of Modern Physical*

Science. In his chapter 7, "The Metaphysics of Newton," Section 4(B), "Space and Time," Burtt writes the following:

> When we come to Newton's remarks on space and time . . . he takes personal leave of his empiricism, and a position partly adopted from others, partly felt to be demanded by his mathematical method, and partly resting on a theological basis, is presented, and that in the main body of his chief work. Newton himself asserts that in "philosophical disquisitions", which apparently means here when offering ultimate characterizations of space, time, and motion, "we ought to abstract from our senses, and consider things themselves, distinct from what are only sensible measures of them."[1]

There are several things wrong with this; but what I call "blatant" is the construction Burtt so casually puts upon Newton's phrase "in philosophical disquisitions." Burtt's target in this entire passage is, of course, the celebrated scholium of Newton on time, space, place, and motion, with its distinction of each of these into "absolute and relative, true and apparent, mathematical and common." Speaking there of time in particular, Newton says:

> Absolute time is distinguished from relative in Astronomy by the equation of the common time. For the natural days are unequal, which are commonly used, as if equal, for a measure of time. The Astronomers correct this inequality, that they may measure the celestial motions by a truer time. . . .The duration or perseverance of the existence of things remains the same, whether the motions are swift or slow, or none at all: and therefore it ought to be distinguished from what are only sensible measures thereof; and out of which we collect it, by means of the astronomical equation. The necessity of which equation in determining phenomena is evinced as well from the experiments of the pendulum clock, as by eclipses of the satellites of Jupiter.[2]

This one might have expected to give Burtt some pause, both in his claim that Newton has taken "personal leave" of his empiricism in asserting a distinction of "absolute, true, and mathematical" from "relative, apparent, and common" quantities—for Newton refers to *experiments* (with the pendulum clock) and *observations* (of the satellites of Jupiter) as providing evidence of the need to distinguish true from apparent times—and in his interpretation of the sort of context in which Newton says this distinction ought to be made (namely, "in philosophical disquisitions"). But what is most ludicrous here is that Burtt has forgotten the very title of the book—rightly called by him "[Newton's] chief work"—in which the discussion he is criticizing occurs. The title is *Mathematical Principles of Natural Philosophy*; and for clarification of the sense attached by Newton to the phrase "philosophical disquisitions," Burtt would have done well to reflect on this fact, rather than to rely on his sense of what that phrase "apparently means."

A second passage that had a significant effect on me as a student is one from Descartes; but I read it first in Norman Kemp Smith's *New Studies in the Philosophy of Descartes*. "Writing to Beeckman," Smith says, "Descartes admits that the whole body of his physics [*totam meam Philosophiam*] would be shaken to its very foundations, should light not travel instantaneously."[3] Although he gives Descartes's own phrase in brackets, Smith's translation implies the clarification: "here by *philosophy* Descartes just means physics." But what Descartes actually says is "my whole philosophy would be utterly overthrown" (or "overthrown from its foundations": *totam meam Philosophiam funditus eversam fore*). This seemed to me at the time rather striking; I had previously assumed that Descartes's physics was of minor interest compared to what I myself then thought of as his "philosophy"—but the passage, even if it is taken as hyperbole, suggested that perhaps the physics ought to be looked at more seriously as part of an organic whole.

Here is a more recent comment—again from a philosopher who, like Burtt and Smith, is aware of and interested in the connection, in the seventeenth century, between what *we* call "physics" and "philosophy"; and who, in this passage, is concerned to emphasize this point.

> The notion that there is an autonomous discipline called "philosophy," distinct from and sitting in judgment upon both religion and science, is of quite recent origin. When Descartes and Hobbes denounced "the philosophy of the schools" they did not think of themselves as substituting a new and better kind of philosophy—a better theory of knowledge, or a better metaphysics, or a better ethics. Such distinctions among "fields of philosophy" were not yet drawn. The idea of "philosophy" itself, in the sense in which it has been understood since the subject became standardized as an academic subject in the nineteenth century, was not yet at hand. Looking backward we see Descartes and Hobbes as "beginning modern philosophy," but they thought of their own cultural role in terms of what Lecky was to call "the warfare between science and theology." They were fighting (albeit discreetly) to make the world safe for Copernicus and Galileo.[4]

But what Rorty tells us here is wrong. Descartes emphatically did think of himself as "substituting a new and better kind of philosophy" for that of the schools: a better metaphysics, a better physics, and in prospect a better ethics—"fields of philosophy" that he explicitly did distinguish. Of course I have included one field—physics—that Rorty does not; Descartes's own list (sufficiently celebrated) contains two more: *mechanics*—that is, more or less what we should call "engineering"—and *medicine*. As to making the world safe for Copernicus and Galileo, there is no doubt that Descartes deeply deplored the condemnation of the latter for teaching the doctrines of the former— a condemnation by which he felt himself very seriously threatened; but his comments on the actual work of Galileo are generally negative and dismissive.

Here is the view of Christiaan Huygens (whose father was a friend of Descartes, and who was himself in his youth an enthusiastic reader of Descartes's works):

> M. Descartes, who seems to me to have been very envious of the renown of Galileo, had that great desire to pass for the author of a new philosophy. Which is apparent from his efforts and his hopes to have it taught in the academies in place of that of Aristotle; from the fact that he wished the society of the Jesuits to embrace it: and lastly because he maintained indiscriminately [*à tort et à travers*] the things he had once put forward, although they were often quite false.[5]

Huygens is indisputably right in saying that Descartes wished to be seen as the author of a new philosophy—that is, of a new science. He wished more than that: he wished to *be*, *in truth*, the author of a new philosophy—and moreover of a true one; or rather, of *the* true one. Further, he thought that he *was so* in truth.

One should understand what this claim of Descartes's meant. I have referred to the fields of philosophy as Descartes names them; the locus is the letter to Picot that serves as a preface to the French edition of the *Principia Philosophiæ*. In that place there occurs his famous metaphor of philosophy as a tree, whose roots are metaphysics, its trunk physics, and its branches—which bear the fruit—the other three fields I have named: medicine, mechanics, and morals. Of these, it is medicine above all that was Descartes's preoccupation. Thus in January 1630 he writes to Mersenne:

> I am distressed over your erysipelas, and over the illness of M. [Montais]; I beg you to preserve yourself [*de vous conserver*], at least until I shall know whether there is a way to find a Medicine founded on infallible demonstrations, which is what I now seek.[6]

And again on April 15 of that year:

> I am now studying in chemistry and anatomy together, and learn something every day that I do not find in the books. I wish I had already come to the point of research into diseases and their remedies, so that I could find one for your erysipelas, with which I am sorry you have been afflicted for so long.[7]

In the concluding paragraph of the *Discours de la Méthode*, he says:

> I have resolved to devote the rest of my life to nothing other than trying to acquire some knowledge of nature from which we may derive rules in medicine which are more reliable than those we have had up to now.

And writing to the Marquess of Newcastle in October 1645 he says, "The preservation of health has always been the principal end of my studies."[8] The sincerity of these professions seems guaranteed above all by the evidence that the medical problem of greatest interest to Descartes was that of longevity—and that this was very much a *personal* concern: a desire for *his own* longevity. On this subject he writes, December 4, 1637, to Constantijn Huygens:

> I have never taken so much care to preserve myself as now, and whereas I used to think that death could snatch from me no more than thirty or forty years at the most, it would not henceforth surprise me if the deprivation could amount to more than a century: for I seem to see very clearly, that if we just beware of certain mistakes that we habitually make in the conduct of our lives, we should be able without any other discoveries [*inventions*] to arrive at a far longer and happier old age than we [now] do; but because I have need of much time and many experiments to investigate everything of use for this subject, I am now working to compose a digest of Medicine, drawn partly from the books and partly from my reasonings, that I hope to be able to use provisionally to obtain some reprieve from nature, the better afterwards to continue with my plan.[9]

Although further evidence of Descartes's optimism in this matter is somewhat variable, there is some indication that at least at times—and to *some* of his followers—Descartes offered even far more favorable estimates, not only of the possibilities, but of what he had actually achieved. One report, deriving from "the chevalier [that is, Sir Kenelm] Digby," has it that Descartes assured the latter that "although to render man immortal is what he would not dare to hope for [*se promettre*], he was quite sure of being able to render his [that is, presumably, "man's"] life equal to that of the Patriarchs."[10] Digby is, to be sure, not an informant who inspires the greatest confidence (one contemporary characterizes him as "the very Pliny of our age for lying");[11] but whether or not Digby exaggerated, there is other testimony in a quite similar vein. Thus Baillet tells us that when the news of Descartes's death in Sweden was received in Paris, the Abbé Picot at first would not believe it; and afterwards said that his death could only have resulted from the derangement of his system by the unhealthy north—since "in the absence of such an alien and violent cause, . . . he would have lived five hundred years, after having discovered the art of living several centuries."[12] Finally, and perhaps most telling, Queen Christina herself—who had been discussing the question of longevity, among other matters, with Descartes just before his final illness—commented wryly when he died that "his oracles have quite deceived him."[13]

But what has this to do with *our own reading* of Descartes's philosophy? For *us*, surely, Descartes's dream of establishing physics (and medicine!) on "infallible demonstrations" is just a quaint delusion of the time and the person, which ought not to prejudice our judgment of the epistemological and metaphysical discussions that *he* thought would serve to found such a science. To this I merely reply: prejudice, no; but influence—yes. In his metaphor of the tree, and his remark that "just as it is not the roots or the trunk of a tree from which one gathers the fruit, but only the ends of the branches, so the principal benefit of philosophy depends on those parts of it which can only be learnt last of all," Descartes clearly had in mind chapter vii, verse 20 of the Gospel of Matthew: this criterion forms, in a certain sense, a part of his own epistemology, and simply to ignore it in our reading of that epistemology is in some degree a distortion. Furthermore, the criterion of judging by the fruits has this merit: that on some matters it really does put an end to dispute. Now, dispute on ticklish questions of Cartesian epistemology and metaphysics has been endless and subtle. But I have been struck by the fact that some of the cruces of Cartesian argumentation and doctrine—for instance, the much-debated issue of the circularity of the main argument of the *Meditations*, or the doctrine of the arbitrary creation of the eternal truths (which Descartes himself advises us not to think about too much)[14]—bear, in the puzzling character of the argumentation or the apparent inconsistency of the principles invoked, a remarkable resemblance to passages of Descartes's physics whose coherence no one could reasonably be tempted to defend; some of these contain even astonishing *mathematical* sophisms.

I do not wish to embark here on a detailed discussion of science and philosophy in Descartes, for although I think him a very important figure and a very interesting one, my conclusions tend (as the remarks already made will suggest) to be rather negative. Such conclusions demand especially careful citation and argument—inappropriate for a brief essay; and they are after all not very attractive or exhilarating. Let me just say summarily that, in the first place, Descartes's philosophical work was devoted to the aim of establishing a new philosophy in the fullest possible sense of that phrase; in the second place, that his view of this new philosophy was *not* that of a collective enterprise to be pursued independently by inquirers like Galileo and himself: rather, it was his ambition—and to the end of his life, his *hope*, in which he continued to express confidence—that he could *by himself*, with the aid only of artisans to make equipment and to carry out experiments under his direction, bring this project to essential *completion*; and, finally, that my own study of this work has led me to conclude that Descartes *as a philosopher in our own sense of the word* fell victim to something he thought he had fundamentally guarded himself against: a deceiver. For in my view, Descartes is an astonishing example of a self-deceiver.

Descartes's influence upon the metaphysics (including epistemology) and physics of the later seventeenth century was of course very great. In the piece from which I have already quoted, Huygens—after remarking that, from having thought in his own mid-teens that in Descartes's *"livre des Principes. . . tout alloit le mieux du monde,"* he had gradually come to change his opinion, until, at age sixty-four, "I find hardly anything that I could approve as true in all his physics or metaphysics"—goes on to deplore that influence:

> He should have proposed his system of physics as an essay in what one might say with probability in that science admitting mechanical principles only, and have invited good minds to investigate on their own part [note: something like what Rorty supposes Descartes to have done]. This would

have been very laudable. But in wishing it to be believed that he had found the truth, as he did in everything, in founding himself upon and glorying in the sequence and the beautiful connection of his expositions, he has done a thing that is of great prejudice to the progress of philosophy.

But at the end of this set of notes on Descartes—made by Huygens during his reading of Baillet's life of that philosopher—he comes to a more generous estimate: that Descartes's grandiose attempt did at any rate show much genius; that his example, and the celebrity he achieved by his work, has inspired others to try to do better; and that if one considers only his achievements in geometry and algebra, he must be accounted a great mind.[15]

I cite this particularly because Huygens himself is one of the two investigators in physics to whom his own remark, that Descartes's example has inspired others to try to do better, most conspicuously applies. The second, of course, is Newton; and a comparison of the two, in point both of their own philosophical achievements and their philosophical relations to Descartes, seems to me a very instructive thing.

Huygens disavows both Descartes's metaphysics and his physics; and he entirely abandons Descartes's epistemology, which demanded that science be based upon principles secured by intuition, and demonstrative reasoning—that is, must be so constituted as to avoid the possibility of error. In contrast, Huygens tells us that the deeper investigations of physics lead to "demonstrations of those kinds which do not produce as great a certitude as those of Geometry, and which even differ much therefrom, since whereas the geometers prove their Propositions by fixed and incontestable Principles, here the Principles are verified by the conclusions to be drawn from them; the nature of these things not allowing of this being done otherwise."[16] (To be sure, this can be seen as influenced by Descartes's own practice; but it contradicts his principles.) However, despite his rejection of the Cartesian identification of the corporeal with the extended, and his embrace of the entirely un-Cartesian doctrine of the atomic constitution of matter, Huygens remained deeply committed to a pair of principles that can be seen as constitutive both of a conception of knowledge and a conception of the world—principles in this sense epistemological and metaphysical—that he did share with Descartes. This is already suggested by a remark in the passage I have quoted above: his reference to the praiseworthiness of "an essay in what one might say with probability [this, of course, is what contrasts with Descartes] in [physics] *admitting mechanical principles only.*"

The phrase "mechanical principles," in the seventeenth century, connoted two things. The first of these is stated with admirable clarity by Huygens at the beginning of his discussion of the nature of light (*Treatise on Light*, p. 3), where he says: "It is inconceivable to doubt that light consists in the motion of some sort of matter"; and then, after listing kinds of phenomena that support this view, adds: "This is assuredly the mark of motion, *at least in the true Philosophy, in which one conceives the causes of all natural effects in terms of mechanical motions.*" And finally: "This, in my opinion, we must necessarily do, or else renounce all hopes of ever comprehending anything in Physics."

It would be interesting to know upon what Huygens conceived this "opinion" of his to be based. In particular, did he believe there was convincing evidence of some kind—and if so, of what kind—supporting what he calls "the true Philosophy": evidence that *in truth all natural effects are caused by mechanical motions?* (The phrase "mechanical motions," here, is to be understood simply to mean *motions of bodies in space* [or—Huygens was in fact a relativist—"with respect to their spatial relations to one another"]: the Aristotelian φορά, as opposed to κίνησις) Of course, if one does have convincing reasons to believe that the world is so, one will *ipso facto* believe it is only by conceiving the causes of natural effects in such a way that one can hope to comprehend them. But it seems to me more likely that Huygens was moved by what I may call a modest transcendental argument: the possibility of advancing physical knowledge along such a path, which had inspired him in his youth, continued to look promising; he saw no alternative possibility that appeared to him worth pursuing; and he concluded—expressly as an "opinion" (although he also characterizes this view as "the true Philosophy"; thus, certainly, a very *firm* opinion)—that *the world must be so, if we are to comprehend it.* In any case, we have, in parallel, a view of the world itself, and an ideal of what science ought to (or *must*) be.

If all phenomena are the effects of motions, then to understand their causes two sorts of knowledge are required: one must know something about the kinds of motion that produce phenomena—appearances—of given types (heat; light—and, in more detail, light of various specific *qualities*, especially colors; weight; smell; etc.)—let us say, one must know the "mechanical" *nature* of such phenomena; and one must know something about how those underlying motions of bodies *affect one another*—that is, one must possess a system of principles of *mechanical interaction.* This latter Descartes had claimed to provide, in his three general "laws of nature," in *Le Monde* and in the *Principles of Philosophy.* The latter work offers also more specific "rules" for determining the effects of impacts of bodies upon one another; and the *Dioptrics* contains examples of reasoning, of a "mechanical" kind (in the sense understood by Descartes), for deducing the laws of light from postulates about its "mechanical nature." It is these detailed principles of interaction, and in the first instance most particularly the rules of impact, that Huygens early became disillusioned with, coming to see them not only as false, but even as unreasonable and absurd. On the other hand, one characteristic of Descartes's conception of the principles of interaction—of what exclusively deserve to be regarded as *conceivable* principles of interaction—continued to dominate the thought of Huygens, and formed an essential part of what he regarded as the "true Philosophy": namely, that (a) motion of itself tends to *persist*—to remain constant in both magnitude and direction; and (b) only when two bodies are in contact, in such a way that the conservation of the motions of both would entail their mutual penetration, can any direct "interaction" occur between them. (For Descartes, from a fundamental metaphysical point of view there is even in that case no "direct" interaction, since "God is the primary cause of motion": he "imparted various motions to the parts of matter when he first created them, and he now preserves all this matter in the same way, *and by the same process* by which he originally created it" [*Principles*, II, §36; emphasis added]. Nevertheless, Descartes admits in a "secondary" sense—so-called "second causes"—what he calls the "force" of a body "to continue its motion or rest," or to "drive or resist another" [ibid., and §§40ff, esp. 43 and 45]; and it is on his estimates of these "forces," these "second causes," that he bases his reasoning in physics.)

This is what I have referred to as the second connotation of the term "mechanical principles": the view that the only intelligible form of interaction is what was called "mechanical action," namely the alteration of the motion of one body by another *through contact*, and more specifically through such contact as *necessitates a change in motion to avoid mutual penetration*. This conception continued to dominate the thought, not only of Huygens, but of all the "mechanical philosophers" of the seventeenth century (most definitely including Leibniz, for instance)—with the single exception of Newton (who for this reason should not be listed, without qualification, among the mechanical philosophers).

Note once more that the mechanical philosophy involves both an ideal of science and a fundamental view of what the constitution of the world is. Of course, in saying that it involves an ideal of science, I do not mean a clear and univocal conception of the *grounding* of knowledge—since, for instance, on this point Descartes and Huygens radically differ—but rather an ideal of the *form*, or "shape," of science. Note too that the view of the constitution of the world is a view of *the nature of causation*. Since the latter involves the notion that all causation among bodies is "mechanical" in the sense I have explained, it follows that for the mechanical philosophy to know *the* fundamental principles of nature—what Descartes calls to "know what a natural power in general is"[17]—is to know *the laws of the communication of motion by contact*. In principle, therefore, in Descartes's own system, his rules of impact ought to be regarded as the laws of fundamental interaction—the foundation of principles of all "second causes." (One striking index of the peculiar texture of Descartes's actual reasoning is Descartes's own assurance to Queen Christina that these rules need not be studied with care, for they are unnecessary to an understanding of the rest of the work.[18])

Huygens—and Leibniz—at any rate did indeed regard the laws of impact as the fundamental laws of physics; the counterpart today would be the elusive unified theory of the so-called four fundamental forces. And it is one of Huygens's very great achievements to have discovered a viable theory of impact. In doing so—by a very beautiful series of theoretical arguments, grounded in part in Galileo's results concerning falling bodies, and in part in two principles of the greatest generality: the principle, borrowed from Galileo but stated for the first time with full clarity by Huygens, that we now call that of Galilean relativity;

and the principle of the impossibility of perpetual motion (or more exactly, of the *creation of energy*)—he contributed essentially to what we ourselves call the science of mechanics. Newton, in his scholium to the laws of motion and their corollaries in his *Principia*, characterizes the laws he has stated as already contained in the work of his predecessors. But, as I have already intimated, "mechanics" as it appears in Newton has been significantly transformed.

In order to explain how transformed, I want first to return to a comparison I have made before: of Newton and Locke. If Descartes is a fascinating figure because of his astoundingly grandiose purpose, his monumental assurance, and (as I think) his grotesque but possibly instructive mistakes, Locke is fascinating for his very serious and honest engagement with problems of knowledge and its advancement, and for his willingness—part of that seriousness and honesty— to *vacillate* over fundamental issues. Locke can be exasperatingly repetitious, and exasperatingly inconsistent; but his repetitions are a sign of his persistent preoccupations, and his inconsistencies are a sign of genuine perplexities. On the other hand, what is surely his most profound vacillation, inconsistency, and—I believe—objectively and in principle his revolutionary *change of mind*, is *not* conspicuous in the text of the *Essay*, but on the contrary is almost undetectable there. The revolutionary change is due directly to Newton.

Among the central motifs of Locke's *Essay concerning Human Understanding* [19] are the doctrine of primary qualities, the problematic "idea" of *substance*, and an altogether skeptical view of the possibility of a science of nature. These themes are deeply interconnected. Locke's view of science in the *Essay* is Cartesian in one important respect that contrasts with Huygens: a body of doctrine has no standing as science unless its constituents are either *intuitive*, or derived from intuitive principles by *demonstrative reasoning*. (To be sure, Locke admits a third degree of knowledge—*sensitive* knowledge; but this not only has a lower degree of certainty than the former two, what is of far greater moment, it does not contribute to what I have called "a body of doctrine"; for it has as object nothing but "the existence of particular external Objects, by that perception and Consciousness we have of the actual entrance of *Ideas* from them" [*Essay*, IV, ii, §14]—that is, it does not extend beyond the objects of *present* sensation.) On the other hand, Locke rejects Descartes's theory of innate ideas and principles; and, in particular, rejects not only Descartes's own view of the nature of corporeal substance, but rejects also—if I may editorialize: quite properly!—the very possibility of knowledge of corporeal nature demonstrated from intuitive principles.

What, then, it might be asked, is the allegedly deep connection of the three motifs I have mentioned? The account I have just given of Locke's skepticism concerning natural philosophy does involve reference to "corporeal nature," and so to "substance"; but the connection is pretty thin and altogether negative, consisting simply in the denial that we have any source of general knowledge thereof; and the motif of "primary qualities" has not occurred at all. Moreover, so far as this point goes, it would seem that Locke's aim in the *Essay*—"to enquire into the Original, Certainty, and Extent of humane

Knowledge; together with the Grounds and Degrees of Belief, Opinion, and Assent" (I, i, §2)—could, as regards the first topic, *Knowledge*, and this main branch, *of corporeal nature*, have been disposed of in very short order indeed. (Of course, Locke himself tells us that when he first put pen to paper, he thought that one sheet would have sufficed for all he had to say.)

The official program of the *Essay*, announced in the immediate sequel to the passage just quoted, is to avoid entirely what Locke calls "the Physical Consideration of the Mind"—including under this any consideration of the dependence of our "Ideas" upon matter—and to study only mental *phenomena*, or "the discerning Faculties of a Man, as they are employ'd about the Objects, which they have to do with." In this program, "speculations" are to be avoided; Locke characterizes his proceeding as "this Historical, plain Method"; and, on his own final teaching, its product can have no standing as "science," or "knowledge."

There may be such an essay embedded in the *Essay*; and if so, the "thin" argument sketched above might—I am not sure: perhaps it could not—find its place there. But the work Locke has given us departs very far from that program, and he is (perhaps somewhat uncomfortably) aware that it does. This is most clearly acknowledged in chapter viii of Book II, which introduces the distinction of "qualities" into primary and secondary: in §22 of that chapter, Locke apologizes for having "been engaged in Physical Enquiries a little farther than, perhaps, I intended"; and asks to be "pardoned this little Excursion into Natural Philosophy," on the grounds that "to distinguish the *primary*, and *real Qualities* of Bodies, which are always in them," from the others, is "necessary in our present Enquiry."

The primary importance of the primary qualities for Locke is twofold, and both aspects bear crucially upon his conception of what natural philosophy should be and what the prospects for it are. In the first place, those qualities, in Locke's express view, constitute the *foundation*, the *principles (principia*; this is the direct connotation of the word "primary"–or, as Locke also calls them, *original*), intrinsic in bodies, of all their other properties; that is, they are *causally fundamental*. Thus, of the "secondary" qualities, Locke says that they are "powers" of a body, "*by* Reason of *its* insensible *primary Qualities*, to operate . . . on any of our Senses" (§23); and of a very important third class of properties, for which Locke subsequently (§26) proposes the designation *"Secondary Qualities, mediately perceivable,"* he says (§23) that they are powers of a body, "*by* Reason of the particular Constitution of *its primary Qualities, to* make such a *change* in the *Bulk, figure, Texture, and Motion of another Body*, as to make it operate on our Senses, differently from what it did before." (Locke's peculiar way of using italics in these passages serves to highlight what he regards as summarizing their essential force. In each case, the phrase *by its. . .primary Qualities* is made to stand out.)

Locke does not tell us how, in his view, we know that the primary qualities are causally fundamental. Indeed, he vacillates on the question *whether* we know this. For instance, he refers (IV. iii. §11) to the secondary qualities of bodies as "depending all (*as has been shewn*) upon the primary Qualities of their minute and insensible parts; *or if not upon them, upon something yet more remote from our Comprehension"* (emphases added). The parenthetic phrase "as has been shewn" seems to imply something like a previous *demonstration*; but the reservation that follows belies this, and raises interesting and puzzling questions. One thing that seems to me unclear is whether Locke considered it as secure knowledge (if so, that knowledge must, I think, be "intuitive") that the primary qualities are *among* the causally fundamental ones, and doubtful only whether there may be other causally fundamental properties as well, that are "more remote from our Comprehension"; or whether he also regarded it as doubtful that the qualities he lists as primary are truly so—whether they themselves may be "secondary"—that is, *derivative* from something "more remote from our Comprehension." (It is worth noting that this last is the teaching of the natural philosophy of the twentieth century.) There are passages in Locke that suggest the latter, more radical, view may have occurred to him as a troubling possibility; but I do not have space to pursue this matter further here.

At any rate, whatever the reservations may be that attach to the position— and the more far-reaching conception of the corpuscular constitution of bodies (with all their properties depending upon the primary qualities of *ultimate indivisible particles*) is explicitly characterized by Locke as a "hypothesis" (IV. iii. §16)—it is certainly the "official" view of the *Essay* that we neither have *nor can have* any clear ideas of causative principles of bodies *beyond* the primary qualities detailed by Locke (all of which, it should be noted, are characterized by him as corresponding to *simple ideas of sensation*). So the twofold importance for Locke's conception of science is this: first, to *have* a natural science would be to have a genuine—intuitively grounded and demonstratively elaborated—causal account of appearances (and, thus, in Locke's conception, *of our "ideas of sensation"*) that should derive the latter exclusively from Lockian primary qualities of the minute parts of bodies. But second, this is (in effect) *demonstrably impossible*—for two reasons: (a) we have no sensory access to the necessary *information* about bodies—"the size, figure, and texture of Parts" on which everything else depends (IV. iii. §11); (b) even worse (Locke says: "a more incurable part of Ignorance"), we have no conception of how the primary qualities of bodies could produce in us any "ideas" whatsoever (IV. iii §§12–13). (In fact, Locke mentions only ideas of *secondary* qualities in this connection; it remained for Berkeley to point out that the same problem arises for our ideas of primary qualities.)

As a summary of this aspect of Locke's position, and at the same time a wonderful example of his characteristic mode of decisive ambivalence, the following passage from his chapter "Of our *Complex Ideas* of Substances" may serve:

> To conclude, Sensation convinces us, that there are solid extended Substances; and Reflection, that there are thinking ones: Experience assures us of the Existence of such Beings; and that the one hath a power to

move Body by impulse, the other by thought; this we cannot doubt of. Experience, I say, every moment furnishes us with the clear *Ideas*, both of the one, and the other. But beyond these *Ideas*, as received from their proper Sources, our Faculties will not reach. . . . From whence it seems probable to me, that the simple *Ideas* we receive from Sensation and Reflection, are the Boundaries of our Thoughts; beyond which, the mind, whatever efforts it would make, is not able to advance one jot; nor can it make any discoveries, when it would prie into the Nature and hidden Causes of those *Ideas*. (II. xxiii. §29.)

That these simple ideas—of which Locke offers several lists, all short—are "the Boundaries of our Thoughts; beyond which the mind. . . is not able to advance one jot" is Locke's central doctrine, stated here in pithy and sinewy prose; but with the qualification: *"it seems probable to me"*! And I say, "Admirable Locke!"

The idea of substance is of course a critical problem in Locke. He himself regards it as a critical problem, and expresses the greatest perplexity about it; and his expressions of perplexity are themselves notoriously perplexed, so that this has become a *crux* for interpreters of the *Essay*. There are a number of issues simultaneously involved in this problem, and as it is out of the question to discuss them here with any claim to adequacy, I shall just summarize those aspects of the theme that are most germane to what I wish to discuss.

One obvious point of difficulty—not for the interpretation of Locke, perhaps not for Locke himself, but for *science*—in Locke's view of science and of substance, is implicit in what has already been said: it is knowledge of the true natures (the "real essences," in Locke's terminology) of substances that he finds barred to us by the limitations of our faculties. (Since Locke says that the "idea of substance in general" is formed *by abstraction* from the ideas of *particular substances* [III. iii. §9], the obscurity—or, rather, "confusion"—that Locke says inescapably attaches to the former idea presumably derives from an imperfection inherent in the latter class; and—although the point seems ticklish—this may somehow concern our ignorance of "real essences.") Very closely related to this—perhaps just an aspect of the very same point—is the fact that, as Locke emphasizes, *"Powers. . . make a great part of our complex Ideas of Substances"*; and, more especially, powers *to affect one another*—that is, Lockian "secondary qualities, mediately perceivable" (II. xxiii. §§9, 10). It is, after all, precisely "qualities" of this type that form the characteristic object of investigation of empirical science: for what else does one study in the laboratory but the "powers of substances to affect one another"?

I have said that perhaps this is not a problem for Locke—having in mind his "official" position that the nature of our faculties makes possible only natural *history*, not natural philosophy (cf. IV. xii. §10). But that depends upon taking a rather narrow view of the "official" position. Let me cite a remark by one of Locke's most penetrating and sympathetic commentators. Michael Ayers writes: "[I]f he let in the possibility that powers or phenomenal properties should belong to things as a matter of brute or miraculous fact not naturally intelligible, Locke's whole carefully constructed philosophy of science and his support for the corpuscularian case against the Aristotelians would collapse."[20] Ayers, therefore, does regard corpuscularianism and the commitment to "mechanical" principles as essential to Locke's doctrine. But Ayers has missed something crucial. The view that interaction consists in the change of motion of bodies in impact makes the laws of impact, as I have said before, the basic laws of physics. Locke points out (in effect) that the process of impact involves a "primary quality" of bodies that Descartes failed to make an explicit part of his conception of body: the property of *resistance to penetration*, which Locke calls "solidity," and which he makes out to be an idea "we receive by touch" (II. iv. §1). But the laws of impact cannot be formulated in terms of the "solidity," or "impenetrability," of bodies alone. They require another concept, first formulated with full clarity by Huygens (in unpublished notes) and by Newton (in the first Definition of his *Principia*): the concept of *mass*. And *this is not "a simple* Idea *we receive from Sensation"* (and of course not one "we receive from. . . Reflection"); it is an attribute we can determine *only* from the study of the actions of bodies upon one another—it is prototypically something to determine in the laboratory—and at the point at which he defines it, Newton tells us that the mass of a body can be determined by weighing it, *as he has found by very accurate experiments* (experiments on pendulums), *which he will detail later in the treatise*. In short, when we juxtapose Locke's claim that our knowledge of the "primary" qualities of bodies is *strictly limited to the simple ideas of sensation* with his claim that the "mechanical" transfer of motion is, in Ayers's phrase, "naturally intelligible," we are confronted with a deep discrepancy.[21] *And it is much to Locke's credit that he recognized this*; although—again his characteristic vacillation—he continued to affirm the principle against his own recognition. For example (II. viii. §11): "The next thing to be consider'd, is how *Bodies* produce *Ideas* in us, and that is manifestly *by impulse*, the only way which we can conceive Bodies operate in"; but (II. xxiii. §28): "Another *idea* we have of Body, is the power of *communication of Motion by impulse*; and of our Souls, the power of *exciting of Motion by Thought*. These *Ideas*, the one of Body, the other of our Minds, every days experience clearly furnishes us with: But if here again we enquire how this is done, *we are equally in the dark*. For in the communication of Motion by impulse, wherein as much Motion is lost to one Body, as is got to the other, . . . we can have no other conception, but of the passing of Motion out of one Body into another; which, I think, is as obscure and unconceivable, as how our Minds move or stop our Bodies by Thought; which we every moment find they do." Clearly, in this latter passage, Locke is recognizing (not, to be sure, the specific role of the concept of mass, but) the general fact that *the "powers" recognized by the mechanical or corpuscularian philosophy do after all belong to bodies* as (in Ayers's words) "a matter of brute or miraculous fact not naturally intelligible."

Now, this is a thing of which Newton was evidently aware, with full clarity, from an early date in his career. Newton, too, in an important investigation, uses the language of "primary qualities"; but in a way that differs very interestingly from that of Locke. In his first published account of an investigation, the "New Theory about Light and Colors," Newton informs us that he has discovered colors to be *"Original* and *connate properties"* of the "Rays of light." He characterizes these properties of the "Rays" more precisely as "their disposition to exhibit this or that particular colour"; and he is led to distinguish, among the colors of light generally, between those he calls "original and simple," or also "[t]he Original or primary colours"—not what *we* call "primary colors," in either of the senses now usual, but the colors of spectrally pure ("homogeneous") light—and those he calls "compounded."[22] The point is that the "original," "simple," "primary" colors are those colors— more exactly, those "dispositions" or (Locke's terminology) *powers*—that are "connate properties" of *individual "rays,"* and so are *causally* "primary"; known to be so *de facto*—"as a matter of brute fact"—through the results of Newton's experiments.

It is important to note, along with the contrast I have made (and those I am about to make), this affinity of Newton with Locke (and thereby also with the "new" natural philosophy of the seventeenth century generally): that the "relational" character of qualities as "powers" was entirely clear to him. This is explicit in the "dispositional" language I have just quoted from his first paper; it is most fully articulated later, in the *Opticks*:

> [T]he Rays to speak properly are not coloured. In them there is nothing else than a certain Power and Disposition to stir up a Sensation of this or that Colour. For as Sound in a Bell or musical String, or other sounding Body, is nothing but a trembling Motion, and in the Air nothing but that Motion propagated from the Object, and in the Sensorium 'tis a Sense of that Motion under the Form of Sound; so Colours in the Object are nothing but a Disposition to reflect this or that sort of Rays more copiously than the rest; in the Rays they are nothing but their Dispositions to propagate this or that Motion into the Sensorium, and in the Sensorium they are Sensations of those Motions under the Forms of Colours.[23]

What distinguishes Newton from Locke—and, ultimately, from the whole orthodox "mechanical philosophy"—is that he does not equate causal primacy *or* "simplicity" with what is in some sense "nonrelational" *in our apprehension*; and, further, that he does not believe that *any* "causally primary" qualities of bodies are manifest to us otherwise than by *inference* from experience.

Such views, I have said, date from early in Newton's career. If the consensus of expert opinion is right in dating the fragmentary treatise known (from its opening phrase) as *De gravitatione et æquipondio fluidorum*) to the 1660s, when Newton was in his twenties, one can (in my opinion) go further and say that the views in question were connected by Newton, very early in his career, with most remarkable metaphysical conceptions. There is not space here to offer a really adequate discussion of the metaphysics expounded in that piece; but let me try to sketch its chief points.[24]

The whole metaphysical discussion is motivated by Newton's remark that his conceptions of place, body, rest, and motion differ fundamentally from those of Descartes. Its first part is devoted to the discussion of motion and space ("extension"), and has a certain affinity with the scholium to the Definitions in the *Principia*; but it is more far-reaching than the latter—or than any other passage in Newton—in its account of his conception of the metaphysical status of space. Newton's doctrine here is that, in the first place, extension has a mode of existence unique to itself. It is not a substance: first, because it does not subsist "absolutely of itself" (*absolute per se*), but "as it were as an emanative effect of God, and a certain affection of every being." These words are rather obscure—what exactly is the meaning of "emanative effect"; and why, if extension is an "affection" of every being (or thing), is it not an accident? As to the latter question, Newton goes on to say that extension is more like a substance than an accident, because it can be conceived as existing without any subject—i.e., we can conceive of void space (or spaces). But both points are clarified in a later passage; and this later passage—something that has not been generally seen, and that I cannot emphasize too strongly—makes it absolutely clear and explicit that, despite Newton's statement that extension is *tanquam Dei effectus emanativus*, he does *not* derive space from his theology. What he says is this:

> Space is an affection of a being just as a being (*Spatium est entis quatenus ens affectio*). No being exists or can exist that does not have relation in some way to space. God is everywhere, created minds are somewhere, and bodies in the spaces that they fill, and whatever is neither everywhere nor anywhere is not. And hence it follows that space is an emanative effect of the first-existing being; for if I posit any being whatever I posit space.

My gloss: space is an "emanative effect" (that is, not something *caused*, but simply "a consequence") of whatever first exists—and this is God; but if God were not the "first-existent being," and something did exist, space would be an "emanative effect" of whatever thing that was: "for if I posit any being whatever, I posit space." For his doctrine that space is not a substance Newton offers a second reason: although it is not part of the received definition of substance that *it can act upon something*, philosophers all tacitly understand this as a distinguishing mark of substances; but space does not *act upon* (interact with) things, it is simply "an affection" of them ("No being exists or can exist that does not have relation in some way to space.")

Newton does not tell us what he conceives to be the grounds of this doctrine—to my own great regret. But he makes a sharp contrast between its epistemic status and that of the next part of the metaphysical passage, which is concerned with the nature of body. He says the following:

> Extension having been described, for the other part corporeal nature remains to be described. Of this, however, as it exists not necessarily

but by the divine will, the explanation will be more uncertain; because it is not at all given us to know the limits of divine power, namely whether matter could be created in one way only, or whether there are several ways by which other beings similar to bodies could be produced.

Thus, by implication, according to Newton, it *is* given to us to know the necessary connection between the existence of anything whatever and that of space, and to know the intrinsic properties of the latter—of which, he emphasizes, we possess "an Idea the clearest of all." Of bodies, however, we have knowledge only from our interactions with them and the sensations that result. This gives us information about properties—and *precisely* about "interactive" properties, i.e., Lockian "powers"—of bodies; but what underlying, metaphysical constitution—what "real essence"—is responsible for these powers, or (as Newton in effect puts it) just "what God would have to do to *make* a body," is something about which we can only conjecture. Nevertheless, in describing at least one possible way, intelligible to us, in which God *could* make beings having all the properties of bodies as we know them, we do as it were clarify for ourselves our own fundamental conception of corporeal nature.

Accordingly, Newton's exposition of a *possible* metaphysical view of body takes the form of a "fable"—undoubtedly modeled upon the creation fable of Descartes—of how God might create a new body in the world. What we may conceive him to do is, first, to choose some definitely delimited region of space, and to prevent any body from entering this region. (If it is asked *how he can do this*, Newton says, we are no more able to answer than we are able to say how we ourselves, by an act of thought and will, can move our own limbs. But we do so; and conceive ourselves able to do so. In this sense [what follows is my gloss], we are able to conceive with full clarity of the *possession and exercise* of a certain power, even though we lack a clear conception of the *intrinsic constitution* of that power—that is, of the *conditions* of its possession.) Next, having willed (as I shall put it) this *distribution of impenetrability*, God may *conserve* this distribution, not in a fixed part of space, but so as (in Newton's words) "to be transferred hither and thither according to certain laws, yet so that the quantity and shape of that impenetrable space are not changed." If these laws are such that, in any *encounter* between the region of impenetrability and an ordinary body—or, supposing God to create more than one such (mobile) region, in any encounter between parts of the distributed impenetrability themselves—the motions of the ordinary bodies *and* the migrations of the regions of impenetrability satisfy the conditions that have been determined to hold between actual bodies in impact, then, Newton says, we shall not only be able to *detect* these regions (by their effects upon ordinary bodies), but we shall find no reason to consider them as anything but ordinary bodies themselves.

It may be objected that there is a gap in this account—that Newton has had to assume the existence of "ordinary" bodies, "ordinarily" detectable, to get his story started, and that without these to use as probes we should remain completely uninformed as to the alleged new bodies. Newton was far too acute a philosopher to be guilty of such a lapse; on the contrary, he not only fills the gap, but does so in a way that simultaneously offers a sharp new criticism of the received view of what is "essential" to matter. Ordinary bodies, too, are—ordinarily—detected by their interactions with other bodies; for instance, we see bodies on account of their interaction with light (and Newton agrees with Huygens that optical phenomena ought to be conceived of as some sort of corporeal motions and interactions). But, of course, the light must affect the eye; and then the optic nerve; and then the brain; and finally—somewhere—in a sensitive region, presumably of the brain, the motions of parts of matter must affect the mind for perception to occur. And contrariwise, in some way mental processes must affect the motions of bodies, if we are able to move our limbs by volition. Therefore, Newton concludes, this property, this power of *interacting with minds*, must be reckoned as no less an essential attribute of matter than its extension, solidity, and mobility. If we suppose God to endow the spatial regions of distributed impenetrability with this power as well (acting according to the laws—unknown to us—by which bodies and minds do interact), our account will be complete.

This fragmentary treatise of Newton's is certainly prior—probably much prior—to Locke's *Essay*. Locke came to know something of the theory it contains, apparently in conversation with Newton. As a result, he made a small revision in the second (1694) edition, changing the statement that the creation out of nothing of a material being and that of a thinking thing are equally beyond our comprehension, to say instead that "we might be able to aim at some dim and seeming conception" of the first, whereas the second is "more inconceivable" (IV. x. §18).[25] But quite apart from the question of the comprehensibility of creation, Newton offers here a solution to—or, at least, a different and interesting way of thinking about—the problem of "the confused [*Idea*] of Substance, or of an unknown Support and Cause of [the] Union" of the "several distinct simple *Ideas*—more properly, the corresponding *qualities*—that make up a particular substance.[26]

If we distinguish between "Support" and "Cause" in Locke's formula and take him (as Berkeley does—see *Principles of Human Knowledge*, I, §11; Ayers, on the other hand, has strongly opposed this reading) to mean by "Support" something that conflates the Aristotelian notion of an "ultimate subject of properties" with the scholastic notion of "prime matter, denuded of all properties," then Newton's analysis may be put thus: The notion of substance as support may be dispensed with entirely. For bodies, in particular, we may employ an ultimate grammar—a set of categories—in which space, or regions of space, are the subjects to which *corporeal substantial attributes* are ascribed. He says this explicitly: "Between extension and the form imposed upon it [impenetrability, laws of motion of the distributed regions, etc.] there is almost the same relation that the Aristotelians posit between the *materia prima* and substantial forms. . . .They differ, however, in that extension has more reality than *materia prima*, and also in that it is intelligible, as likewise is the form that I have assigned to bodies"; and "[S]ubstantial reality [is] rather [to be] ascribed to these kinds of Attributes which are real and intelligible in themselves." What makes the attributes "substantial" is precisely that they are what determine *interactions*: they are "*powers*." (I remind you that Newton has identified the power to act as the mark of substantiality; the metaphysics of this work is one of "*substantiality* without *substance*.") As to the *cause* of the coexistence of attributes, that coexistence is in effect *simply an ultimate fact*—thus, for Newton's way of thinking, a fact of the arbitrary exercise of God's will—or in Ayers's phrase, "a matter of brute or miraculous fact not naturally intelligible": like, I should add, *any* postulate regarded as ultimate by the science of one's day.[27]

The metaphysics of *De gravitatione* is discernible beneath the surface both of Newton's *Principia* and his *Opticks*—as fundamentally compatible with, and (in my opinion) illuminating, the teachings of those works; although not fully *implied* by them—and although those works entail an important amendment of that metaphysics. In the terminology of these later treatises, the "primary qualities" (as *I* have here used that term, in explicating both Locke and Newton) of corporeal nature are called the "natural powers" (a Cartesian term, it should be recalled) or *forces of nature*. In the *Opticks* (p. 401) Newton tells us that these are to be conceived of as—or characterized through—"general Laws of Nature, by which the Things themselves are form'd" (cf. the laws of the migration of regions of impenetrability, which constitute *a part of the nature itself of the bodies created* in Newton's "fable"). One such force of nature is the *Vis inertiæ,* a passive Principle by which Bodies persist in their Motion or Rest, receive Motion in proportion to the Force impressing it, and resist as much as they are resisted" (p. 397): the law characterizing this principle is, thus, the three Laws of Motion of the *Principia* taken together. But there are in addition, as Newton says "it seems to him," "active Principles," by which bodies are moved (p. 401); the *action upon a body* of a *vis naturæ* of this latter kind Newton in his *Principia* calls *vis impressa*: "impressed force."

The *Principia* of Newton, taken as a whole, is described by him in the Author's Preface—accurately, as is his wont—as presenting a particular (in fact, although he does not explicitly say this, a *new*) "method of Philosophy." It is a method based upon the concept of causes or principles of motion to be expressed in the form of laws of interaction, themselves subject to the three laws of motion that constitute the Axioms of the *Principia*. This departs from the Cartesian, and more generally from the orthodox "mechanical," philosophy, by substituting for *pressure or impact* as the basic causal mode the far more general notion of a Newtonian "force of nature." Of this method, he says, he offers an example in the third book of the *Principia*, in which he derives, by a very beautiful, subtle, and—as I have elsewhere argued (in disagreement, on this point, with Newton's own view)—*bold and risky* (and, thus, "hypothetical") argument, the first case ever known of what we still regard as a "fundamental force": the force (and law) of universal gravitation.

It is in response to that achievement of Newton that Locke made what I have called his revolutionary change of mind—not reflected in the *Essay*.

Again, I have discussed this elsewhere; but let me cite two striking passages to illustrate the point. In his *Thoughts concerning Education*,[28] published in the same year (1690) as the first edition of the *Essay*, Locke first repeats his decisively skeptical view about physics as a science: "Natural philosophy, as a speculative [that is, of course, *theoretical*] science, I imagine, we have none; and perhaps I may think I have reason to say, we never shall be able to make science of it" (§190). But a few pages later (§194) he qualifies this significantly. Having referred first to what has been achieved, "in the knowledge of some, as I may so call them, particular provinces of the incomprehensible universe" by "the incomparable Mr. Newton," he adds:

> And if others could give so good and clear an account of other parts of nature, as he has of this our planetary world, in his admirable book "Philosophiæ naturalis Principia Mathematica," we might in time hope to be furnished with more *true and certain knowledge* in several parts of this stupendous machine, than hitherto we could have expected. [Emphasis added.]

The second passage I wish to call attention to occurs in Locke's posthumously published treatise *The Conduct of the Understanding*, which he originally intended to form a new chapter of the *Essay* but decided to omit because of its length.[29] It is in a section (§43) titled: "Fundamental Verities." These fundamental verities, Locke says, are "teeming truths, . . . and, like the lights of heaven, are not only beautiful and entertaining in themselves, but give light and evidence to other things that without them could not be seen or known." Of such pregnant principles he gives two examples. The first of these is "the discovery of Mr. Newton, that all bodies gravitate to one another, which may be counted as the basis of natural philosophy"; the second is "our Saviours great rule, that *we should love our neighbor as ourselves*": by this alone, Locke says, he thinks "one might without difficulty determine all the cases and doubts in social morality." Setting aside any questions raised by the audacity of the personal parallel, we ought to remember that according to the "official" doctrine of the *Essay*, morals, unlike natural philosophy, *is* capable of being made a science. The passage I have just cited drastically narrows the difference between the two.

Of course it might be argued that Locke in this is simply untrue to his own philosophical principles and arguments—that he has succumbed to *enthusiasm*, that disease of the mind which in the *Essay* he so earnestly deprecates. For there is no sense in which Newton's achievement can be seen as establishing natural philosophy as a *demonstrative science grounded in intuitive principles*. I do not wish to dismiss such an argument as simply wrong but only to suggest the interest and importance of another point of view. What I emphasized at the outset was that the concern of seventeenth-century philosophers with the *foundations* of knowledge was most intimately connected with a concern for the *advancement* of knowledge, and in particular, with finding a new and more fruitful way for that advancement. If one views the *Essay*, and the change of mind I have described, in this light, one can describe this change as follows: In the *Essay*, Locke adopts a conception of knowledge that in point of its *grounds* is a kind of modified Cartesianism (knowledge must be based upon intuition and deduction), and that makes the objective ideal of knowledge of bodies also a modified Cartesianism (only mechanical principles are intelligible, and therefore only such are admissible). Unlike Descartes, Locke came to a pessimistic conclusion: no such knowledge of corporeal nature is attainable. But Newton convinced him that something else *was* attainable; and that this was something whose objective content was so surprising and so important, whose evidence was after all so convincing, and whose systematic structure was so deserving of appellations like "speculative, demonstrative science," that a sterile consideration of the old, unattainable ideal ought to give way to the adoption of the new method and new ideal whose fruitfulness had been proved.

To be sure, this overstates the matter so far as concerns Locke: he did *not* revise or renounce the *Essay*, and he expressed himself (as he knew so well how to do) at once eloquently and emphatically and yet guardedly. So we may conclude that he was tempted by conversion to the new faith, but remained in a state of interesting hesitation.

In speaking of Huygens's commitment to the mechanical philosophy, I referred to the likelihood that he was motivated by what I called a "modest transcendental argument." It has been the usual fate of transcendental arguments to be defeated by *modus tollens*: one defines *science* in a certain way; demonstrates that things must be such-and-such, or science will be impossible; and it is then found that things are not such-and-such, so that *in this sense of the word* there is no science. It is not the *obvious* duty of a philosopher then to continue to insist on the definition.

I should like to conclude with two points that have a bearing upon the philosophy of Hume—who, as we all know, took Newton's work as a paradigm of science.

The first of these concerns the notion of *cause*. There has been a tendency to read Newton as if "force" equals "cause," more particularly "cause of change of motion"—that is, Newton's "impressed force" (and when the concept is made quantitative, what he calls "motive force"). On this reading, basic Newtonian causes are the analogue of traditional "efficient" causes.

That is a misattribution to Newton of the point of view of the orthodox mechanical philosophy—where, however, pressure and percussive forces rather than the more general Newtonian forces are the only ones allowed. (In his very early days, when he shared the "mechanical" viewpoint, Newton himself wrote: "force is ye pressure or crouding of one body upon another."[30] When the *Principia* appeared, Newton was tasked with having reintroduced Aristotelian "occult qualities" into physics. There is something quite correct about this: the fundamental *causes* in Newton's developed philosophy of science are not *vires impressæ* or *vires motrices* but *vires naturæ*—"forces of nature"; and these Newton himself, in the *De gravitatione*, likens to Aristotelian "substantial forms" (or, in the *Opticks*: "general Laws of Nature, *by which the Things themselves are form'd*)—but with this crucial difference, that the substantial forms of the scholastics are obscure and useless, whereas the forms invoked by Newton are clear, his principles not occult but made manifest by phenomena.

But Hume has certainly not understood this. Claiming to follow Newton, he tells us that three relations together make up the essential conditions of the relation of cause to effect: contiguity, temporal priority of the cause, and constant conjunction. A crucial proposition of Newton's *Principia* states that the behavior of the moon in its orbit and that of falling bodies on the earth have *the same cause*: namely, *gravity*, or *weight, towards the earth*. It is very hard to see how Newton's proposition makes sense by Hume's analysis of the idea of cause.

The second point I shall cite is not unrelated to the first, and again suggests that Hume has not understood the nature of Newton's real achievement. He offers his own theory of human nature as the analogue of Newton's theory of corporeal nature, and his own principle of the association of ideas as "a kind of *Attraction*, which in the mental world will be found to have as extraordinary effects as in the natural" (*A Treatise on Human Nature*, I. i. §iv). Now, Hume's theory of *belief*, most intimately connected with his theory of the causal relation, is that belief is a kind of "feeling" or "sentiment," which he describes in terms of the "force" or "vivacity"—one might say, the compellingness—that an "idea" has for us. He employs this analysis in a most striking way in the *Enquiry concerning Human Understanding* to dispose of a far-fetched metaphysical theory (that of the universal and exclusive efficacy of the Supreme Being). Here is what he says (§vii, part 1):

> It seems to me, that this theory . . . is too bold ever to carry conviction with it to a man, sufficiently apprized of the weakness of human reason, and the narrow limits, to which it is confined in all its operations. Though the chain of arguments, which conduct to it, were ever so logical, there must arise a strong suspicion, if not an absolute assurance, that it has carried us quite beyond the reach of our faculties, when it leads to conclusions so extraordinary, and so remote from common life and experience. We are got into fairy land, long ere we have reached the last steps of our theory; and *there* we have no reason to trust our common methods of argument, or to think that our usual analogies and probabilities have any authority. Our line is too short to fathom such immense abysses. And however we may flatter ourselves, that we are guided, in every step which we take, by a kind of verisimilitude and experience; we may be assured, that this fancied experience has no authority, when we thus apply it to subjects, that lie entirely out of the sphere of experience.[31]

How then, I ask, can Hume himself believe Newton's theory of gravitation— the theory, that is, that *every particle of matter in the universe attracts every other particle*, and does so according to a precisely stated quantitative law? By what line does Hume suppose Newton was able to fathom *that* abyss? If one just reflects upon the enormous scope of the law Newton propounds, how far it extends, not only beyond "common life and experience," but beyond *any* observation or experiment in any way accessible to Newton as evidence, one may in fact come to sympathize with such scientifically well-informed judges as Huygens and Leibniz, who rejected the law as unproved and incredible. Hume has failed to appreciate the depth and subtlety both of the issue and of Newton's actual reasoning; and this has had a damaging effect both on his epistemology generally, and on his conception of what is *required* to establish a science of human nature.

To return in conclusion to the two theses I said I wished to present, they may be rephrased as follows: If epistemology is to be concerned with knowledge in the sense—or *a* sense—of that term that characterizes what history has shown us capable of *actually acquiring* about the world we inhabit; and if metaphysics is to be concerned with fundamental aspects *of that world, as we have come to know it*; then both of these fields of inquiry must have some serious relation to the disciplines we call "scientific," which have produced that "knowledge." Philosophers of the seventeenth and eighteenth centuries took it for granted that there was such a serious relationship (and employed the word "philosophy" itself accordingly). To pay attention to this fact is therefore deeply pertinent to our historical understanding of what those philosophers were about; and deeply pertinent, also, to the possibility that we ourselves may learn cogent philosophical lessons from their work and their experience.

NOTES

The material here presented is based in part on work supported by the National Science Foundation under Grant No. DIR-8808575. A version of this paper was delivered as part of a symposium on science and philosophy in the early modern period at the annual meeting of the Central Division of the American Philosophical Association, April, 1991.

1. E. A. Burtt, *The Metaphysical Foundations of Modern Physical Science* (reprinted Garden City, N.Y., 1955), 244. (The work dates from 1924; rev. ed., 1932.)

2. Isaac Newton, *Philosophiæ Naturalis Principia Mathematica*, edited by Alexandre Koyré and I. Bernard Cohen, the third edition (1726), with variant readings from the first two (1686 and 1713), 2 vols. (Cambridge, Mass., 1972) 1:48. (The translation in the text is my own; it corrects some inaccuracies in the published translations.)

3. Norman Kemp Smith, *New Studies in the Philosophy of Descartes* (London, Macmillan, 1952), 120, n.2.

4. Richard Rorty, *Philosophy and the Mirror of Nature* (Princeton, N.J., 1980), 131.

5. *Œuvres complètes de Christiaan Huygens*, published by the Société hollandaise des sciences, vol. 10 (La Haye, The Hague, 1905), 404.

6. *Œuvres de Descartes*, edited by Charles Adam and Paul Tannery, vol. 1, rev. ed. copublished by the Centre National de la Recherche Scientifique (Paris, 1969), 1056. This edition will henceforth be referred to as "AT *Œuvres.*"

7. Ibid., 137.

8. AT *Œuvres*, vol. 4, rev. ed. (Paris, 1972), 329.

9. AT *Œuvres*, 1:649.

10. The story is quoted in AT *Œuvres*, vol. II, rev. ed. (Paris, 1967), 670–71, from *La Vie de Saint-Evremond* by Des Maizeaux (that is, Pierre Desmaiseaux). Des Maizeaux reports it as having been told him by the subject of his biography, Charles de Marguetel de Saint-Denis, Seigneur de Saint-Évremond, who heard it from Digby.

11. See the article "Digby, Sir Kenelm," *Encyclopædia Britannica*; 11th ed., 8:262.

12. Des Maizeaux, who evidently had some reservations about the reliability of Digby, continues at the place cited in n. 11 above by indicating the existence of further evidence of Descartes's view; and cites, among other things, this passage from Baillet, *La Vie de Monsieur Descartes* (2 vols.) 2: 452–53. See also Charles Adam, *Vie et Œuvres de Descartes*, supplement to the Adam and Tannery edition, republished with the assistance of the Centre National de la Recherche Scientifique (Paris, 1957), 552, note a.

13. Cited by Adam, *Vie et Œuvres*, in the same footnote to which reference has just been made (p. 552, note a.): "This [namely, the indefinite prolongation of human life] was still the subject of his conversations during the last months of his life; or at least it was this that most struck Christina: but then, '*ses oracles l'ont bien trompé*,'she says of Descartes, in a letter to Saumaise."

14. AT *Œuvres*, 4:119 (letter to Mesland, May 2 1644).

15. Huygens, *Œuvres*, 10:403, 405–6.

16. Huygens, *Treatise on Light*, translated by Silvanus P. Thompson (Chicago, 1945), p. vi (Note: this translation was originally published in 1912.)

17. Descartes, *Rules for the Direction of the Mind*, under Rule VIII; e.g., in *The Philosophical Writings of Descartes*, translated and edited by John Cottingham, Robert Stoothoff, and Dugald Murdoch, vol. 1 (London, 1985), 29.

18. Letter to Chanut, Feb. 26 1649; AT *Œuvres*, vol. 5 (Paris, 1974), 291.

19. I shall cite the *Essay* by book, chapter, and section, for the sake of invariance over editions. My own quotations are taken from John Locke, *An Essay concerning Human Understanding*, edited by Peter H. Nidditch (Oxford, 1975; reprinted with corrections 1979). (It should be noted that the chapter referred to below as 'I,i' is the Introduction, placed by some editors before Book I—so the chapters that Locke numbered ii–iv of Book I are, by those editors, called i–iii of that Book.)

20. M. R. Ayers, "The Ideas of Power and Substance in Locke's Philosophy," *Philosophical Quarterly* 25 (1975):22.

21. R. S. Woolhouse, *Locke's Philosophy of Science and Knowledge* (Oxford, 1971), 111, without any comment, names mass first in a list of Lockian primary qualities. This has no justification whatever.

22. See *Isaac Newton's Papers and Letters on Natural Philosophy*, edited by I. B. Cohen (Cambridge, Mass., 1958), p. 53 and p. 54 ¶5.

23. Isaac Newton, *Opticks* (reprint, following the 4th ed. [1730]; New York, 1952), Book I, Part II, pp. 124-5 ("Definition," preceding Proposition III of that Part).

24. The fragment has been published in *Unpublished Scientific Papers of Isaac Newton*, edited by A. Rupert Hall and Marie Boas Hall (Cambridge, 1962), 90–121, with an English version following; pp. 121–156. That version has serious defects; the translations used here are my own.

25. For an account of some of the circumstances of this revision, and of the source of our knowledge of what lay behind it, see the edition of A. C. Fraser, *An Essay Concerning Human Understanding* by John Locke, collated and annotated [etc.] by Alexander Campbell Fraser, 2 vols. (Oxford, 1894; reprinted New York, 1959) 2:321–22, n.2.

26. Cf. Locke, *Essay*, III, vi, §21: "[I]n Substances, besides the several distinct simple *Ideas* that make them up, the confused one of Substance, or of an unknown Support and Cause of their Union, is always a part."

27. Cf. the comments on this point in my article "Yes, but. . . Some Skeptical Remarks on Realism and Anti-Realism," *Dialectica*, 43:64.

28. Locke, *Some Thoughts concerning Education*, in *The Works of John Locke: a New Edition, Corrected*, 10 vols. (London: Thomas Tegg *et al.*, 1823), 9: iii–v, 6–205.

29. Locke, *Of the Conduct of the Understanding*, in *The Works of John Locke*, 9 vols., 12th ed. (London, 1824) 2: 323–401.

30. In the extracts from Newton's so-called *"Waste Book"* published by John Herivel,

The Background to Newton's Principia: *A Study of Newton's Dynamical Researches in the Years 1664–84* (Oxford, 1965), 138.

31. Here quoted, for the sake of Hume's original orthography, from the edition of T. H. Green and T. H. Grose, David Hume, *The Philosophical Works*, 4 vols., "reprint of the new edition London 1882" (Darmstadt, 1964) 4:59–60. (To avoid a possible confusion, it should be noted that this is vol. 4 of the entire edition, but at the same time vol. 2 of Hume's *Essays: Moral, Political, and Literary*, within that edition.) The passage is to be found in the more easily accessible edition of Selby-Bigge/Nidditch—David Hume, *Enquiries Concerning Human Understanding and Concerning the Principles of Morals*, edited by L. A. Selby-Bigge; 3rd ed. revised by P. H. Nidditch (Oxford, 1975)—on p. 72.

There's a Hole and a Bucket, Dear Leibniz

MARK WILSON

The long duration of a belief. . . is at least proof of an adaptation in it to some portion or other of the human mind; and if, on digging down to the root, we do not find, as is generally the case, some truth, we shall find some natural want or requirement of human nature which the doctrine in question is intended to satisfy.

S. T. Coleridge

I

Recastings, in modern mathematical idiom, of the dispute between Newton and Leibniz over the nature of space have become so commonplace in philosophy that the art of anachronistic evaluation can claim its own traditions of scholarship. Before I prolong this odd lineage further, I should offer some rationale for our endeavors. Obviously, we should not expect to reach an *interpretation*, in any straightforward sense, of the often mysterious words of the historical Leibniz and Newton. What kind of "interpretation" could wisely appeal to subtle varieties of differential manifolds of which our seventeenth-century savants never dreamed? Alternatively, our mission might be conceived as one of illustrating methodological morals, e.g., "The modern settings articulate crisply the scientific choices that Leibniz and Newton faced and allow us to adjudicate their dispute properly." There is much "hurray for X; phooey on Y" cheerleading in the anachronistic literature that partakes of this dubious orientation.

In truth, the historical dispute over space has neither winners nor losers. To suppose otherwise is to overlook the deep conceptual tensions native to so-called "classical mechanics." But such overlooking is a frequent occurrence, because virtually everything in our philosophical and physical training conspires to assure us that the grounds of "classical mechanics" are conceptually transparent. The scoutmasters of my youth felt impelled to treat every gnat-ridden slog as if it were a pleasant stroll unter-den-linden; clinical optimism of an allied kind infects most instruction in "classical mechanics." The boggy parts of the subject are breezed over as if they constitute firm pavement.

To be sure, we know what kinds of objects belong in the world of the classical physicist: extended bodies capable of distortion, fracture, and recombination. They can terminate in sharp boundaries and interact through impact, steady pressure, sliding and, arguably, from a distance. They can mix or alter chemical nature. How they respond to, e.g., a fixed pressure will depend both upon their composition and their history—if a steel is tempered, it becomes harder, but if repeatedly flexed, it softens. The most doggedly unscientific observer is familiar with all of these behaviors; they are the primal stuff of which the classical world is composed.

But knowing which horses belong to a corral does not mean that they can be easily persuaded to assemble there. Finding a mathematical framework to accommodate even a small portion of the expected range of classical behaviors can be very difficult. In our century, prodded by industrial needs to deal with paints, rubbers, plastics, and steels, applied mathematicians have formulated theories that capture wide swatches of the classical picture. Insofar as these recent theories approach the classical ideal, they tend to be subtle and rather intricate mathematically. Relative to such complexities, one expects that Leibniz and Newton, armed with more primitive mathematical tools, will have focused primarily upon smallish regions of classical behavior, with the hope that the remainder can be made to conform to the selected paradigms. In this essay, I will try to show how Newton's and Leibniz's differing philosophies of space and time correlate, to a reasonable extent, with distinct areas of classical behavior. In particular, there is an important notion (which I will dub "Leibniz structure") whose classical salience seems to have been missed in most of the philosophical and historical literature.

The world is much with us, early and late, but, during working hours when we sit in our armchairs or next to our accelerators, it is easy to forget how beams bend or water flows. The sophisticated "classical mechanics" of modern times, oddly enough, can remind us of the ordinary yet demanding realities that confronted Leibniz and Newton just as they confront us (after hours). Did Leibniz *understand* the concept of "Leibniz structure" that I will extract from modern formulations of mechanics? No. But he might have *felt* the distinction, in the sense that his work stretches towards the notion. One has the frequent impression that "there's something right in what Leibniz says"; the reader may judge how well "Leibniz structure" captures that notion.

As I stated, standard educations in philosophy and physics rarely indicate a need for sophistication within classical mechanics (except insofar as irrelevant trifles as the alleged "nonobservationality" of force are mentioned). Most physics texts encourage the pretense that "classical mechanics" ultimately boils down to Newton's three laws of motion, supplemented by specific force laws such as the law of gravitation. Certainly, a sharp theory[1] can be associated with these ideas, based on ordinary differential equations, whose postulates are easy to state. But this theory, whose proper name ought to be "point particle mechanics," treats classical objects as unextended centers of force in the manner of Boscovitch. What, then, of the continuously distributed bodies found in the intuitive classical picture? Beware of assuming automatically that continuum mechanics "reduces" to point particle mechanics. At best, one's old physics primer showed that a few special varieties of continuum behavior can be approximated within point particle models. Any more systematic survey of expected continuum behaviors reveals many lacuna; how can the Boscovitchian "central force" model supply a force parallel to stream flow in a fluid under pure shear? (This can be answered, but at a cost I will discuss later.) Nineteenth-century physicists articulated *molecular* (not point particle) theories that account for the gross behavior of a variety of macroscopically continuous substances, but these models usually rely upon "molecules" that are continuously distributed themselves. In truth, few physicists saw the Boscovitchian model as a serious foundation for classical physics until the twentieth century (although its associated framework of ordinary differential equations has always been extensively studied, especially within celestial mechanics). The twentieth-century Boscovitchian revival was sparked largely by a desire to erect an easy bridge between "classical" behavior and quantum mechanics.

A nineteenth-century classic such as Thompson and Tait's *Principles of Natural Philosophy* also claims to be based on "Newton's three laws," but their "particle" represents a complicated kind of infinitesimal designed to support the *partial* differential equations needed in a continuum. Concern over the shaky mathematics involved in such books prompted Hilbert to place the project of rendering mechanics mathematically rational upon his famous 1900 list of unsolved problems. Workers in Hilbert's school subsequently developed many of the tools—integro-differential equations, measure theory—utilized in the modern treatments of classical mechanics, but, by the time these tools were available, Hilbert's worries about classical mechanics had drifted from the attention of professional physics (and, with it, philosophy) to less publicized academic arenas.

Philosophers in their own right have encouraged classical simplemindedness through continual appeal to loose contrasts between "the mechanistic worldview" and "the field conception of electromagnetism," etc. But what, exactly, *is* the "mechanistic worldview"? A plausible contender can be formulated (see "kinematics of mechanism" below), but it falls embarrassingly short of accommodating the gamut of classically expected "mechanical" behaviors.

One of the major problems in framing a unified view of classical mechanics stems from the fact that our proposed list of "classical" behaviors combines, in a disconcerting way, *continuous* actions (material flow, states of stress and strain) with *discontinuous* actions (fracture, impact). It is hard to get such activities to work in harness; the transition from "smooth" to "broken" continua usually requires fairly sophisticated mathematical tools. Historically, this tension leads to such apparent paradoxes as the fact that Leibniz, who hoped to clear a ground for differential equations by postulating that "nature does not make jumps," also tried to analyze differential interactions using the model of impulsive impact! In this paper, we shall deal only with the distinctions that arise in a treatment of smooth continua; our framework is not adequate to fracture or impact.

As I stated, the bulk of this essay will be concerned with isolating a special variety of spatial structure to be called "Leibniz structure." It is a notion that has passed unnoticed in the standard philosophical literature on space-time, presumably because it naturally arises only in a continuum physics context. Without it, however, I do not believe our understanding of the roots of space-time "relationalism" can be complete. In saying this, do not expect "relationalism" to be ratified here. In fact, "relationalism" seems a rather vague doctrine with two distinguishable aspects: (i) the sense that all physically important spatial relationships "travel with" a material as it flows; (ii) the idea that spatial structure independent of material flow is intrinsically incoherent or incapable of performing the jobs for which it is intended. The first is an ill-formulated physical claim; the second is a philosophical argument, of a verificationist stripe, intended to shore up claim (i). "Leibniz structure" helps codify the possible content of (i) directly, without need of operationalist underpinnings. In my opinion, "verificationism's" primary historical virtue is that it has often served as a stop-gap defense for physical distinctions that are otherwise hard to articulate, as "Leibniz structure" indubitably is. Verificationism seems a credible doctrine only because it has fronted for so many worthy notions over the years.

And this brings me to the background motivations for my project. Contemporary metaphysics and philosophy of science are riddled with all sorts of unfortunate impulses to instrumentalism, anti-realism, and worse. Historically, the headwaters of these philosophical streams originate in nineteenth-century struggles with classical mechanics, rather than the later confrontations with quantum mechanics and relativity. (Leibniz, of course, foreshadowed many of the nineteenth-century concerns.) In the mouth of a Leibniz or a Heinrich Hertz, a bit of philosophizing may seem especially weighty because of its coupling to physical circumstance. But once "classical mechanics" is read as "point particle physics," all the complications inherent in continuum mechanics are forgotten and one tends to credit the weightiness of the philosophizing to the philosophy itself. Let's face it; much of the impetus behind philosophical debate is driven by tradition, even among those philosophers who most disdain historical study (indeed, these folks are most apt to be suckers for a silly problem that has broken loose of its substantive moorings). I do not pretend to genuine historical study here, but I hope to tug "relationalism" back to physical shore.

It happens that a very clever argument, also of anachronistic demeanor, has recently emerged that seems to bolster exactly the "verificationist" aspects of Leibniz's thought that I deplore. I have in mind the so-called "hole argument" proposed by John Earman and John Norton.[2] Since the road to "Leibniz structure" passes by so many of the concerns that spark Earman and Norton's argument, I will address these matters as well.

Everything I say about continuum mechanics I have learned from the writings of Clifford Truesdell[3] and his school (especially Walter Noll). Truesdell, who is not only a key contributor to the modern articulations of mechanics but, arguably, our greatest historian of the subject, has long complained of the misunderstandings that arise when "classical mechanics" is read as "point particle physics." It has seemed to me that the philosophers and historians who could most benefit from his advice have shunned it. Perhaps the topics raised here, which constitute but a small piece of the full story, might help reverse this trend.

II

Let us review some standard considerations in respect to the notion of space-time structure. In general, we take space-time to be the locus of physical events, presumed to be topologically four-dimensional. In "De Gravitatione,"[4] Newton presented an argument, directed against Descartes, that attempted to show that the space-time for any reasonable physics ought to embody a structure we will call "Newtonian." Specifically, "Newtonian structure" requires that the space-time $\mathcal{N}$ break into a "space" consisting of orthodox Euclidean three-space and an one-dimensional "time" carrying a Euclidean metric (i.e., $\mathcal{N} = E^3 \times E^1$). Under this decomposition, $\mathcal{N}$ also splits into (a) slices of events that are simultaneous according to the temporal metric and (b) filaments of events that occupy the same "spatial place." $\mathcal{N}$ shall be called "Newtonian space-time."[5]

Newton's argument in favor of this structure, reconstructed anachronistically, runs as follows. Any reasonable physics will rely upon an appeal to **velocity**. By definition, **velocity** represents the time derivative of **displacement**. Accordingly, we must rely upon the "same place as" filaments in a space-time so that (i) **displacements** can be unambiguously defined upon each temporal slice and (ii) a path exists whereby a **displacement** within the time t_1 slice can be pulled back into the t_0 slice in such a way that a *change* of **displacement** can be calculated below. Without Newtonian structure, **velocity** will not be meaningful in the space-time. See figure 1.

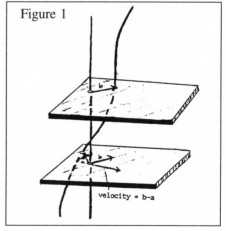

Figure 1

velocity = b-a

As Newton recognized, such considerations leave some wiggle room whereby a physical theory might utilize a weaker space-time structure if a *relativized* notion of **displacement** is consistently used. I will return to such considerations in section VIII. What Newton did not anticipate, of course, is that a physical theory might utilize a space-time that does not split into unique slices, that it might work with intrinsically 4-vectors instead of the 3-vectors within slices presumed in Newton's picture of theory (indeed, Newton's point-of-view was, understandably, so 3-vector oriented that the notion that space-time might be usefully conceived as a unified object never occurred to him). Special and general relativity evade Newtonian structure through this four-dimensional ploy, yet otherwise conform to Newton's observation that geometrical structure is needed to define the derivative operations that carry a quantity like **displacement** into **velocity**.

The well-known problem with Newton's conclusion is that the laws Newton himself placed in $\mathcal{N}$ do not seem to utilize all of the scaffolding placed there by Newtonian structure (by "Newton's laws," I intend the standard "$\mathbf{f} = m\ \mathbf{a}$" principles for *point particles*[6]). In particular, none of his laws distinguish between states of spatial rest and states of constant velocity (in the jargon, Newton's laws are invariant under a Galilean transformation: $t \to t + a; x \to Rx + t\mathbf{v} + \mathbf{b}$; $\mathbf{v}$ is the relative velocity of the frame; R is a constant rotation). This symmetry suggests that Newton's filaments of "constant spatial position" represent an unwanted construct within a space-time designed for his laws. It would clearly be preferable if these laws could be set in a weaker space-time. Indeed, it is easy to describe the space-time $\mathcal{G}$ wanted, which will be called "Galilean" here. $\mathcal{G}$ continues to split into unique temporal slices carrying a unique one-dimensional temporal separation, but not into unique "same place as" filaments. Instead, a richer family of straight lines, interrelated by Galilean transformations, pierce the space-time in time-like directions. These so-called "affine lines" stitch the temporal slices together, but none of them is selected as privileged in the manner of Newton's filaments. (See figure 2.) A parallel family of such lines correlates with the standard notion of an "inertial frame"—in the diagram, A and B represent two such inertial families. Here the A-stitching looks as if it is "shorter" and more "perpendicular" than B, but this appearance is simply an artifact of the way $\mathcal{G}$ happens to have been pictured within the two-dimensional Euclidean structure of the printed page. Our Euclidean picture contains a "rigidity" that is not present in $\mathcal{G}$'s time-like directions. An analogy: a gelatine mold of the Eiffel Tower lacks a structural backbone to tie its layers firmly together, but a photo of the stuff may look as if it possesses the rigidity of the true Tower. A picture imprints a rigidity upon its image whether or not that rigidity is present in the photographic subject. The A-stitches seem as if they support the t_1 layer better than the B-stitches in our picture of $\mathcal{G}$, but in reality the A-lines are no more "underneath" the t_1 layer than any other inertial family ("underneath" is a Euclidean angular concept excluded from $\mathcal{G}$).

III

Having described $\mathcal{G}$ intuitively, let us enter upon a digression that affects the hole argument. The fact that coherent geometries can exist that *lack* ingredients present in standard Euclidean space became almost universally accepted by

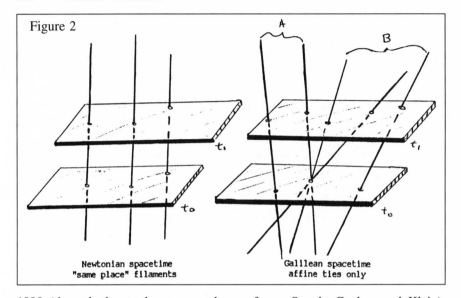

Figure 2

Newtonian spacetime
"same place" filaments

Galilean spacetime
affine ties only

1890 (through the work, among others, of von Staudt, Cayley, and Klein). Galilean space-time is weaker than Newtonian space-time in that the former stretches only an "affine" structure between the temporal slices, whereas $\mathcal{N}$ is closely related to Euclidean 4-space.[7] Klein's "projective spaces" lack metric structure as well, although none of them are marbled with the odd metrics-along-slices that distinguishes $\mathcal{G}$. But the acceptance of such non-traditional structures poses a delicate dilemma for philosophy of mathematics, viz., how can the novel structures be brought under the umbrella of *safe* mathematics? Certainly, we rightly feel, after sufficient doodles have been deposited on coffee shop napkins, that we understand the structure desired in $\mathcal{G}$. A modern approach to the foundations of mathematics would be quite remiss if $\mathcal{G}$ were not included in its menagerie. But it is hard to find a fully satisfactory way that permits a smooth integration of non-standard structures into mathematics. The dilemma is this. Ideally, Galilean space-time should stand as autonomous, *qua* mathematical structure, of orthodox Euclidean space. Newton overlooked the possibility of using $\mathcal{G}$, we would like to say, because his sense of "mathematical possibility" was too narrow. We would hope that "any coherent structure we can dream up is worthy of mathematical study and potential physical application." The rub comes when we try to determine whether a proposed structure is "coherent" or not. Raw "intuition" cannot always be trusted; even the great Riemann accepted structures as coherent that later turned out to be impossible. *Existence principles* beyond "it seems okay to me" are needed to decide whether a proposed novel structure is genuinely coherent. With some sadness, late nineteenth-century mathematicians recognized that, despite our hope that $\mathcal{G}$ stands as a self-contained, coherent whole, existence principles for such structures need to piggyback eventually upon some accepted range of more traditional mathematical structure, such as the ontological frames of arithmetic or Euclidean geometry. In Klein's era, Euclidean geometry constituted the fiducial point for the coherence of other geometrical structures; in our century, set theory has become the canonical backdrop to which questions of structural existence are referred. So when we go beyond napkins and set up a structure like $\mathcal{G}$ formally, we need both: (a) to mark the "intrinsic" elements proper to the internal workings of the structure, without extraneous fat; (b) to provide sufficient "extrinsic" links to set theory so that the existence of our structure can be established. If one is not careful, "linkage" elements needed for purpose (b) can become confused with internal features of type (a). Sundry conventions govern the notion of an "embedding into set theory" that must be treated properly if nonsense is not to ensue. We shall return to this topic in section XVI.

IV

Returning to our main theme, although Galilean space-time $\mathcal{G}$ seems like a better setting for Newton's laws, we have not shown that the "De Graviatione" argument in favor of Newtonian structure can be evaded. Recall that Newton's argument assumed that the laws of any "reasonable theory" will rely upon a notion of **velocity**. Insofar as $\mathbf{f} = m\mathbf{a}$ goes, none of Newton's satellite laws for f (e.g., universal gravitation) actually evoke **velocity**. "$m\mathbf{a}$" requires **acceleration**, rather than **velocity**, but this seems grist for Newton's mill: doesn't **acceleration** definitionally represent an interslice change in **velocity**? Isn't Newtonian structure needed to make sense of **acceleration**?

Although a somewhat hazy awareness that a structure like $\mathcal{G}$ would be better for Newton's laws arose earlier, no crisp reply to the "De Graviatione" argument seems to have been supplied until the 1920s, by E. Cartan. His reply consists of two observations. First, a manifold may admit a notion of derivative without relying upon the close-to-metrical structures of Newtonian space-time. What are the weakest conditions that can be asked of a derivative operation upon vectors? Assume a vector $\mathbf{x}$ runs along a curve upon some manifold. Suppose $\mathbf{y}$ is another kind of vector smoothly defined in the neighborhood of the point p through which $\mathbf{x}$ runs. A "directional derivative," at the very least, should tell us how $\mathbf{y}$ changes as one moves along $\mathbf{x}$'s curve. The answer should be some third vector $\mathbf{w}$. Intuitively, our directional derivative hauls the $\mathbf{y}$ vector situated just ahead on the $\mathbf{x}$ curve back to p for a comparison with the $\mathbf{y}$ vector that actually sits at p. $\mathbf{w}$ is then the local difference between these vectors.[8] However, unless some additional structure lies upon our manifold, we have no particular criterion for how $\mathbf{y}$ should be "properly" hauled back to p. A rule that settles such matters of vectorial transportation is dubbed a "connection," written as Γ, and its associated "covariant derivative" is written as $\nabla_{\mathbf{x}}\mathbf{y}$. When the manifold contains a richer structure such as a metric, a natural choice of covariant derivative is automatically determined. But a Γ can stand alone without the support of richer structure—it is then the minimum structure needed to support a "derivative" operation in the space-time.

Cartan's second observation exploits the four-dimensional character of space-time. Heretofore, we have worked with 3-vectors, written as boldfaced lower case letters. But now consider 4-vectors in $\mathcal{G}$, written in upper case. Assume that $\mathcal{G}$ bears a connection Γ. If $\mathbf{x}$ and $\mathbf{y}$ lie within a common spatial slice, they can be naturally identified with their 3-vector counterparts $\mathbf{x}$ and $\mathbf{y}$. For example, in formulating Newton's laws, we will introduce a vector $\mathbf{F}$ that is called a "4-force." But we also require $\mathbf{F}$ to always lie within single temporal slices. This demand (which is not made in relativistic theories) ensures that $\mathbf{F}$ secretly acts as if it were simply a familiar **force** 3-vector $\mathbf{f}$. If $\mathbf{X}$ and $\mathbf{Y}$ are vectors confined to the same slice, the $\nabla_{\mathbf{x}}\mathbf{Y}$ determined by Γ must conform to the usual notion of "directional derivative" defined by the Euclidean geometry within the slice. But if $\mathbf{X}$ and $\mathbf{Y}$ stretch across slices, where no metric reigns, $\nabla_{\mathbf{x}}\mathbf{Y}$ enjoys a greater freedom to behave as we might wish. In particular, let $\mathbf{V}$ be the 4-"velocity" vector tangent to the curve traced by a moving particle. Such $\mathbf{V}$'s exist on any 4-manifold, but (in the present context) $\mathbf{V}$ does not bear any intimate connection to **velocity** in our usual 3-vector, Euclidean sense ($\mathbf{V}$ is tied to our particle's motion in a much weaker way). Using our freedom to fuss with Γ in time-like directions, we can require $\nabla_{\mathbf{v}}\mathbf{V}$ to always lie within temporal slices. We force $\nabla_{\mathbf{v}}\mathbf{V}$ to act as a 3-vector, although V itself does not. Cartan now suggests that a slice-restricted $\nabla_{\mathbf{v}}\mathbf{V}$ be employed as a replacement for Newton's 3-acceleration $\mathbf{a}$ in $\mathbf{f} = m\mathbf{a}$. We can thus construe $\mathbf{f} = m\mathbf{a}$ as $\mathbf{F} = m\nabla_{\mathbf{v}}\mathbf{V}$, retaining $\mathbf{a}$'s vital link to particle motion, without needing to accept absolute notions of either **displacement** or **velocity** as the "De Gravitatione" argument presumed—truly a beautiful response to Newton.

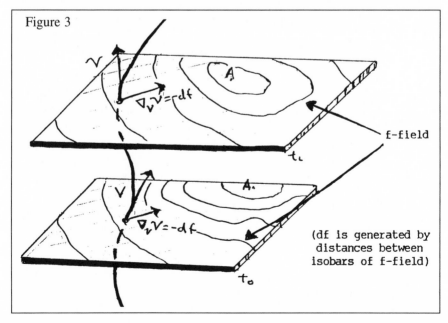

Figure 3

f-field

(df is generated by distances between isobars of f-field)

Incidently, in figure 3 I have shown how a force like gravitation can be extracted from the local potential field g, an operation that depends solely upon the internal geometry of the slices. This is the simplest way that forces can arise in classical mechanics.

Our informal discussion makes it seem as if the Cartan connection needs to be added to Galilean space-time as an additional ingredient. Actually, $\mathcal{G}$'s "affine stitching" is simply the curves whose 4-"velocities" satisfy the "geodesic" condition $\nabla_V V = 0$. *Modulo* a few global/local technicalities,[9] Γ and our affine stitchings determine one another (in the literature, $\mathcal{G}$ is often described "locally" in terms of Γ, rather than "globally" in terms of stitching as we have done).

Cartan's treatment finesses the "De Gravitatione" argument neatly— Newton's laws of motion are set up without getting near **displacement** or **velocity**. But the reader should be warned that many texts (including virtually everything written in continuum mechanics) proceed in a less tidy way. They typically define *frame-dependent* notions of 3-**displacement** and 3-**velocity** and then show that the 3-**acceleration** defined from these frame dependent vectors is not itself frame-dependent (to be more precise, if the derivatives of the components of 3-**velocity** in a coordinate system adopted to the frame are calculated, the resulting array of numbers will transform as a 3-vector to any other frame in the allowed group). A vector (or, more generally, tensor) that is independent of the choice of *inertial* frame in this sense is called "absolute"[10] (later we shall encounter a distinct class of vectors dubbed "objective" whose formulation rests upon a different family of frames). If only "absolute" quantities are allowed to appear in statements of physical law, this "untidy" way of proceeding gives the same net results as Cartan's approach. However, the "untidy" treatment seems to run the danger of falling into Newton's clutches: haven't we once again made **acceleration** definitionally. dependent upon **displacement**?

In fact, tidiness has merely encouraged, as it so often does, an excessive intolerance. Don't presume that notions like 3-**displacement** do not exist in Galilean space-time; they are present as *frame-dependent* objects. A frame, after all, is a well-defined geometrical gizmo within $\mathcal{G}$ and from it, a relativized notion of **displacement** can be immediately defined. One would hardly wish to say that normal vectors do not exist in Euclidean space, although they can only be defined relative to a choice of surface. Like normals, frame-dependent vectors are often important objects in physics. Why, then, do we expect that the laws of nature should not depend upon frame-dependent quantities? Their problem is not that they do not exist, but that they are insufficiently *local* to appear in differential equation laws.[11] That is, if a law seems to need a frame dependent quantity, we should expect that there is a local space-time element from which the relevant frame dependent notion can be grown on a global scale. It is the local element that belongs in the law. In short, a hidden presumption of "locality" lies behind the misleading claim, often found in philosophical discussions of Cartan's approach, that "**velocity** doesn't exist in $\mathcal{G}$." In any case, the "untidy" methodology is just as defensible as Cartan's, albeit not so prettily presented.

This reconciliation of "tidy" and "messy" descriptions of space-time structure will prove important when we encounter "Leibniz structure."

V

Leibniz and Descartes seemed to believe that classical physics ought to be set within a weaker structure that has been dubbed "Leibnizian space-time" (our anachronistic muse allows us to ignore the fact that neither historical figure wished to place his physics "in" anything!). Leibnizian space-time $\mathcal{L}$ is exactly like $\mathcal{G}$ except that $\mathcal{G}$'s Γ is eschewed. The universe still carves uniquely into slices carrying Euclidean 3-metrics, with a temporal metric betwixt the slices, but nothing privileged stitches the slices together any longer.

Lacking a Γ, Cartan's trick for making sense of "$\mathbf{f} = m\mathbf{a}$" seems to fail. More generally, the prospects for setting up *any* kind of orthodox deterministic physics within $\mathcal{L}$'s barren landscape seems dubious, at least if we expect our physics to be based on differential equations. Generally we expect the data along a given slice of space-time to determine the conditions on the immediately ensuing slices. But if no Γ exists to make interslice comparisons, it seems impossible to formulate law-like connections between the slices. As Earman would express it, Leibnizian space-time looks like "an inhospitable setting" for any kind of deterministic physics. Let's dub these worries about $\mathcal{L}$, the "Γ-argument." Clearly, they represent the salient residue of Newton's "De Graviatione" argument.

Actually, the Γ-argument can be evaded in a rather coarse way. Nothing forces us to install all of the ingredients needed in a Γ within the space-time proper; we are allowed to use a Γ-surrogate as long as it is "non-geometrical." On the face of it, such a retort sounds completely stupid—the purest form of philosophical pettifoggery. In fact, continuum mechanics has deep, and instructive, reasons for wanting to treat its Γ in this "stupid" manner. The chief burden of this paper is to explicate these reasons. We shall also find that our "stupid" reply also cuts to the heart of what is wrong in Earman and Norton's "hole argument."

VI

Before we examine continuum seriously, let us inspect a second argument that also claims that $\mathcal{L}$ cannot accommodate a deterministic physics. This argument, which I will dub the "twist argument," was first proposed by Howard Stein[12] and has been endorsed by John Earman (whose formulation I will follow). A symmetry of a space-time is a mapping ϕ of its points that preserves the structural features of the space-time. For example, every space-time we have discussed declares that if ϕ maps point p to $\phi(p)$ a fixed displacement $\mathbf{b}$ away, then space-time will look exactly the same at $\phi(p)$ as at p. $\mathcal{L}$ has the special symmetry that every point looks exactly the same if $\mathcal{L}$'s slices are mapped through differing degrees of rotation (the transition must be smooth enough from layer to layer to not tear $\mathcal{L}$'s differentiable structure). In the jargon, $p \rightarrow R(t)p$ is a symmetry of $\mathcal{L}$, where R is a rotational matrix. $\mathcal{G}$ does not obey this symmetry because a slice-dependent rotation map will carry the affine line stitching over to curves that are not straight, hence the conditions at $\phi(p)$ will not look like those at p.

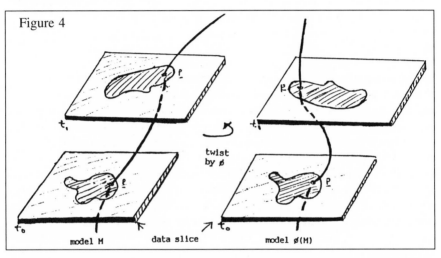

Figure 4

twist by ϕ

model M   data slice   model $\phi(M)$

The twist argument now proceeds as follows. Let T be a theory that determines how matter is supposed to evolve within $\mathcal{L}$. Let the condition of the matter up to time t_0 be fixed and assume that T declares that the matter should evolve past t_0 in manner M. Map the matter beyond t_0 by a rotation map ϕ that leaves the stuff before t_0 fixed. Obviously the ϕ distribution of matter will be different from M (except in degenerate cases where M is unusually uniform). But, the argument runs, a theory will be deterministic only if there is only one model of the theory that continues a fixed set of initial conditions beyond the data slice. But our $\phi(M)$ distribution must satisfy T as well as the M-model, for we have already conceded that space-time at $\phi(p)$ acts just the same as at p. Since we have arranged everything else at $\phi(p)$ to be like that at p, the distinct models M and $\phi(M)$ show that T cannot be deterministic. Note the argument fails for $\mathcal{G}$.

The common thread between the twist and the Γ-arguments is that, lacking stitching between its layers, a space-time cannot accommodate a deterministic theory.

I am somewhat uncertain how Earman and Stein view the conclusion of the twist argument—does it prove that determinism is truly impossible within Leibnizian space-time or simply that some further story needs to be told on this score (e.g., about Γ's non-geometrical nature)? Many passages[13] in Earman suggest the second, guarded position, yet he seems to accept the closely related "hole argument" in a more definitive spirit. Yet if the hole argument works, the twist argument surely does.

My own evaluation is the twist and hole arguments both turn upon defective uses of their characteristic ingredients ("determinism"; "symmetries"). The Γ-argument, I believe, isolates the central worry about the physical utility of $\mathcal{L}$ in the most blunt and straightforward way. Most of our discussion will be devoted to showing how the worries about Γ are set to rest in continuum physics. We will then return to the other arguments.

VII

By the standards of the "plausibility arguments" they regularly employ, most physicists will accept the Γ and twist arguments as correctly codifying unhappy features of Leibnizian space-time, without regarding these infelicities as irredeemable. Recently John Earman and John Norton have developed a variation of the twist argument that runs contrary to conventional physical opinion. This argument—the "hole argument" mentioned earlier—tries to show that many theories standardly regarded by physicists as deterministic cannot be unless the notion of "determinism" is altered in an unexpected way. Although consideration of the hole argument carries us slightly afield of our central concerns within classical mechanics, it is worth rehearsing because its problems are so close to those of the twist argument.

It runs as follows. Consider any theory, such as general relativity, whose collection of symmetries accepts any smooth mapping ϕ of the manifold to itself (the general symmetry group is the full set of diffeomorphisms). As in the twist argument, fix all data below a selected spacelike slice. Let M be a predicted manner of evolving beyond the slice. Unlike $\mathcal{L}$, the manifolds of general relativity have a Γ that stretches across these slices (indeed, a metric is present). However, unlike the Γ in $\mathcal{G}$ general relativity's Γ is not immutable; it is coupled with the matter (more exactly, the mass-energy) within a model. If the matter in M is transported by ϕ, the geodesics et al. determined by Γ will follow along with $\phi(M)$. The symmetries of general relativity declare, that once this is done, p will not retain any geometrical features that prevent $\phi(M)$ from acting as a model of the theory (whereas p's retention of stitching structure in $\mathcal{G}$ keeps $\phi(M)$ from serving as a proper model). But we now have two models that continue the same set of initial data in different ways. Hence, general relativity, contrary to conventional opinion, is never deterministic. Virtually the same argument will work for any theory that accepts a suitably generous collection of symmetries.

This is called "the hole argument" because Einstein once dubbed the region where the ϕ mucking around occurs a "hole." His early tergiversations about general covariance, later abandoned, inspired Earman and Norton's argument.

I have presented the hole argument in the restricted manner favored by Earman, where its effectiveness is limited to theories accepting generous space-time symmetries. Norton believes that the argument applies almost universally. Consider $\mathcal{G}$, which seems to resist the twist argument. Norton suggests that if the stitching is moved under ϕ as well as the rest of M, a model distinct from M is generated although ϕ does not count as a conventional symmetry of the space-time. If Earman's hole argument is correct, it is hard to see what is wrong with Norton's.

Earman and Norton cheerfully concede that something has gone dreadfully wrong somewhere. Point particle mechanics and general relativity ought to count as deterministic (*modulo* some specific cause of failure). Earman attributes the source of the error to the presumption that $\phi(M)$ is a different model from M. M, after all, is "observationally equivalent" to $\phi(M)$ in the sense that all events "look the same" to observers within the two universes. The logical positivists claimed that such models should be identified and Leibniz himself struck similar verificationist chords. Accordingly, Earman forms equivalence classes of models according to whether they can be isomorphically mapped to one another, a relation Earman dubs "Leibniz-equivalence." A genuine "model" for general relativity, Earman thinks, should be an equivalence class of the old models cited in the hole argument. He writes:

> [T]he intended interpretation of $M \equiv_L M'$ [M and M' belong to equivalence class E] is that M and M' are different modes of presentation of the same state of affairs; that is, at base, physical states are what underlie a Leibniz-equivalence class of absolutist models.[14]

By collapsing multiple models into one, Earman hopes to restore determinism to deserving theories. Earman quotes several physics texts that seem to support this point of view.

As I understand him, Earman thinks that matters cannot be left here. Individual models M and M' error in attributing too much "spatial structure" to a world properly represented by E, just as Newton runs unnecessary filaments through Galilean space-time. In the latter case, we would feel that $\mathcal{G}$ had not been described properly if we could characterize it only as "the space-time behind the class of all observationally equivalent Newtonian structures." We want to delineate the features of $\mathcal{G}$ *directly*, as we did above. The unwanted structure revealed by the hole argument seems to turn upon the notion of "matter being located at point p." We should seek a direct articulation of a structure that is able to support the usual mathematical operations permitted on a manifold while eschewing "being located at p," a notion that is unfortunately built into the customary mathematical definition of "manifold." Earman makes a few tentative suggestions about what the structure might look like, but eventually leaves the quest as a project for future research.

On my diagnosis, this enterprise is ill-conceived. There is nothing wrong with the usual notion of manifold and Earman's difficulties arise from a misreading of the conventions that underwrite mathematical notions such as this. These topics will be briefly discussed in section XVI. In any case, the idea that "Leibniz-equivalent" structures ought to be identified for purposes of determinism seems wrong. Consider a standard case of failure of determinism:[15]

a column with a weight on top. At low values of the weight, the condition of the column below is completely determined. At a critical value of the weight, however, the column suddenly sags. But this collapse is not unique; although the *shape* of the buckling is the same, the column is free to bulge in any 360° direction. In a universe consisting of a solitary weighted column, "Leibniz equivalence" would identify these different responses, contrary to conventional mathematical opinion that the governing theory does *not* determine unique solutions. A proper resolution of the hole argument should *not* identify "Leibniz-equivalent solutions."

As I stated, critical discussion of the hole puzzle will be reserved until the concluding sections of the paper.

VIII

We are almost ready to discuss continuum mechanics! It is a useful warmup to consider how an anachronistic Descartes might have responded to the worry about Γ (as well as Newton's "De Gravitatione" argument). Despite his trumpeting of "clarity and distinctness," Cartesian physics seems a rather confused assemblage of doctrines. However, an important component of the Cartesian picture is still studied today under the rubric "the kinematics of mechanism"— viz., the study of how systems of gears and other forms of rigid mechanical linkage behave. Roughly described, the subject worries about how machines can move, without bothering about the forces that induce them to do so. Remember those old I.Q. tests: if gear A turns a certain amount, then the movements of the parts connected to A can (usually) be calculated by geometrical reasoning alone. If A's angular velocity is known, the velocities of the other parts are fixed, and so on. What kind of space-time does the "kinematics of mechanism" require? The theory relies upon the following ingredients:

1. A time lapse function that carves space-time into simultaneity slices.

2. Euclidean 3-geometry upon each slice.

3. A tracking of material points between slices—that is, a way of knowing when the material at p on slice A is the same as on slice B. We render this precise by setting up a *reference set* of machine parts with a map Ψ that carries the points in the reference set into each slice.

4. Under Ψ, the metrical relations between material points belonging to same machine part are always the same—i.e., our machine parts are rigid.

5. Let a and b be contacting material points on two machine parts at time t_0. Let d_a and d_b be the arc length displacements of a and b from their former contact point at time t (as long as contact between the parts is maintained). As $\lim \Delta t \to$

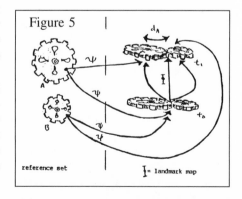

Figure 5

reference set $\mathbf{\int}$ = landmark map

0, $d_a/\Delta t$ varies smoothly and is equal to $d_b/\Delta t$. That is, gears do not slip at their points of contact.

6. On each time slice, all parts connect with one another through some chain of connection.

Note that in condition (5), **velocity** arises as a comparison of d_a-**displacements** between slices. Does this mean that our space-time contains a Γ-like structure that makes sense of these comparisons? No, "places" on different slices are correlated through an identification of geometrical *landmarks* in the slices, viz., we track the continuing "place" on each slice where gears A and B contact. Tracing a few of these "landmarks" through time defines a interslice map Φ that engineers call a "fixed space." Relative to Φ, most material points and other potential landmarks will be in motion. Clearly, if Φ serves as a suitable "fixed space," then so might any other Φ' linked to Φ by a "Leibnizian transformation" (i.e., $\mathbf{x} \to R(t)\mathbf{x} + \mathbf{c}(t)$). We now have two structures—the material flow Ψ and the family of "fixed spaces" Φ—but both seem attached to the matter rather than the space-time. The only features in our list required of the event space itself are those of Leibnizian space-time. Thus, the geometrical requirements of the kinematics of mechanism seem fully in accord with Cartesian presumptions about the nature of "space." The definition of velocity relies upon Ψ and landmark linkages across the slices, rather than a Γ planted within the space-time backdrop.

Machine movements of the type contemplated are not always deterministic, but, insofar as they are, they exemplify a different sort of "determinism" than the "initial value" format presumed in the twist and hole arguments. Rather than demanding that data upon a slice should determine the universe's later evolution, our theory of machines supports only "here versus there" determinism. That is, we take as "data" a specification of the *full* temporal history of a particular gear relative to its immediate surroundings. We hope that this local history will determine the movements of the remainder of the machine (note that such "data" lie on a time-like surface in the manner of "boundary data" for differential equations). But, without supplementation, our theory has no means for determining how a gear will move in the future given that it turns in a particular manner now.

In other words, our approach to machines lacks resources to support claims like "the watch spring is wound up" or "the flywheel carries inertia." As is well known, Descartes wanted to use **velocity** (or **momentum**) as a repository of "motive power" that can sometimes transfer from one machine part to another in a time-like direction. But we have no assurance that our Ψ and Φ-bound **velocity** can be used in this way—there is a great danger that a stronger Γ-like structure has been surreptitiously smuggled in. If the "De Gravitatione" argument is examined carefully, one finds that Newton recognizes that a "kinematics of mechanism" notion of **velocity** can be coherently introduced. He takes pains to show that Descartes speaks of **velocity** in circumstances where landmark identifications break down (e.g., the machine parts separate or distort). Descartes's main hope of rebuttal is to claim that a gear can retain a "memory" of the landmark-based **velocity** it bore when it was last in contact with another machine part. This "memory" will then determine the rate at which the gear will turn a new part, once a fresh contact has been struck. If this kind of appeal to "memory" can be legitimated, then Descartes will have successfully adhered to his landmark based conception of **velocity**. However, it is obvious that the rules of such a theory will need to be ridiculously *ad hoc* (as Descartes's rules of contact in fact were). Hence we will not pursue this line of inquiry further. In the next section, we will see that continuum mechanics relies upon a more sophisticated notion of "memory" that is quite viable.

IX

One advantage of walking through the kinematics of mechanism is that it provides an easy introduction to some of the constructions used in continuum mechanics. However, the latter is a considerably more sophisticated subject and its response to the problem of **velocity** is quite different from that we have just seen. My plan is to make a few observations on the ingredients wanted in continuum mechanics and then examine the requirements that these constructions demand of the event space.

Observation #1: *Need for an Inherent Notion of Material Point Identity.* Once again a function Ψ mapping a "reference set" of objects into each space-time slice is introduced to track the flow of material points through space-time. Here Ψ's role grows more serious because it is used to capture the idea that a bit of matter remains the same although its volume has dilated. In smooth continuum mechanics, a "reference set" becomes a "body manifold" whose purpose is to ensure that the topological and differential features of a continuous substance remain fixed over time.[16] At each time t, Ψ maps the body manifold B into a slice of space-time, the image under Ψ being called "B's configuration at t." Only the image, $\Psi(B, t)$, carries any metrical attributes; B itself lacks lengths (it carries "volume" however). Although "distance" is not assigned to B, "mass" is (as a measure over suitable subsets of B). Intuitively, conservation of mass claims that if a portion of matter is tracked through time, no matter how that matter distorts, its mass will remain constant. By assigning mass to the non-temporal body manifold, $\Psi(P, t)$ automatically retains its mass, where P is a portion of B.

These humble considerations give rise to the distinction familiar to all workers in continuum mechanics between "material" and "spatial" pictures of a process (the more traditional terms are "Lagrangian" and "Eulerian," but these labels are less evocative and historically misleading). Since matter can flow and distort, two natural ways of studying a material process arise: (a) we watch where a given portion of B flows over time; (b) we watch the behavior of the material found within some constant *spatial* volume, where "constant volume" is hooked to a coordinate system centered, say, in an inertial frame. When we study fluids, for example, we tend to think of "same volume" in the second, "spatial" manner because it is easy to observe what flows in or out of a fixed cube, but it is quite hard to determine where the stuff flowing out

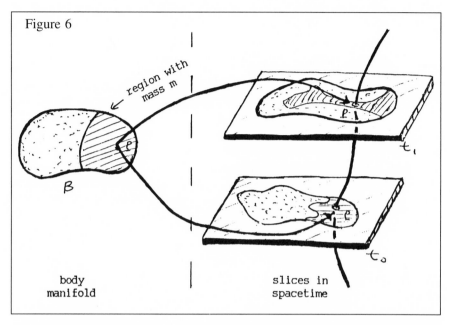

Figure 6

region with
mass m

P

B

P

t₁

t₀

body
manifold

slices in
spacetime

goes. When we distort a piece of iron, however, tracking the flow of material is easy. Independently of difficulties of observation, however, the material picture is better suited to the formulation of physical laws. Conservation of mass is very simple there: $\int_{V(t)} \rho dx = \int_{V(t*)} \rho dx$, where $V(t)$ and $V(t*)$ are the same "material volume" at different times and ρ is the mass density. But in the same spatial picture, we have to worry about how matter flows in and out of the volume we are investigating (so the Jacobian of the flow must be cited). Likewise, conservation of charge, balance of momentum, etc., are all expressed more simply within the material picture. Loosely stated, physical laws "go with the flow," rather than lingering over the structures fixed within the space-time. This overall preference for the material picture, of course, does not prove that laws do not also pay attention to fixed space-time structure as their materials waft by. The chief salience of "Leibniz structure" is that a surprisingly large amount of the structure required by the laws of continuum mechanics can be viewed as "carried along with the matter." Recognition of this fact seems to lie at the heart of the Leibnizian distrust of an "independent" space-time in the spirit of Newton. We miss the source of distrust if we regard point particle mechanics as the full repository of "classical physics," for the material-spatial dichotomy is almost invisible there.

While on the topic of the Ψ mapping, I will mention a consideration that helps explain why continuum mechanics cannot be viewed as a trivial extension of point particle physics. In the latter, each point particle retains its mass and identity through time. However, when we try to build a point particle model of a classical fluid, the Ψ wanted in the continuum description will not correlate with the temporal identifications natural to the point particles. In a point particle model of a viscous fluid, the particles must regularly *cross* the boundaries of the tubes of macroscopic Ψ-flow. Indeed, an arbitrary swarm of point particles is not organized enough to support the macroscopic continuity of "material particle flow" that lies at the basis of continuum mechanics. In consequence, the principle of "conservation of mass" in continuum mechanics cannot be viewed as a simple reiteration of "conservation of mass" within particle mechanics, despite the similarity in nomenclature, because the continuum principle requires some mechanism for ensuring that the mass which flows out of a tube of Ψ flow is replaced, on the average, by an incoming mass. Similar caveats apply to most continuum and point particle principles that "sound the same." As C. Truesdell is fond of pointing out, it is difficult to "reduce" continuum mechanics to particle mechanics in any rigorous way, whereas the latter stands as simply a degenerate case of the former.

Observation #2: *Need for a "Referential Configuration."* Thus far no spatial notions besides volume have encroached upon our laws—physics can live solely on **B** for all that conservation of mass cares. In continuum mechanics, reliance upon spatial ideas arrives in several grades of increasing "spatial commitment." The most minimally committed laws are those that correlate forces upon a material with its *current* spatial configuration. Thus the gravitational law looks at the distances within a distribution of mass density upon a slice and calculates, through Poisson's equation, the force each material point will feel because of the gravity produced in its environs (see figure 3). Here the 3-metric on each slice is used to calculate a distance between material points and to define directional derivatives in sundry spatial directions.[17] But no appeal has yet been made to any four-dimensional gizmo (like Γ) that compares conditions upon distinct temporal slices. Forces in point particle physics are generally of this modestly committed variety.

A principle that relates forces (or, more accurately, stresses) to other features of the material is called a "constitutive law." Continuum mechanics is distinguished by a rich variety of these, most of which require a higher grade of "spatial commitment" than the gravitational law. *It is here that "Leibniz structure" enters the picture.* In thinking about materials intuitively, the metaphor of a "material memory" seems quite apt. Elastic-like substances, for example, retain a "memory" of the state they would assume if they were not strained. When the strain is released, elastic materials snap or creep back to their natural configuration. Hooke's law for a spring is a "constitutive equation" that relates the spring's stress to its distortion from its "remembered" relaxed state. But if the spring "sets" under prolonged strain, its "memory" shows signs of "fading" over time and Hooke's law needs to be modified to take account of this forgetfulness. A typical fluid, however, only possesses "short-term memory"— the stress it feels is determined solely by the rate at which its configuration is presently changing. More complicated materials such as paints display both types of "memory." Obviously everyday materials display a huge variety of distinct types of "constitutive behavior." Rather than trying to construct molecular mechanisms for the myriad ways in which a macroscopic material might develop stress, modern continuum mechanics tries to organize the possible "constitutive laws" into tractable classes[18]—i.e., first study purely elastic and purely viscous (= fluid like) responses independently, then "mix or match" to suit more complicated behaviors and so on. Certain general laws oversee the construction of this taxonomy, one of which will be crucial to "Leibniz structure."

How might we make this idea of "memory" precise?[19] The standard procedure is to introduce on each slice an *imaginary* portrait of the material in some fixed condition, say, the relaxed configuration of an elastic material. Formally, one defines a second function σ that maps B into space-time with the restriction that the images under σ are isometric (= have the same metrical properties on each slice). However, it should be clearly irrelevant how σ sets the "referential configuration"

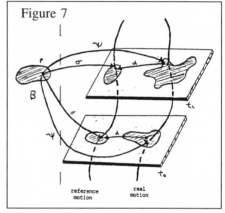

Figure 7

$(= \sigma(B, t))$ within each slice. Therefore we want to introduce the full family Σ of allowable σ in our formalization of "memory" (i.e., Σ should be closed under the maps $\sigma'(p, t) = \mathbf{R}(t)\sigma(p, t) + \mathbf{d}(t)$). Such a Σ, I dub a "Leibniz structure."

Σ is used in the following way: relying upon a selected σ, we can define various σ-dependent (tensorial) quantities. For example, once σ is fixed, the current **displacement** of a material point from its σ-dependent "memory" can be defined. However, we are interested only in the quantities that are independent of σ's placement in the space-time, in the sense that an "arrow" defined from a σ should transform as a "vector" within the full array of Σ-maps. Objects that are independent of a specific σ in this fashion are called "objective" in the continuum literature. **Displacement** is not "objective" in this sense. However, there are various (tensorial) measures of "stretching" away from the reference state (e.g., the Cauchy-Green strain tensor $\mathbf{C}$) that turn out to be "objective." That is, $\mathbf{C}$ captures a kind of distortion within a material that is independent of how the "memory" of the reference state has been placed in the space-time, how it has been rotated, etc. Intuitively, the idea is this: suppose one has a photograph of one's formerly thin self. The increase in bulge, the change in posture, the history of dietetic failure all represent "objective" changes from the condition pictured. But the fact that the photo is placed 16" from one's nose signifies nothing. It is remarkable that we can capture the "objective" changes by "factoring out" the differences among the possible photo placements (= Σ). In this same manner, an elastic material can "remember" its internal deviations from its relaxed configuration, but it has no conception of where it would now be if it had been kept constantly relaxed. It has only $\mathbf{C}$-memories; not

displacement memories. Fluids, on the other hand, are too forgetful to maintain C-memories. Relative to a σ, **velocity** arrows are defined on the fluid; these arrows will not be "objective." However, the degree to which such **velocity**'s shift in the neighborhood of a material point (= the velocity gradient tensor **L**) is, and **L** correctly captures a fluid's memory of its immediate past. In short, the Leibniz structure Σ ratifies as "objective" a variety of quantities that are then allowed to appear in the "constitutive laws" of sundry materials: the law for an elastic body takes the general form **stress**$(p, t) = F(\mathbf{C}, p, t)$ for some function F; for a fluid, **stress**$(p, t) = F(\mathbf{L}, p, t)$.

In setting matters up this way (including our intuitive talk of material "memory", σ-placements, etc.), I have followed the standard literature closely. Notice that a family of "objective" quantities has been introduced by first constructing a bunch of "non-objective," σ-dependent notions. In so doing, the construction resembles the literature, discussed in section (iv), that constructs 3-**acceleration** from an inertial frame-dependent 3-**velocity** and then proves that the 3-**acceleration** is "absolute" (= independent of choice of "inertial frame"). Note that "absolute" ≠ "objective," because the latter relies upon a larger class of frames (among tensors as well as human beings, it is easier to be "absolute" than "objective"). Cartan, as we noted, obtained 3-**acceleration** through direct calculations from a (suitably restricted) 4-dimensional Γ. It is presumably possible (although I have never seen matters presented in this way) to set up an algebra that calculates "objective" quantities directly from an Σ (rather as the Cartan-Grassmann calculus of forms isolates the subset of tensorial objects natural to integration). However, such a presentation might easily obscure the fact that "Leibniz structure" represents a formalization of an intuitive conception of "material memory." We shall return to the importance of "Leibniz structure" later. For the moment, let us note two facts: (1) a large group of laws rely upon a Leibniz structure Σ for their proper formulation; (2) Σ represents a *weaker* degree of "spatial commitment" than Γ supplies,[20] in the sense that there are fewer "objective" quantities than "absolute" ones. In the simpler world of point particle mechanics, the notion of Leibniz structure never arises.

"Leibniz structure" should be of interest to philosophers who like to defend modality through appeal to the dignity of physics.

Observation #3: *Surface and Volume Forces.* Another crucial difference between continuum mechanics and point particle mechanics is that in the former, forces fall into two classes, depending upon where they act. Isolate an imaginary volume V within a material and let δV be its boundary. Volume (or "body") forces like gravitation act *directly* upon the material points of V. The total volume force on V can be found by integrating these individ-

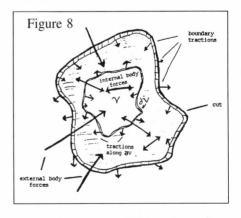

Figure 8

ual point actions over all of V ($= \int_V \rho \mathbf{b} dv$). However, much of the influence that V feels arrives *through contact along its boundary* δV. The total effect of these "traction" forces is found by integrating along the *surface* δV ($= \int_{\delta v} \mathbf{t} ds$). Cauchy showed that a tensor field $\mathbf{T}$ could be introduced that permits a reconstruction of the tractions across any selected δV within the material. This $\mathbf{T}$ is called the **stress**, and, compared to a volume force, represents a relatively sophisticated concept.

Observation #4: *The Need for a "Cut" in the Formulation of Laws.* Although, in general, forces between objects are mutual, it is seldom that all forces acting upon a system can be treated in a fully reciprocal manner. Instead, a "cut" must be introduced between the system under study and its external environment, where the "external" forces arriving from the exterior are treated as less variable than they would be if they were handled in a fully reciprocal manner. In dealing with a body of fluid, the gravitational force exerted by the earth is typically set up as a constant force arriving from outside. Newton's gravitational law, strictly speaking, requires that the water, in its sloshing around, should influence reciprocally the external gravitating bodies to some extent. When we select a "cut" along the fluid's surface, we decide to ignore this back effect upon the outside world, allowing the force upon the fluid to be treated as constant. Likewise, the "boundary conditions" that specify the forces acting on the surface of a material (where the usual choice of a "cut" is placed)[21] are almost always specified in a more static manner than could be honestly accomplished if the boundary forces act in the reciprocating manner of traction forces inside a "cut." Designation of a force as "external" simply means that it is being treated in an approximation, hopefully of a minimal nature (in particular, the approximated force should not be too different from the result one would obtain if it had been treated as "internal" in a setup where the "external/internal" cut had been placed further away). If all matter in the universe were concentrated in a local island, the need for a "cut" could be eventually evaded by moving outward far enough. But if the universe is filled with matter, as Leibniz supposed, it is impossible to remove the "cut" dependence from the formulation of physical law completely; a residue of the universe always remains whose workings are not described wholly accurately. Indeed, since "external forces" are merely averaged approximations for real forces, they should not be regarded as really residing in space-time.

We shall exploit the need for a "residue" in our later treatment of inertia.

X

Which laws of continuum mechanics seem to require a stronger degree of spatial commitment in the form of a Γ? The so-called "balance of (linear) momentum" law comes to mind, along with its angular momentum companion.[22] The former represents the continuum analog of $\mathbf{f} = m\mathbf{a}$, which requires a Γ in Cartan's treatment to formulate the 3-**acceleration** part. However, the continuum balance law is not a local principle, since it integrates over extended material. If the universe is smooth enough, Cauchy showed how to reexpress the balance principle in local form, where the potential role of a Γ can be surveyed more easily. To do this, the total force **F** acting upon a volume V is decomposed into its surface and volume contributions, viz., $\int_v \rho \mathbf{b} dv + \int_{\delta v} \mathbf{t} ds$, where **b** is the total body force and **t** the total traction force acting on V. By appeal to Stokes's rule and Cauchy's association between **t** and the stress **T**, we obtain a point-based formulation of momentum balance called Cauchy's law of motion: $\rho D\mathbf{v}/Dt = \rho \mathbf{b} + \text{div}\mathbf{T}$ ("D/Dt" is the so-called "material derivative"). But what is **v** here? It is a 3-**velocity** field based upon some selection of σ-referential configuration. Unfortunately, neither **v** nor $D\mathbf{v}/Dt$ count as "objective" under the full Σ, so Leibniz structure alone cannot support "$D\mathbf{v}/Dt$"'s appearance within a physical law. It seems that Γ has finally forced its space-time appearance here, converting our background space-time from $\mathcal{L}$ to $\mathcal{G}$. So the Γ-argument seems finally vindicated by our inspection of continuum mechanics.

Once a Γ is available, perhaps we can discard the quaint apparatus of "reference configurations" in favor of the more familiar notion of inertial frame. Assume (by an axiom) that the geodesics of Γ can be globally assembled into inertial frames. Why not let the isometric regions that track inertially through our space-time replace the reference configurations? A "reference configuration," on this reading, is simply a space-time worm that holds metrically constant on slices, rather than the "ghost" of an idealized memory, as our earlier treatment suggested.

In fact, it is perfectly possible to set up continuum mechanics within $\mathcal{G}$ in the fashion suggested—Marsden and Hughes essentially follow this course in their excellent textbook. In doing so, they make a choice which carries certain mildly awkward consequences. By adhering to $\mathcal{L}$, these drawbacks can be avoided (although the $\mathcal{L}$-treatment carries infelicities of its own). Walter Noll first presented continuum mechanics in the $\mathcal{L}$-based vein, an approach that has been generally adopted by Truesdell and his school.

XI

In selecting $\mathcal{L}$ or $\mathcal{G}$, we confront a "Scylla or Charybdis" choice, which is revealed if we reorganize Cauchy's law in a trivial, but quite traditional, way. Group the terms using ρ as a coupling constant on the same side of the equation: $\rho(D\mathbf{v}/Dt - \mathbf{b}) = \text{div}\mathbf{T}$. Think of $D\mathbf{v}/Dt$ as an "inertial force" that, in concert with **b**, balances the divergence of the material's internal stress. Cauchy's equation by itself does not determine how a material moves—the stress **T** must be tied to quantities that somehow derive from the material's previous history of motion. Recall that an elastic material obeys a "constitutive law" of the form $\mathbf{T} = F(\mathbf{G})$, where **G** is the "stretch" tensor recording the material's deviation from its rest state. The material's law of motion is formed by a substitution into Cauchy's law: $\rho(D\mathbf{v}/dt - \mathbf{b}) = \text{div}F(\mathbf{G})$.

"Scylla or Charybdis" arises as follows: the two sides of the expanded Cauchy's law seem to "see" different underlying space-times—the left side "sees" $\mathcal{G}$; the right side, $\mathcal{L}$. In our rearrangement, the "constitutive nature" of the material—its manner of response to outside influence—is isolated on the right. We have seen that an incredible variety of different manners of response can be found in the world's materials. Despite this heterogeneity, the constitutive principles for any material obey certain basic limitations. The restriction that is salient for our concerns is standardly called "material frame indifference" or "objectivity."

A paper clip should not be flexed too much or it will weaken. Even when restored to its original shape, it winds up weaker (= develops less stress against bending), unlike a fully elastic substance that reclaims its original fortitude. Suppose we try to gauge, through experimental test, the constitutive behavior of a metal destined for some industrial application. We will twist or pull the material in sundry ways and measure the stresses produced. Our paper clip experiences warn us that it will be foolish to test only *fresh* blocks of material, for we will not be able to see, from experiments that do not probe the role of a material's history, whether the metal behaves elastically (good) or is subject to fatigue (dangerous). Spectacular engineering disasters have resulted because materials were accepted on the basis of good behavior within a overly narrow spectrum of tests.

In this context, it is striking that we instinctively dismiss as pointless experiments that propose investigating whether paper clips on slowly rotating turntables are subject to a special kind of rotational fatigue. That kind of motion, we feel certain, will not affect the material. Or, more exactly, although paper clips *can* be slightly weakened within a briskly moving centrifuge or by accelerating them incredibly, these are not novel phenomena that mandate special experiments. Why? Accelerating a substance induces "inertial forces" upon it, which may stretch it, but materials do not react to a history of inertial influences differently from the way they respond to an analogous schedule of regular forces. The right hand side of Cauchy's law "sees" the entire left side merely as a combined "outside influence"; it is unable to respond differentially to its decomposition into "$D\mathbf{v}/Dt$" and "$\mathbf{b}$" pieces. Accordingly, a material's response to an inertial history, however idiosyncratic it may otherwise be, must display itself in experiments that involve only orthodox "true" forces. We take this "equivalence of response" for granted in everyday life. If spring A stretches twice as much as spring B when loaded at rest, who doubts that A will extend twice as far as B when spun on a turntable that duplicates the load through centripetal effects? If an arbitrary fluid is placed in Newton's famous rotating bucket, its response can be hard to predict—a paint piles up much differently than water. But we can be certain that, however it reacts, the fluid will assume the same shape if it is pulled at by suitable forces in a stationary setting. It is striking that we accept this inertial/regular force equivalency readily, when the auguring of other aspects of an unknown material's response seems risky. Newton claimed that the differences between a stationary and a rotating bucket show that space-time must incorporate inertial structure; in reading this, one has the uneasy feeling that some salient fact about the situation has been slighted.

The missing consideration is "material frame indifference"—the fact that the *character* of a material's response to outside influence is the same in both kinds of bucket.

Accordingly, although a material's current behavior may reflect its past in very intricate ways (e.g., it becomes fatigued, displays set, loses viscosity, etc.), it will ignore the purely rotational aspects of that history and attend only to the schedule of stretching and shear. Inertial influences are noticed only insofar as they contribute to this distortional history. And we rely upon this fact when we decide that it is patently stupid to test paper clips for a differential response on turntables. Our certitude in these matters seems almost *a priori* in character. Of course, we recognize that this appearance is mistaken; paper clips may surprise us some day.[23] Its proper source can only be the tacit wisdom derived from extensive experience with the materials of our world (many years of paper clip manipulations in childhood). But suppose one believes, as Leibniz did, that the nature of space and time is fixed *a priori*. As we will soon detail, "material frame indifference" falls out of the structure of $\mathcal{L}$ as a free gift. "Frame indifference's" strong intuitive appeal can be taken as strong evidence that the true *a priori* space-time is Leibnizian, rather than Newtonian or Galilean. Accordingly, the sources of Leibniz's space-time convictions may trace partially to the tacit knowledge of genuine experimental fact.

Let us look at a loose argument on behalf of "material frame indifference" that will help us see how the principle should be precisely formulated. Intuitively, all materials can be forced to maintain a *steady state* by a suitable choice of outside influences. Strong pressure will hold a putty still; a suitable schedule of continual shearing will set a liquid into steady flow. Their steady response indicates that the materials have found internal means of adjusting that are able to match the outside influences. "Internal adjustment" suggests the internal stresses are generated by a disparity between the material's current condition and a "memory" of a less constrained state. But a mere translation or rotation of the material cannot be expected to kick any additional internal response into play; such movements affect the material only insofar as inertial effects alter the *external* constraints that have been placed upon the material.

These considerations suggest that our Leibniz structure Σ of "material memories" be used to formulate "material frame indifference" precisely.[24] Recall the definition of "objective" tensor, based upon the reference configurations σ in Σ. "Material frame indifference" is the demand that only objective quantities can appear in the constitutive equations of a material. We see that this requirement captures our intuitive sense that a material only remembers "internal" adjustments from ideal states.

XII

With these ideas in hand, which are completely absent in point-particle mechanics, let us ask once again about the proper space-time for continuum mechanics. When we left the issue, the subject had been placed in $\mathcal{G}$. We had asked whether "reference configuration" could be defined in terms of Γ and its confederate "inertial structure." Recall that an object that is "tensorial under all changes of inertial frame" is called "absolute." However, "absoluteness" is too weak as a condition to satisfy "material frame indifference"—we demand "objectivity," i.e., "tensorial under all changes within the Leibnizian group Σ." After all, the rotational condition of a turntable is "absolute," but not "objective." So our suggested treatment of "referential configuration" will not work. If we remain in $\mathcal{G}$, we must *add* an additional axiom the principle that, although absolute quantities exist (locally) within the space-time, no material is smart enough to "see" them—their constitutive behavior responds only to Σ-objective quantities.

Recall, from section II, our reasons for abandoning Newton's $\mathcal{N}$ in favor of $\mathcal{G}$: $\mathcal{N}$ contains an absolute notion of "place," but no material is "smart enough to see it," in the sense that no appeal to "place" is permitted in physical law. As long as physics is set in $\mathcal{N}$, we must postulate that force laws are always "place-blind." That is, Newtonians must assume an axiom of "Galilean frame indifference" to capture what they know about the general behavior of force laws. But if the same physics is set in $\mathcal{G}$, we get "Galilean frame indifference" for free; since 3-**velocity** does not exist locally within this space-time, no axiom needs to claim that it is ignored. If I tell you that UFO's do not exist, I do not have to add that no one has actually seen one.

Walter Noll's idea is that a special axiom of "material frame indifference" can be avoided by never installing a Γ within the space-time. Instead, we keep the space-time Leibnizian or, conceivably, as $\mathcal{L} + \Sigma$, which is weaker than $\mathcal{G}$ (and still subject to the twist argument). The $\mathcal{L} + \Sigma$ choice seems less natural, however, because Σ seems to formulate the "memory" a material carries with it as it flows, rather than a feature of the background space-time. I might add that, although I have generally eschewed genuine Leibniz interpretation in this essay, it is striking that Leibniz constantly credits matter with "personality traits" such as memory and appetite ($\approx$ hunger for ideal state). What we have learned is that there is a definable level of "spatial commitment," viz., Σ, that ratifies many of these metaphors. Σ, moreover, represents a higher degree of commitment than the Φ utilized by Descartes, yet falling short of Γ (let alone Newton's "same place" filaments). Leibniz's "relational view of space" reads as if it is more tolerant than Descartes's; perhaps these remarks about Φ, Σ and Γ explain why.

In any event, the Charybdis of setting continuum mechanics in $\mathcal{G}$ is that a special axiom is needed to ensure the universal blindness of materials. The Scylla affecting Noll's program is that the *left* side of Cauchy's law, $\rho(D\mathbf{v}/Dt - \mathbf{b})$, seems to see inertial structure. To complete Noll's program, we must make sense of $D\mathbf{v}/Dt$ without inserting a richer structure in the space-time. The basic trick is to sneak in an equivalent of Γ, but keep it out of $\mathcal{L}$. We begin by regarding the two terms containing the coupling constant ρ, $\rho D\mathbf{v}/Dt$ and $\rho\mathbf{b}$, as the "same kind of animal." This is quite traditional; even Newton regarded $\rho D\mathbf{v}/Dt$ as an "inertial force." In fact, let's abbreviate $D\mathbf{v}/Dt$ simply as $\mathbf{i}$. Recall, from section IX, the need for a "cut" and the subsequent division of forces into "external" and "internal." Let us view **i** as a "force" that reaches from outside the "cut" in the same manner as any external **b** such as gravitation. We have learned that the material inside the "cut" only feels the sum **e** = **i** − **b** and, indeed, **e** itself counts as an "objective" vector under Σ. It is only **e**'s decomposition into **i** and **b** components that is not "objective." Now look at the "cuts" that surround our reference configurations. The **e** upon these boundaries are able to hold the reference configuration steady over time. That is, some mixture of **b** and **i** "forces" holds the material fixed. We make the following postulate: if σ is an acceptable reference configuration with fields **b** and **i** assigned to its "cut," there will be another σ′ in Σ that carries **b** ′ = **b** − **i** and **i** ′ = **0** upon its cut. We dub the subclass of reference configurations with **i** = **0**, "inertial." Relying upon this special group, the **e** around the *real* material can be invariantly dissected into **i** and **b** parts. Under any choice of inertial frame, **i** winds up identical to the frame-dependent quantity Dv/Dt. By this route, we reach the full law of motion containing the differential operator that the Γ-argument is concerned with, but the trick has been turned without installing our Γ-surrogate in the space-time itself. Where is it then? Actually, nowhere; it hovers on the outside of the "cut," generated by a "memory" of the "cut" conditions needed to hold the material steady. Like all external body forces, its effects are felt at definite places within the material, but its places of origin are nebulous, reflecting the "averaged" character common to all outside-the-"cut" influences. In any case, the decomposition of **e** into **i** and **b** cannot be seen locally by any material point, allowing their constituitive behaviors to respond to only Σ-certified quantities. Thus we evade a special postulate of "material frame indifference."

We might note that this "forcifizing" of inertia is different than Mach's famous suggestion.

XIII

The Scylla of Noll's approach, of course, is that the decomposition of **e** is treated in such an odd "out–of–space-time" way. Which approach supplies the proper space-time for continuum physics? Temperamentally, I prefer the Galilean formulation, but the arguments in favor of each seem to stack up rather symmetrically. There are philosophers who will see in our two approaches distinct, yet "observationally equivalent," theories—a viewpoint to which I am grumpily unsympathetic. Clearly, I think, we have only one theory here; the discrepancy between our two settings lies only in which component structure has been *honored* with the title "space-time." The features constructable from the Galilean Γ (e.g., geodesics) are certainly present in the Leibnizian setup, but they are labeled as "material"-dependent objects there. The same lines are singled out whether they are called "geodesics" or not. The distinct demands of the two sides of Cauchy's law show that continuum mechanics contains two substructures that can be regarded as equally plausible pretenders to the throne of "space-time" (perhaps $\mathcal{L} + \Sigma$ has its claims as well). In my opinion, these multiple space-times coexist in continuum mechanics. No *a priori* rule dictates that a physics must carry a single "geometry," especially when different groupings of its laws "see" different underlying structures. Greater legal agreement is required before a particular space-time can claim dominion over the others.

Neither Newton nor Leibniz was prepared to be as tolerantly accepting of coexistent space-times as I. The two sides of Cauchy's law help explain why they saw space-time so differently. Their differing viewpoints can then be correlated with the patterns of everyday phenomena that each selected as paradigmatic. Although the second book of the *Principia* deals, in a somewhat haphazard way, with terrestrial continua, Newton's main emphasis is clearly upon celestial mechanics. In that setting, the distribution of stress within a body is of little concern—planets get treated, in Truesdell's phrase, as "singularities of inertia" (i.e., the right side of Cauchy's equation is set to 0, allowing the dynamics of a planetary system to be treated as a balance between $\rho D\mathbf{v}/Dt$ and $\rho\mathbf{b}$). A more terrestrially oriented thinker, as Leibniz surely was, looks first to the behavior of materials under differing schedules of static loading. In such circumstances, the right side of Cauchy's equation usually dominates the left. Since it is this right hand side that requires most of the technical trickiness, it is little wonder that Leibniz's articulated physics was always hazy and incomplete. His philosophy of matter seems generally informed by an awareness that fairly complicated distinctions will be needed if the behavior of continuous materials is to be properly captured (although his own suggestions seem to head in the direction of a complex philosophy of infinitesimals, rather than paralleling the apparatus used here). We miss all of this if we pretend, in the manner criticized in section I, that "classical mechanics" is captured entirely in point particle mechanics.

XIV

The arguments that Leibniz supplies for his view of space-time in, say, *The Leibniz-Clarke Correspondence* generally carry a "verificationist" character that some still find inspirational. As we all know from political disputes, the grounds of a conviction are not always the same as the arguments advanced upon its behalf. A lousy argument will be dearly cherished if it seems the only way to capture a deeper concern that cannot be otherwise articulated. I see "material frame indifference" in this light: it constitutes the objective ground that supports an unwillingness to capitulate to a wholly Newtonian picture of space-time structure. I have claimed that "material frame indifference," or some facsimile thereof, seems to be deeply internalized within the "commonsense physics" of everyday life. If so, it rationalizes our uneasy sense that "there's something right in what Leibniz says." Paraphrasing Coleridge, then, "the long duration" of space-time verificationism gives "proof of an adaptation in it to some portion or other of the human mind."

As I announced at the outset, this suggestion in no way counts as an "interpretation" of Leibniz. I have provided something more akin to a psycho-analytic explanation of belief fixation, except that the beliefs we have studied are fixed by unconscious memories of paper clip wavings, rather than Oedipal urges.

It is interesting that in most contemporary textbook presentations, "material frame indifference" is presented as if it were an innocent remark about observation. Thus, Eringen's excellent text:

> The physical properties of materials are not dependent on the coordinate frame selected. It is intuitively clear that whether the observer is at rest or in motion, the material properties he observes should be the same. If this viewpoint is accepted, then the measurements made in one frame of reference are sufficient to determine the material properties in all other frames that are in rigid motion with respect to one another.
>
> In the formulation of physical laws, it is desirable to employ, as far as possible, quantities that are independent of the motion of the observer. Such quantities are called *objective* or *material frame indifferent*.[25]

In this paper I have avoided this kind of language. Material frame indifference is not an expression of the trivial principle that paper clips do not care how we describe them. Indeed, once one recognizes that "continuum flow" and "point particle flow" do not track together, it is hardly obvious that material frame indifference will be ratified in the continuum limit of point particle mechanics (despite the fact that the latter puts stronger restrictions upon its force laws than frame indifference does).

XV

The patient reader would be gratified if, having "physicoanalyzed" verification-ism away, I might fold my tent and depart. However, the current consensus in philosophy of science, derived from Earman and Norton's hole argument, is that there is "something right" *of a verificationist stripe* in Leibniz's writings. Namely, space-time points, as orthodoxly conceived, are too structureless to do proper work in physics, as exemplified by the way they foul up questions of "determinism." As noted earlier, Earman thinks that we must abstract from their "Leibniz equivalence" if we ever hope to reach a setting in which a universe can be properly run. Earman sees Leibniz's verificationist strands as anticipating the concerns captured in the hole argument. I believe that there is absolutely nothing wrong with the usual conception of "space-time" and "space-time point." The hole worries are generated by a complicated variety of factors resting primarily upon: (i) a defective approach to the notion of "determinism"; (ii) a misreading of the conventions of mathematical specification. I will concentrate largely upon (i), because its resolution falls naturally out of our prior discussion. Since the resolution is the same for the twist argument, we shall discuss that as well.

My treatment of (ii) will be less complete, since it hinges on matters properly belonging to philosophy of mathematics.

One of the diabolical features of the hole argument is that it internally offers up a phoney card as a key to its solution. The twist and hole arguments both operate with a "two-model" approach to determinism: a theory is deterministic if there is no pair of models M and M' that agree on all facts up to some appropriately selected data slice t_0, but disagree beyond t_0. Yet when we look at the "new model" that, say, the twist arguments builds, we are inclined to complain: "But $\phi(M)$ is the same structure as M". The impression that $\phi(M)$ should be different from M only arises, we say, because speaking of "twists" conjures up an irrelevant picture where $\mathcal{L}$ has been embedded in a richer Euclidean frame and literally twisted. It is only this *extrinsic* embedding that supplies a rationale for differentiating M and $\phi(M)$, just as it is the "extrinsic" embeddings within the integers that distinguishes the structure $1,3,5,7,\ldots$ from $2,4,6,8,\ldots$. When a physical theory is assigned a space-time model, that model should be treated *intrinsically*, in terms of its own internal features. Within $\mathcal{L}$'s viewpoint, M and $\phi(M)$ are identical structures.

All of these remarks are perfectly unexceptionable. Here is where the phoney card is forced: judging from the orthodox textbook definition of "manifold", M and $\phi(M)$ count as nominally different, albeit isomorphic, structures. To actually identify them (and thus resolve the hole problem), Earman thinks that the textbook definition should be corrected by moving to equivalence classes. The problem then becomes one of gaining a better handle on the structures that stand behind this equivalence class account.

The strong temptation to move in these directions is an example of a misunderstanding that I will call "Dedekind overkill" later. However, as the column example of section VII shows, a theory whose competing models are all isomorphic need not be deterministic. Why do we judge the twist and the column cases so differently with respect to determinism? We are inclined to say something like, "In the second case, we begin with the same basic column, but find that it doesn't contain a mechanism sufficient to force a unique reaction to the weight. But in the twist argument, we switch to a completely disconnected second model whose standards of comparison with the original structure are quite hazy. This lack of connection between M and $\phi(M)$ generates all of the pseudo-worries about model identification that the twist and hole arguments exploit."[26] Taken literally, however, this intuitive reaction seems incoherent; it seems to claim, "Inside *the* model for the column, there are many ways it can sag." But how can a single model "contain" multiple models?

Our intuitions can be redeemed if "determinism" is construed as a matter of "the unique completion of a partial model" (as we will see, this treatment does not arise simply as a way out of the hole problem, but as a consequence of the basic meanings of differential equation). A simple prototype for "partial model completion" is Beth's notion of "implicit definability" with respect to, say, the propositional calculus. We first divide the propositional variables into two classes: the *defining* variables ($a, b, c, \ldots$) and the variables *to be defined*

$(p, q, r, \ldots)$. Let T be a collection of sentences asserting relations between the independent and dependent variables. Assign a model to $a, b, c, \ldots$ (i.e., supply them with truth-values). Consider all possible ways of extending that model to the $p, q, r, \ldots$ that make the T sentences true. If only one possible extension exists, Beth says that the defining variables implicitly define the variables to be defined. This condition is a primitive analog of "existence and uniqueness" for differential equations.

Let T now be a concrete set of differential equations whose "determinism" is to be investigated. Such equations partition their variables into "independent" and "dependent" groups. For example, in the wave equation $(\delta^2 h / \delta t^2 = -k(\delta^2 h / \delta x^2 + \delta^2 h / \delta y^2)$ for a membrane, the independent variables are x and y (spatial) and t (time); the dependent variable is h (height above the x axis). With more terms and equations, temperature, density, etc. could be added as further dependent variables. Insofar as the relevant physics is concerned, we enjoy a lot of freedom to select other choices of independent or dependent variables, e.g., we might use polar quantities instead of the Cartesian x and y for our membrane. Except in rare cases, dependent variables cannot be exchanged for independent variables.

The problem of determinism for T should be seen as roughly that of filling in an array like table 1.

Table 1

	x_0	x_1	x_2	x_3	x_4
t_0	1	0	$-1/2$	-1	$-1/2$
t_1	?	?	?	?	?
t_2	?	?	?	?	?

Here we might have a string governed by the one-dimensional wave equation and the numerical values along the t_0 row capture the string's "initial condition" on the data slice at t_0. When we ask about "determinism" (= existence and uniqueness) in this kind of case, we are wondering whether our equations are powerful enough to allow us to fill in the *dependent* variable question marks on the chart (with differential equations, all of our chart boxes must be "infinitesimally separated," a factor we will ignore for the time being). If the dependent variable squares cannot be completed at all or can be completed in a variety of ways consistent with the equations and the data line, determinism fails. Note how similar this treatment is to that of implicit definability—the independent variables form the "frame" that underlies our intuition in the column case that "we want to stay in the same model and see what is forced by the underlying mechanism." But note—and this is crucial to what follows—that the differential equations do not pretend to advance the *independent* variables; their job is only to push the dependent variables forward on rails that have already been laid down by the independent variables.

Here we have introduced the least complicated case of determinism—a simple "initial value" (or "Cauchy") problem admitting globally specifiable independent variables. All cases of "determinism" among differential equations follow the basic "completion of a partial model" format displayed here, but other cases introduce some important subtleties that complicate the setup. The equations of general relativity, unfortunately, participate in all these complications, which is why "hole"-type worries about "determinism" first arose there.

First complication: Suppose we are dealing with a so-called "boundary value problem," say, a membrane fixed at its edges where one wonders whether some equilibrium configuration is uniquely determined. In the relevant equation (the wave equation minus the time term), the independent variables are x and y (spatial) and the dependent variable is h (height). We now wonder whether the *interior* of table 2 can be filled in uniquely.

Table 2

	x_0	x_1	x_2	x_3	x_4
y_0	1	0	$-1/2$	-1	$-1/2$
y_1	$1/2$	?	?	?	0
y_2	$1/3$	?	?	?	$1/2$
y_3	0	$1/2$	1	1	1

Here we do not expect to find a solution as before, "marching forward" from the data line. Rather, one works out a solution as in a crossword puzzle and hopes that it is the only one that works. Uniqueness is usually harder to establish in cases like this. The typical situation in physics relying upon partial differential equations (including continuum mechanics and general relativity) is that one is faced with a *mixed* initial condition–boundary value problem.

Second complication: in both of these cases, we know choices of acceptable independent variables whose global behavior can be specified in advance. At the base of the membrane problem, the Cartesian coordinates x and y are used as independent variables. Their global behavior is completely transparent—we know exactly how they will lay down in advance. But other quantities can be adopted as independent variables quantities whose long-range behavior is *not* so easily foreseeable. Such choices are often advantageous. Suppose, in the membrane problem, we *adapt* our independent variables to the guess we have just made of the membrane's shape. We can let our new independent variable u point continually in the direction where the membrane seems to be changing height most rapidly. A complementary independent coordinate v can be made to run perpendicular to u. Based upon the differential equations (which need to be recast in terms of u and v), calculate[27] the height of the membrane one (infinitesimal) u-step ahead. That is, we run u and v one step ahead of where we are, and then use the differential equation to catch up with the dependent variable h. After this is done, reappraise the current climb of the membrane and push u and v a second step forward in this new direction. Catch up with h, reorient u and v and continue until the possible extent of the membrane is exhausted (i.e., when the projection of the u/v grid fills the appropriate region of the x/y plane below). Viewed as an approximation scheme, such "adapted" ($=$ u and v reflect the *prior* condition of h) variables often behave better than non-adapted ones, such as our Cartesian x and y which pay no attention to h whatsoever.

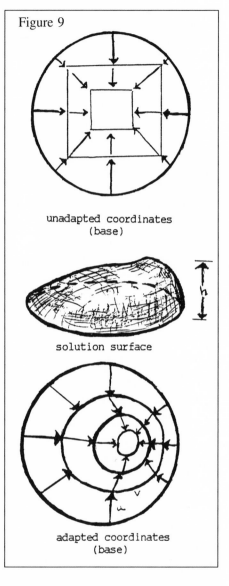

Figure 9

unadapted coordinates
(base)

solution surface

adapted coordinates
(base)

The salient fact about this technique is that we can no longer foresee *globally* how the u/v grid will form; its twists and turns reflect the h-values that have been previously deposited upon the grid. The independent variables stay "independent" of the dependent variables only in the sense that at each stage they run one (infinitesimal) step ahead of them. In a pure initial value problem of this type, we can accurately claim that, as each stage of marching the data forward is completed, the manner in which the adapted independent variables run forward to the next time slice is completely fixed. Moreover, the differential equations themselves play no role in determining how these run ahead; they merely help catch up with the dependent variables. This unequal rate of marching forward gives the right general picture of the asymmetrical manner in which independent and dependent variables interrelate. However, if boundary conditions are also involved, the nature of the interrelationship must be weakened to that described in our first complication. Use of adapted variables forces our formulation of determinism to be more delicate, because if our membrane problem is not determinate, the u/v grids form differently on the different structures tolerated. What we must say is this: pick a rigid *recipe* for building a u/v grid. The system is deterministic only if an identical grid with an identical completion of dependent variables is always built from the original data. Note that the recipe for the independent variables must be held constant; the fact that two grids fill in differently when different schemes for independent variables are employed shows nothing about determinism.

Third complication: In the last case, we relied on our membrane's embedding within Euclidean 3-space to decide when our u/v grid is complete. In this sense, we have not totally avoided knowledge of global behavior of some set of variables. However, we can set up a problem quite similar to our membrane case where these Euclidean underpinnings are removed. Suppose we look at the membrane *intrinsically*, from the point-of-view of a bug living on its two-dimensional surface. Instead of utilizing membrane height as a dependent variable, our bug uses intrinsic curvature, metric, and stress (as in the classical "theory of shells"). Our bug no longer has the option of using globally foreseeable independent coordinates; he *must* use adapted ones of u/v type. Furthermore, he must find some intrinsic criterion for deciding when his u/v grid closes—say, by correlating possible stretching of the membrane with the curvature it displays. If he can do this, he has assembled enough of a frame in which the question of determinism of the dependent quantities can be raised.

Fourth complication: Unfortunately, many surfaces do not admit truly global independent variables; every recipe for a coordinate scheme goes bad somewhere on the surface. Our bug may need a criterion for patching patches of independent variables together so that his world can be completely described.

It is obvious that the complications general relativity presents for determinism are essentially those of our bug. In the absence of global knowledge of a universe's geometry, one is forced to find schemes for constructing patches of adapted independent coordinates. In simpler branches of physics, one expects to learn a suitable selection of independent variables directly from the "equations." In general relativity, the "equations" are usually written in so-called "covariant forms" (i.e., in terms of relationships among tensors). Covariant equations, strictly speaking, are not differential equations; they are merely *recipes* for writing differential equations once a suitable set of independent variables has been selected. This project is left completely to the user; the covariant equations tell how to advance the dependent variables once independent variables have been found. The latter problem is left to a separate analysis of what kind of independent variables will be supported, in their infinitesimal stretches, within any model of the theory.

The complications that our four problems raise for the proper formulation of determinism became widely appreciated only after the birth of general relativity. The fact that earlier theories generally accepted globally foreseeable coordinates ensured that no one worried too much about their essential character. Usually a crude method of variable counting told a physicist whether her equations were likely to be deterministic or not. In fact, Einstein seems to have fallen upon the worries that generated his version of the "hole problem" by relying upon such techniques. But the covariant presentation of any theory will come up short on traditional equation counting, because covariant presentations purposely run away from the facts needed to set up a coordinate scheme. Without the latter, one does not have true differential equations. But until such equations are on the table, the problem of determinism cannot be properly formulated.

Throughout all of our sundry complications, determinism remains a partial model extension problem, albeit of an intricate sort. Once determinism is correctly set up, it becomes easy to see what goes wrong in the twist and hole arguments. In the final analysis, the resolution traces to the same "stupid" reply we have given to the Γ-argument for $\mathcal{L}$: a Γ-surrogate does not have to be entirely "geometrical" in character. Similarly, *the independent variables that underwrite a theory's differential equations neither have to be "purely geometrical" nor "globally specifiable" in character*, although, within the simpler theories of physics, the standard choices of independent variables happen to possess both of these traits. Consider the twist argument first. When continuum mechanics is set in $\mathcal{L}$, the Γ-surrogate is viewed as hooked to the "cut" rather than the space-time. But we still rely upon this Γ when we eventually get around to writing down definite ($=$ expressed in a coordinate system) laws of motion. In particular, Γ helps form a working grid of independent variables. Here the coordinates do not have to be adapted; indeed, we can imitate any scheme that can be set up in $\mathcal{G}$. The only difference is that we cannot claim that any of $\mathcal{L}$'s independent variables are wholly "geometrical" in character (because they rely upon all of the "cut"/Σ apparatus that backs up Noll's Γ-surrogate). But so what? The use of such "geometrically impure" independent variables does not keep the physics from working. After all, the u/v coordinates in our bug problem are not "purely geometrical" either.

The twist argument appeals to symmetries of $\mathcal{L}$ to claim that, under a twist map, $\mathcal{L}$ winds up looking exactly the same as before. True, but this claim would be relevant to determinism *only if we could assume that the independent variables of our theory's equations are laid down in a purely geometrical manner*. We have seen that determinism takes the form: the universes tolerated by this theory have enough structure to lay down fully specified independent variables over which the theory's differential equations can infinitesimally advance the dependent variables in a unique manner. All that the twist argument really shows is that we cannot expect to found suitable independent variables solely upon $\mathcal{L}$'s structure, but the Γ argument taught us this moral in a more direct way. Arguments from purely geometrical symmetries are irrelevant to theories whose natural choices of independent variables rely upon other than purely geometrical considerations.

The appeals to space-time symmetry employed in the "hole argument" fail for a closely related reason. Here the class of symmetries cited are not specific to the geometry of the particular space-time utilized, but represent generic features of any space-time accepted by the theory. The hole argument shows merely that generic space-time geometry does not supply enough strength to support a working set of independent variables. Once again this fact should come as no surprise, because the standard choices of independent variables used in the existence proofs for relativity (e.g., Gaussian coordinates) are always adapted to structures existing in the particular space-time considered (specifically, they reflect the geometry of the data surface from which they depart). But it is possible, though difficult, to show that most data slices allow one to build grids of suitable adapted coordinates that are filled in uniquely by the equations.

A final comment: in this discussion, I have engaged freely in speaking of "infinitesimally close slices" and the like. This, I think, is the best way to see what is intuitively going on with determinism. To make it rigorous, one retreats to "the limits of finite difference approximations" and the like utilized in the textbooks. Roughly speaking, a family of discrete structures is built within a richer space that, as slice differences go to 0, tends towards a limit that can be read, when its richer trappings are thrown away, as an intrinsic description of the unique space-time model sought.

XVI

It would be a mistake to leave the hole argument without a brief remark on what I called "its internal phoney card." Recall, from section III, the dilemma that mathematics faces in trying to present novel structures "intrinsically," yet preserving enough links to previously tolerated structures that existence of the new objects can be proved. Commonly, new structures are built by "killing" and modifying familiar ones. By "killing," I mean dropping unwanted features of the richer frames, as $\mathcal{G}$ is built by "killing" some of the metric in Euclidean E^4. But most ways of forming new structures modify the old in more extreme ways. Following Dedekind, the rational numbers can be built from the integers by forming equivalence classes of integer pairs (a, b) that cross-multiply correctly (i.e., $(a, b) \sim (c, d)$ if $ad = cb$). Note that rationals wind up structured differently from the integers. A certain amount of "structure killing" is also tacit in this technique—the rationals are supposed to carry only the structural properties they possess at the equivalence class level (one should not think that $1/4$ bears the property "has infinitely many members"). Unfortunately, the temptation sometimes arises to use equivalence classes to kill structure that has already been killed—a form of mathematical sadism I will dub "Dedekind overkill." For example, suppose one wishes to isolate the structure common to $1,3,5,7,\ldots$ and $2,4,6,8,\ldots$ by placing them in a common equivalence class ω. As soon as ω has been built, one realizes that the structure $\omega,1,2,3,\ldots$ deserves to be in ω as well! Obviously, completing the desired equivalence class is impossible, a basic lesson from the early days of logicism. One has to swallow the fact that, no matter how generously an equivalence class structure is constructed, it is likely to carry features that distinguish it extrinsically from structures that, by all rights, embody the intrinsic core that the construction is designed to isolate. Moral: the role of equivalence class constructions needs to be construed more modestly; in pulling a novel structure out of Dedekind-style construction, one obeys basic *conventions* that determine which features belong and which do not belong to the structure produced (e.g., which features should be considered "killed"). If the sequence 1,3,5,7, . . . is presented intrinsically—that is, not regarded as a subsequence of the integers—then, by that very stipulation, its unwanted "extrinsic" trappings are suppressed. An attempt to suppress the extrinsic trappings a second time over by forming equivalence classes is Dedekind overkill. What good does it do?—the idiot who does not catch our intrinsic drift is likely to focus upon the unwanted extrinsic aspects of the Dedekind construction as well. In general, equivalence class constructions are useful only when intrinsic structure gets altered (as in integers → rationals); otherwise one is simply beating a dead structure.

Dedekind overkill, although a crime against mathematical convention, frequently pops up in many otherwise impeccable quarters. The temptation towards overkill commonly arises in cases like this. As noted at the end of last section, existence proofs for space-time theories often construct their models from "finite difference approximations" within what amounts to a Euclidean space. After the building process is completed, the Euclidean scaffolding should be discarded in favor of the features that are germane to the space-time in which our final model is set. From the point-of-view of the richer space, however, there may be a wide variety of ways to build a product that, when stripped of its richer space trappings, supplies the intrinsic structure desired. Suppose our students are confused by the need to ignore the richer space's point-of-view. Properly speaking, they require a mini-lecture on the philosophy of mathematical structure. Being fonder of technique than philosophy, many texts finesse their pedagogical problem by blithely declaring, "We now derive the structure we seek by forming the equivalence class of workable constructions." Because of Dedekind overkill, our hapless students have been encouraged to waste their time puzzling out the mysterious structure that underlies the equivalence classes. In fact, Hawking and Ellis,[28] in their existence proof for general relativity, use equivalence classes in very much this way. Such passages encourage Earman in his quixotic search for a "manifold" substitute.

Indeed, the concept of "differentiable manifold" happens to particularly invite Dedekind overkill. Placing a "differentiable structure" on a topological manifold is simply a way of classifying spots on the manifold according to a rather abstract notion of "degree of smoothness" (roughly speaking, it settles whether the surface contains analogs of sharp mountain peaks). Specification of differentiable structure relies upon a variant convention (use of extrinsic charts) that often confuses. In particular, this convention allows *two* or more differentiable structures to be placed upon a common manifold, which are incompatible in the sense that a motion classified by one structure as smooth is regarded by the other as rough. One naturally expects that if only *one* of these "different" structures is set upon a manifold, the results will be different than if the other structure had been selected. This is usually not the case; the resulting manifolds can be completely isomorphic. If we attend to the intended meaning of "differentiable structure" as a grading of smoothness, we see that our two "one-structure" manifolds are clearly the same intrinsic object; we have just misunderstood the conventions of "chart" specification. If our students are confused, we should say, "Think of the features we want in a differentiable manifold," not "The *true* differentiable manifold is the equivalence class of all isomorphic manifolds." Unfortunately, well-regarded texts replace common sense by formal construction at this particular pedagogical juncture.

So the "phoney card" of the hole argument tempts us, through imitation of our scientific betters, to commit Dedekind overkill. Once we have fallen for the trick, we find ourselves in a terrain where the unexceptionable notion of "differentiable manifold" looks strange and in want of improvement.

XVII

To summarize a long paper brutally: Leibniz's "relationalism" seems attractive because it contains a nugget of physical truth, not because its verificationist supports are sound.[29]

NOTES

1. A good presentation is G. Galavotti, *The Elements of Mechanics*, (New York, 1976).

2. Throughout this paper, I shall use John Earman, *World Enough and Space-Time* (Cambridge, Mass., 1989) as a canonical reference. John Norton's variant reading of the argument is found in "The Hole Argument" in *PSA 1988*, edited by A. Fine and J. Leplin, Philosophy of Science Association (East Lansing, Mich., 1989).

3. A good start for philosophers is "Suppsian Stews" in *An Idiot's Fugitive Essays in Science* (New York, 1982). The reader should be warned that Truesdell's prose is sometimes, to put it gently, acerbic.

4. In *The Unpublished Scientific Papers of Isaac Newton*, edited by A. and M. Hall (Cambridge, 1962).

5. I will rely upon Earman's book, *op. cit.,* for more formal descriptions of the space-times I shall discuss. With respect to continuum mechanics, I have relied primarily upon C. Truesdell, *A First Course in Rational Mechanics* (New York, 1977) (set in $\mathcal{L}$) and J. Marsden and T. Hughes, *Mathematical Foundations of Elasticity* (Englewood Cliffs, 1983) (set in a generalization of $\mathcal{G}$ with an eye towards relativity).

6. The historical Newton probably did not read his laws in the "ordinary differential equation" manner intended, but the anachronistic literature invariably treats his laws this way.

7. $\mathcal{N}$'s temporal and spatial metrics can be added to supply the Euclidean 4-metric on $\mathcal{N}$. Or they can be subtracted to give the Minkowski's metric—the important difference between $\mathcal{N}$ and special relativity's space-time lies in the slicings, not the metrics.

8. In my account, I leave the distinction between the manifold and its tangent bundle a bit blurred. This identification is legitimate in the particular kinds of manifolds we wish to discuss.

9. I.e., a local treatment allows space-times to grow that may be cylindrical in the large, etc. We shall ignore these issues. Also a "connection" also settles the behavior of the "torsion" on the manifold. In the present context, this is usually set to 0. However, there are an interesting class of approaches to continuum physics where the torsion quantifies the residual strain trapped in the material through dislocations. The spatial slices are "flat," but nonetheless not fully Euclidean.

10. This is not the notion of "absolute object" employed by J. Anderson and Michael Friedman.

11. In continuum mechanics, important constitutive principles are often integro-differential in form and a special axiom declares that the "non-locality" implied by the integral is confined to a neighborhood of a material particle's past history.

12. Howard Stein, "Some Pre-History of General Relativity" in *Minnesota Studies in the Philosophy of Science*, vol. 8, edited by J. Earman, C. Glymour and J. Stachel (Minneapolis, 1977).

13. For example, Earman feels that the theories of J. B. Barbour, of which he apparently approves, sit in $\mathcal{L}$, without comment upon the twist problem. Actually, a volume element has presumably been added to $\mathcal{L}$.

14. Earman, *World Enough*, 171.

15. For simplicity, I cite the non-uniqueness of a boundary value problem rather than an initial value problem. I am also ignoring the unstable response to the post-critical weight, which is not isomorphic to the others. In any event, "Leibniz equivalence" clearly invites us to count solutions wrongly.

16. Preservation of these features entails that the material is not allowed to combine, rip, or even slip over time. These restrictions seem far too strong but, unless they are maintained, we cannot expect our material's behavior to be governed by differential equations alone. How the setup is generalized will not be pursued here.

17. More exactly, Poisson's equation defines a distribution of a scalar ϕ. We use the metric to push the $d\phi$ 1-form into a vector.

18. For a good introduction to this plan, cf., L. E. Malvern, *Introduction to the Mechanics of a Continuous Medium* (Englewood Cliffs, 1969).

19. A material need not have actually assumed its "reference configuration" in the past; in this sense, it may represent as "idealized memory" of what is possible.

20. I mean "Γ supplemented with a unique slicing." Once "unique slicing" is given up, as in relativity, our degrees of "spatial commitment" become harder to define. The fact that most laws of continuum mechanics rely upon unique slicings and long-term "memory" makes it hard to find a simple recipe for transferring classical laws into the relativistic setting.

21. Remember we are only working in smooth continuum physics, which cannot accommodate boundary discontinuities without supplementation. Except for a few artificially smoothed cosmological settings, the "cuts" of practical continuum mechanics are very unambitiously placed. The issues raised here, which are necessary if a comparison with the Newtonian point-of-view is to be made, are lightly touched upon in even the most rigorous texts (which often ignore the internal aspects of gravitation, presumably on the grounds that, in practical applications, such forces will be negligible). I discuss "cuts" further in "Law Along the Frontier" in *PSA 1990*, Vol. 2, edited by M. Forbes and A. Fine, Philosophy of Science Association (East Lansing, Mich., 1991).

22. Balance of angular momentum is also a basic law in this framework; it relies upon Γ. I should add that continuum mechanics does not reach happy closure unless thermodynamic ideas are introduced into the framework. Such matters do not affect the issues discussed here.

23. There is a controversy in the literature whether the doctrine fails in delicate cases.

24. For a history of its explicit formulations, see C. Truesdell and W. Noll, *The Non-linear Field Theories of Mechanics*, Vol 3 of *Handbuch der Physik*, edited by S. Flugge (New York, 1965), 45–47.

25. A. C. Eringen, *Mechanics of Continua* (Huntington, New York, 1980), 91.

26. In an interesting article, J. Butterfield attempts to tackle the hole problem by a direct assault on such counterfactuals. Cf., "Albert Einstein Meets David Lewis" in *PSA 1988*, edited by A. Fine and J. Leplin (East Lansing, Mich., 1988).

27. Since we are working in a boundary value problem, we are faced with our first

complication and we must begin with a guess, consistent with the strictures of the differential equation, of the height we seek. The solution offered is ratified only after the whole grid has been filled in consistently.

28. S. W. Hawking and B. Ellis, *The Large Scale Structure of Space-Time*, (Cambridge, 1973), cited by Earman, *World Enough*, 183.

29. I would like to thank Michael Friedman, Paul Teller, Ed Slowik, and Bob Batterman for useful discussions; I'm also grateful for John Earman's long litany of inspiring work.

MIDWEST STUDIES IN PHILOSOPHY, XVIII (1993)

The Transcendental
Character of Determinism

PATRICK SUPPES

DETERMINISM

In philosophical discussions of determinism I believe it is fair to say that there is not ordinarily a sharp separation of a process being deterministic and a process being predictable. Much of the philosophical talk about determinism proceeds as if it is understood that a deterministic process is necessarily predictable. Here is a typical quotation, taken from A. J. Ayer's essay "Freedom and Necessity:" "Nevertheless, it may be said, if the postulate of determinism is valid, then the future can be explained in terms of the past: and this means that if one knew enough about the past one would be able to predict the future." Without attempting quite general definitions for arbitrary systems, it will still be useful to examine in somewhat more detail how we ordinarily think about these two related but different concepts of determinism in prediction.

One formulation of determinism grows naturally out of the theory of differential equations. If a set of differential equations is given for a phenomenon, then we say that the phenomenon is deterministic if there is exactly one solution as a function of time of the differential equations satisfying the given initial and boundary conditions. There is no general conceptual reason, of course, for restricting ourselves to differential equations. We can easily say that a system of equations for discrete time intervals is deterministic in the same sense.

Sometimes in formulating what we mean by deterministic systems we put the emphasis rather differently. For example, in asserting the claim that classical mechanics is deterministic we may formulate the condition along the following lines. The history of an isolated system of particle mechanics is determined by the masses and forces acting on the particles, together with appropriate initial conditions. As is well known, these appropriate initial conditions give for some particular instant of time the position and velocity of each of the particles. When we approach determinism from this more general system standpoint there are difficulties, however, that are immediately encountered. For example, the proof of the deterministic character of the system will rest on additional assumptions. In the case of mechanics, for example, in ordinary formulations we will need to assume no collisions occur, or we will on the other hand need to introduce a number of special assumptions about the nature of the collisions. The kinds of differential equations or systems just discussed might be said to be ones that are completely deterministic in character.

I introduce the modifying adjective *completely* because systems that we do not think of as deterministic still have many deterministic features, and it has been an important part of their analysis to clarify the nature of these features. A simple sort of example avoiding any complexity is provided by elementary stochastic processes. We certainly think of such processes as not being deterministic because the sample path, as we ordinarily talk about the path of a particle in a stochastic process, is certainly not uniquely determined by the initial and boundary conditions together with a set of equations. It is just the feature of such processes that many different sample paths can result from exactly the same initial circumstances. On the other hand, many features of the process do have a deterministic character. The simplest example of all is coin flipping with a fair coin, which has a probability half of heads on each flip, with the flips being conducted in such a way that they are independent. If we flip a coin a thousand times then we have with equal possibility 2^{1000} sample paths. To give some idea of the magnitude of this number indicating the diversity of possible sample paths in such a simple process, consider the following: If we were to complete the equivalent of 2^{1000} flips, with each thousand flips taking one hour, then to have a sample equal to the number of possibilities (not that, of course, each possibility would show up in the sample) would itself take a time about 10^{285} times longer than the present universe has been in existence, which I am assuming roughly to be thirty billion years. So, at first thought, nothing could be further removed from determinism than such a simple stochastic process of coin flipping. On the other hand, a major thrust of theoretical analysis of such processes has been to establish what results hold with certainty, i.e., with probability one, which in our present discussion is equivalent to being deterministic. The various central limit theorems of probability theory are a good example of the kinds of results that can be obtained.

I am not for a moment suggesting that the whole theory is deterministic. What is important is that any standard stochastic theory, simple or complex, has important deterministic elements and a major effort of theory is to find them. Note that I have widened the use of *deterministic* to some extent in this discussion, for I am treating a probabilistic law that we know with certainty, or with probability one, as being in a general sense deterministic. What I want to claim is that such laws are just as deterministic as are laws that arise from theories that are fully or completely deterministic.

It is important also to separate the statistical data that we use to test the correctness of such a deterministic distribution law and the deterministic character of the law itself. In no sense do the finite sample data satisfy the law with probability one. The data are finite, and a statistical analysis is appropriate to determine the degree to which the law is satisfied. But there is nothing special about this because the underlying theory is stochastic. We test in the same general way completely deterministic theories, for the data here too are finite, subject to experimental errors of measurement and the inevitable finiteness of observation. I see, in principle, no strong difference between completely deterministic theories and stochastic theories that are only partially deterministic, from the standpoint of testing particular deterministic laws when they arise from either of the two kinds of general theories.

Of course, I am not attempting to amalgamate under a common heading completely deterministic and partially deterministic theories. The great thrust of classical physics for the development of completely deterministic theories has been and continues to be of importance. My point is rather to emphasize that we often have deterministic laws arising from partially deterministic theories.[1]

PREDICTION

As already mentioned, in much general talk, including much by philosophers, there is a confounding of determinism and prediction. If a theory is completely deterministic, it is often talked about as if phenomena governed by the theory are completely predictable. It is of course important and fundamental to modern science that many aspects of phenomena are predictable. It is also fundamental that in restricted cases of deterministic theories we think of phenomena as completely predictable, except for observational errors of initial conditions or other parameters. But the existence of predictable phenomena, which constitute some of the most important results of science, by no means guarantees anything like universal predictability. In fact, the conceptual separation of determinism and predictability is one of the fundamental themes of this article.

A good example to illustrate this separation is the famous three-body problem of classical mechanics—or more generally the N-body problem, which is easy to describe. It is that of determining the trajectories of three (or more) bodies interacting only under the force of gravity, and with given initial conditions. It is of course assumed that the three-body system is isolated, that is, no forces external to the system are affecting the motions of the three bodies. Already in the nineteenth century the problem of making long-term predictions of the three bodies was investigated in a deep way by Poincaré. It had been shown by Bruns in 1877 that essentially quantitative methods other than series expansions could not settle the three-body problem. There is no closed analytical solution in general. Poincaré then showed that the series expansions developed earlier in the work of Laplace, Lagrange, and others diverged rather than converged, as required in order to have a proper long-term solution. Methods of numerical approximation must be used, and when the solution, which may in principle be proved to exist, is sufficiently pathological in character, no extrapolations based on various methods of series expansion will give accurate predictions for extended periods of time. I shall return later to still other fundamental problems of prediction for simple three-body systems, problems which exist quite apart from any question of methods of numerical approximation in the construction of solutions.

It is of great importance to emphasize that the three-body problem is not at all special in the sense of the limited results one can obtain. For most mechanical problems that one would naturally write down, it is fair to say that we can neither solve the resulting system of differential equations nor give a sufficiently detailed description of the solutions to claim that we have complete predictability of behavior. It is a fact not sufficiently emphasized in foundational discussions of these matters that we actually can obtain explicit solutions or very accurate approximations for only a very small number of types of problems. To be more specific about these matters, it is important in mechanics to analyze problems in terms of a potential or a potential energy function. In a mechanical system with one degree of freedom it is always possible to introduce the potential energy of the solution of the differential equation. For a system with more than one degree of freedom, this is not always possible, and in general our analysis by current mathematical methods of potential systems with more than one degree of freedom is quite incomplete.

A skeptical reader may want to reply that of course I have not proved any of these results about the absence of predictability. I have been discussing general classes of models, for example, classical mechanical systems with a potential of two degrees or more of freedom. I shall shortly turn to more formal results that can be stated in a clear mathematical form about predictability. But it is important to emphasize as well the broad scientific experience with the use of mathematical methods in the sciences. Only for special classes of problems are solutions possible. Predictability of general system behavior is limited. Numerical computations can be carried out for particular values of parameters for a certain distance, but such methods will not provide a satisfactory general approach to predictability and will not even be practical for many kinds of applications. What I consider important to stress here is the chasm that separates many deterministic theories from being theories that are satisfactorily predictive in character.

Turing Machines

I turn now to some more specific results, first about Turing machines and then about cellular automata. A Turing machine, like any computing machine, can only really compute when it has a physical embodiment. It is therefore natural to think of Turing machines as finite physical processes or, in other terminology, the embodiment of a Turing machine is a physical process of computation. Note that I have said "embodiment of a Turing machine," because there is a good argument for treating Turing machines themselves as abstract objects, but it is not pertinent here to belabor the distinction. The important point is rather that physical processes that embody Turing machines provide excellent examples of physical processes that are not predictable when initial conditions are given.

The important objective of our analysis is to show that such simple discrete elementary mechanical devices as Turing machines already have behavior in general that is unpredictable. What I have in mind, of course, is the well-known unsolvability of the halting problem for Turing machines. Given a Turing machine in an arbitrary configuration with a finite length string of nonblank tape symbols, will the Turing machine eventually halt? The well-known result is that the answer is not decidable. In terminology useful here, the following important theorem is provable.

Theorem 1: *(Halting problem): There is no algorithm to determine if an arbitrary Turing machine in an arbitrary configuration will eventually halt.*

Put in other words, given the finite nonblank string and the configuration, which correspond to initial conditions of a mechanical system, the theorem says that there is no predictive algorithm as a function of these initial conditions such that it can be predicted whether or not the Turing machine will eventually halt. Notice, of course, that a particular configuration and a particular string of nonblank tape symbols may possibly be analyzed and a prediction made. But as in the case of Newton's celebrated solution of the two-body problem, what we are interested in is a general solution predicting the behavior of systems in closed form, that is, algorithmically, as a function of initial conditions. What is conceptually surprising is that such an elementary and simple physical machine as a Turing machine has such fundamentally unpredictable behavior. Note that there is no question of determinism here, or errors of measurement of continuous quantities, because the setup is essentially discrete. The behavior of the machine is deterministic.

It is worth mentioning in this connection that if we do not require that a Turing machine be deterministic but permit the machine to have several choices for the next move, then we have what is called a nondeterministic Turing machine. It is natural to ask if this weakening of the restrictions on the moves of the machine increases its power. The answer is negative, as formulated in the following theorem:

Theorem 2: *Any nondeterministic Turing machine can be simulated by a deterministic Turing machine.*

By the way, I am not suggesting that the notion of a nondeterministic automaton of a given class catches, in any sense, the philosophical sense of nondeterminism.

Cellular Automata

The formal definition of Turing machines is rather complicated. I want now to turn to some discrete systems that are closer to a wide range of physical systems and that are, above all, easy to characterize.

To illustrate ideas, here is a simple example of a one-dimensional cellular automaton. In the initial state, only the discrete position represented by the coordinate 0 has the value 1. All other integer coordinates, positive or negative, have the value 0. The automaton can in successive "moves" or steps produce only the values 0 or 1 at any given location. The rule of change is given by a function that depends on the value of the given position and the values for the simple case to be considered here of the adjacent values on either side. Thus, using a_i' as the value at site i, after supplying the function for change, we can represent the rule of change as follows:

$$a_i' = f(a_{i-1}, a_i, a_{i+1}). \tag{1}$$

Note that because of the restriction in the values at any site this function takes just eight arguments, the eight possibilities for strings of three 0's and 1's. The automata being discussed here are at the very lowest rank on the number k of possible values: $k = 2$ is the number of possible values and $r = 1$ is the distance away from a_i. The characterization here in terms of the infinite line can be replaced by a finite characterization, for example, in terms of a discrete set of points on a circle. The same kind of function as that of equation (1) applies.

Cellular automata have the following obvious characteristics: discrete in space; discrete in time; discrete state values; homogeneous—in the sense all cells are identical; synchronous updating; deterministic rule of change corresponding to a deterministic differential equation of motion in the case of classical mechanics; locality of change rule: the rule of change at a site depends only on a local neighborhood of the site; temporal locality: the rule of change depends only on values for a fixed number of preceding steps—in the present example just one step.

Notice that one characteristic of ordinary mechanical systems has not been mentioned, and that is reversibility. It is also easy to characterize reversible automata but I shall not enter into the details here. (For information on these and many other aspects of cellular automata, see Wolfram 1986a.)

The basic problem of prediction of interest here for cellular automata can be given a simple formulation. Knowing the transition rule for the cellular automata and given an initial configuration, can one predict in closed form the configuration after n steps? Once again I emphasize that determinism is not an issue. Clearly the process is deterministic. Predictability is another matter. By putting stress on closed-form expressions, I do not mean to imply that this is the only form of prediction. It is just a good criterion of our understanding of a phenomenon.

In the case of cellular automaton we can, of course, get predictability by directly simulating the automaton, but this is not what is ordinarily meant by predictability. There is, in fact, a concept that is useful to introduce at this point in these discussions. This is the concept of *computationally irreducible* (Wolfram 1985). A system is computationally irreducible if a prediction about the system cannot be made essentially by shorter methods than simulating the system or running the system itself. Wolfram (1986b) has shown that the following $k = 2\, r = 1$ cellular automaton generates highly complex sequences that pass many tests for randomness, in spite of its totally elementary character. The automaton is defined by the equivalent equations, one in terms of exclusive *or* the other in terms of mod 2 arithmetic.

$$a_i' = a_{i-1}\text{XOR}(a_i\text{OR}a_{i+1}) \tag{2}$$

$$a_i' = (a_{i-1} + a_i + a_{i+1} + a_i a_{i+1})\bmod 2 \tag{3}$$

Already the 256 cellular automata with $k = 2$ and $r = 1$ fall into four natural classes: (1) the pattern becomes homogenous, (2) the pattern degenerates into a simple periodic structure, (3) the pattern is aperiodic, and (4) the structure is complicated and localized.

Another way of thinking about the complexity of cellular automata is that it seems intuitively clear that cellular automata can be constructed which simulate universal Turing machines, and thus can compute any partial recursive function. In fact, a quite reasonable one-dimensional cellular automaton with only fourteen states has been shown by Albert and Culik (1987) to be a universal computational device, that is, it can simulate a universal Turing machine. The interesting conjecture, due to Wolfram, is that we have naturally occurring physical processes that are universal computing devices, so that a way to think about the complexity of much natural phenomena is that it is identical with the complexity we associate with computation. Such ideas, which are still to some extent speculative, help to drive a further wedge between determinism and predictability.

Chaos

I now turn to another closely related topic which provides many illustrations of systems that are deterministic but not in any practical sense predictable in their behavior. Such systems possess the property of chaos, a concept now much studied for physical systems of a great variety.

To illustrate ideas, let us begin with a simple example that is not really physical in content, but shows how a really simple case can still go a long way toward illustrating the basic ideas. Let f be the doubling function mapping the unit interval into itself.

$$x_{n+1} = f(x_n) = 2x_n(\bmod 1) \tag{4}$$

where mod 1 means taking away the integer part so that x_{n+1} lies in the unit interval. So if $x_1 = 2/3, x_2 = 1/3, x_3 = 2/3, x_4 = 1/3$ and so on periodically. The explicit solution of equation (4) is immediate:

$$x_{n+1} = 2^n x_1(\bmod 1) \tag{5}$$

With random sequences in mind, let us represent x_1 in binary decimal notation, i.e., as a sequence of 1's and 0's. Equation (4) now can be expressed as the rule: for each iteration from n to $n + 1$ move the decimal point one position to the right, and drop whatever is to the left of the decimal point:

$$.1011 \ldots \rightarrow .0111 \ldots .$$

We think of each x_n as a point in the discrete trajectory of this apparently simple system. The remarks just made show immediately that the distance between successive discrete points of the trajectory cannot be predicted in general without complete knowledge of x_1. If x_1 is a random number, i.e., a number between 0 and 1 whose binary decimal expansion is a random sequence, then nontrivial prediction will be out of the question. Moreover, any error in knowing x_1 spreads exponentially—the doubling system defined by equation (4) is highly unstable.

Surprisingly simple natural physical systems can exhibit chaos. Lighthill (1986) gives a very nice example of a pendulum which is forced to move back and forth in a sinusoidal oscillation with a period just slightly greater than the natural period of the pendulum. The pattern of motion followed by the pendulum is quite unpredictable by standard methods. It may be shown that the dependence on initial conditions is even deeper in chaotic systems than it is in the classical unstable systems that have been studied for a very long time. Perhaps one way of describing the characteristic feature of chaotic systems is the existence of what has been termed a *predictability horizon*. Such a horizon is the time after which solutions of very close initial conditions, close to the accuracy of specification being used, become quite remote from one another, but even more, the solutions vary in a discontinuous fashion in response to last decimal place rather than in any smooth continuous fashion. In short, chaotic systems exhibit the property that solutions that are close to each other for given initial conditions of a given number of decimal places diverge exponentially from one another.

One of the earliest and most influential chaotic models has been the Lorenz model, which was drawn from the theory of the atmosphere and was developed by Lorenz to persuade meteorologists that simple deterministic models of the atmosphere could lead to chaotic behavior, that is, to the weather being unpredictable even under the most accurate set of observations. Lorenz's model was drawn more particularly from the theory of convection in a fluid heated from below (Lorenz 1963). What is most surprising about Lorenz's system is that he abstracted and simplified so much that he ended up with a system of three ordinary differential equations with no more than a polynomial of second degree in the unknowns. A full-scale study of the Lorenz equations is to be found in Sparrow (1982).

Lorenz equations depend on just three real positive parameters. Variation in parameters, as is to be expected, changes the behavior of the solutions of the equations. In particular, for some parameter values, numerically computed solutions oscillate, apparently forever, in a way that appears random and which is now called chaotic. I emphasize that the oscillation is not random but seems extremely complex and is what is sometimes termed pseudorandom. For other parameter values there is what has been "preturbulence," which is a phenomenon in which trajectories behave chaotically for long periods of time but finally settle down to stable behavior. There are still other values of parameters in which intermittent chaos is observed. In these cases the trajectories alternate between chaotic and apparently stable behavior. It is not appropriate here to enter into technical details concerning the solutions of the Lorenz equations. What is important is that the equations represent an extremely good example of a simple physical system that is obviously completely deterministic in character but is, for all practical purposes, unpredictable for large sets of values of the parameters. To the best of our knowledge, the system represents for such values of the parameters a computationally irreducible system in the sense defined earlier.

It is important to emphasize that I have only touched the surface of the now enormous literature on chaotic systems. What is central here is simply to cite them as examples of deterministic systems that are unpredictable in behavior.

RANDOMNESS AND DETERMINISM

It is characteristic of chaotic systems and also of the deterministic behavior of cellular automata that actual randomness in the behavior has not been proved, where I have in mind using a very strict definition of random behavior in the sense, for example, of Kolmogorov for infinite sequences. Remember that Kolmogorov's sense of randomness is that of sequences of maximum complexity, where complexity of a sequence is defined in terms of the length of the program required to describe the sequence.

It might be thought that strict randomness is inconsistent, in a formal sense, with determinism. It is a major point of this article to emphasize that this is not the case. Strict randomness and strict determinism are mutually compatible, contrary to much philosophical and even physical thought of the past. I shall use an example that I have used previously in discussions of randomness of deterministic systems, namely, a certain special case of the three-body problem, and I shall end by indicating what deeper questions it would be nice to have answers to concerning the relationship between randomness and determinism (cf. Suppes 1987).

Our special case is this. There are two particles of equal mass m_1 and m_2 moving according to Newton's inverse-square law of gravitation in an elliptic orbit relative to their common center of mass, which is at rest. The third particle has a nearly negligible mass, so it does not affect the motion of the other two particles, but they affect its motion. This third particle is moving along a line perpendicular to the plane of motion of the first two particles and intersecting the plane at the center of their mass—let this be the z axis. From symmetry considerations, we can see that the third particle will not move off the line. The restricted problem is to describe the motion of the third particle.

With these restrictive assumptions it is easy to derive an ordinary differential equation governing the motion of the third particle. The analysis of this easily described situation is quite complicated and technical, but some of the results are simple to state in informal terms. Near the escape velocity for the third particle—the velocity at which it leaves and does not periodically return, the periodic motion is very irregular. In particular, the following remarkable theorem can be proved. Let $t_1, t_2, \ldots$ be the times at which the particle intersects the plane of motion of the other two particles. Let s_k be the largest integer equal to or less than the difference between t_{k+1} and t_k. Variation in the s_k's obviously measures the irregularity in the periodic motion. The theorem, due to the Russian mathematicians Sitnikov (1960) and Alekseev (1968a, b, 1969a, b), as formulated in Moser (1973), is this:

Theorem 3: *Given that the eccentricity of the elliptic orbits is positive but not too large, there exists an integer, say α, such that any infinite sequence of terms s_k with $s_k \geq \alpha$, corresponds to a solution of the deterministic differential equation governing the motion of the third particle.*

The correspondence between a solution of the differential equation and a sequence of integers is the source of the term *symbolic dynamics*. The idea of such a correspondence originated with G. D. Birkhoff in the 1930s. A corollary about random sequences immediately follows. Consider any random sequence of heads and tails—for this purpose we can use any of the several variant definitions—Church, Kolmogorov, Martin Löf, etc. We pick two integers greater than α to represent the random sequence, the lesser of the two representing heads, say, and the other tails. We then have:

Corollary 1: *Any random sequence of heads and tails corresponds to a solution of the deterministic differential equation governing the motion of the third particle.*

In other words, for each random sequence there exists a set of initial conditions that determines the corresponding solution. Notice that in essential ways the motion of the particle is completely unpredictable even though deterministic. This is a consequence at once of the associated sequence being random. In this case the notion of unpredictable is a sharp one, for surely the strongest sense of unpredictability is that of randomness.

Of course, we also have the following corollary, which some may find disturbing.

Corollary 2: *Let the books of the Library of Congress and the Bibliothéque Nationale be encoded according to the current sequential arrangement of the books, with the letter "a" being encoded by the first integer larger than α, "b" the next integer, etc. Then the sequence of integers whose initial segment encodes the contents of these libraries corresponds to a solution of the deterministic differential equation governing the motion of the third particle.*

Upon reflection, this second corollary, given the first one, is not too surprising. Any information desired can be expressed by the continuum of initial conditions possible for this restricted three-body problem. It just took time and analysis for the idea to surface that indeed any sequence could be obtained in this fashion with only the mild restriction stated in the theorem.

It certainly seems right that there are a large number of other simple physical systems that are deterministic in character and that can generate random sequences, but I know of no other examples that have been worked out in sufficient detail to lead to Corollary 1 stated above. It is to be emphasized that the proof of the theorem stated above by Sitnikov, Alekseev, and Moser is quite long, technical, and difficult. Almost all the results in the theory of chaos have not yet received a similar intense scrutiny. I do emphasize that there is a definite step requiring specific mathematical argument to pass from a system's being chaotic to its exhibiting strict randomness, a step very likely not possible for many familiar chaotic systems.

A striking feature of randomness is complexity. So what are random sequences? Under one view, they are the limiting case of increasingly complex deterministic sequences. And the most complex deterministic systems are completely unpredictable in their behavior.

INDETERMINISM

The richness and complexity of deterministic systems suggest that any phenomena can be accounted for by a suitable choice of system type and suitable parameters. But there is a long scientific and philosophical tradition of urging a role for indeterminism. One of the problems has been deciding what indeterminism should mean, a problem made more difficult by the inclusion of chaos and randomness within the framework of determinism.

There is, I am sure, no definitive analysis of indeterminism, even for the limited purpose of this article. For a variety of reasons, the most prominent concept of indeterminism is that derived from quantum mechanics. But exactly what this view is is not transparent. There are at least three different ideas advanced in the literature.

(i) Quantum mechanics is indeterministic because it implies the existence of objective probabilistic phenomena in nature. But, as we have already seen, this argument by itself is not decisive because such phenomena are also implied by classical mechanics.

(ii) Quantum mechanics is indeterministic because it implies the Heisenberg uncertainty principle, which does not have an analogue in classical physics. But, it may be argued, this is misleading, because the classical theory of measurement, embodying a theory of accidental or random errors, also yields an uncertainty principle. The difference is that the theory of measurement is not part of classical physical theory but stands alongside it as a necessary supplement to provide a theoretical basis for the actual measurement of classical physical quantities.

(iii) Quantum mechanics is indeterministic because it is inconsistent with the existence of local hidden variable theories. Beginning in the 1930s there has been a series of proofs that quantum mechanics is in principle inconsistent with any deterministic theory satisfying what appear to be plausible classical assumptions about the locality of common causes—there should be no instantaneous action at a distance. The latest arguments have centered on the inequalities first formulated in 1964 by John Bell. The inequalities formulate a simple condition that must be satisfied by any local hidden variable theory, i.e., any satisfying locality, for a wide class of experiments involving simultaneous measurement of two spin-1/2 particles from a single source, or similar phenomena. Quantum mechanics is inconsistent with the Bell inequalities. This seems straightforward and a clear triumph for indeterminism, given the strong experimental support of quantum mechanics.

In fact, from the standpoint of the most natural intuition that indeterminism implies probabilistic phenomena, the Bell results are ironic rather than straightforward. For the nonexistence of a local hidden variable, i.e., local common cause, is exactly equivalent to the nonexistence of a joint probability distribution of the spin or other experimental observables. As Laplace would have appreciated, causes exist if and only if probabilities exist.

As I have argued elsewhere (Suppes 1990), quantum mechanics in general only generates average probabilities. Hidden-variable extensions, possibly non-local in nature, can be made that are either deterministic or indeterministic in character. Quantum mechanics by itself does not provide a knockdown argument for indeterminism, contrary to much popular thought.

On the other hand, the general case for indeterministic theories of much natural phenomena seems highly persuasive, in view of the apparent hopelessness of formulating detailed deterministic theories that have serious scientific support. Examples abound, ranging from the communication of neurons in animal nervous systems to the motion of clouds and the formation of weather disturbances. It is not mere fashion or fancy that accounts for the use of probabilistic models, without deterministic underpinnings, in essentially every area of science. Deterministic models or theories cannot be made to work. But, as I shall now argue, the philosophical consequences of this robust fact about current science is not support for the absolutely indeterministic structure of nature, but rather the transcendental character of the issue.[2]

TRANSCENDENTAL ISSUE

To begin with, there is a relevant fundamental point to be made about both determinism and indeterminism. If we could make a case for the universe consisting entirely of stable deterministic or indeterministic systems, it would be easy to show that there could be no place for any biological species in such a universe. If the universe consisted of something like the eternal heavens of ancient Greek astronomy with only a touch here and there of radioactive decay or some similar phenomena, it would indeed be a lifeless, mostly very orderly place. But in fact nothing could be farther from the truth. The extension of the standard ideas of determinism or indeterminism, as embodied in current scientific theories, to the entire universe in its full blooming, buzzing confusion is a metaphysical fantasy as extreme as any that can be retrieved from the archives of past philosophical thought.

For a great variety of empirical phenomena there is no clear scientific way of deciding whether the appropriate "ultimate" theory should be deterministic or indeterministic. Philosophy would like a general answer, but fortunately science is opportunistic—going for limited but highly constructive results. The metaphysics of either determinism or indeterminism is transcendental, in the sense that any general thesis about the nature of the universe must transcend available scientific facts and theories by a very wide mark. I now turn to a strong argument for this thesis.

The modern research on dynamical systems, whose lineage extends back to the deep analysis of Poincaré of motion in celestial mechanics in the nineteenth century, has produced a variety of philosophically significant results, but none more so than that expressed in the following theorem.

Theorem 4: *(Ornstein): There are processes which can equally well be analyzed as deterministic systems of classical mechanics or as indeterministic semi-Markov processes, no matter how many observations are made.*

The theorem is due to Donald Ornstein. It depends on earlier work of Kolmogorov, T. G. Sinai, and others; an excellent detailed overview is provided in Ornstein and Weiss (1991). It is this theorem that justifies the title of this paper. The existence of physically realistic models of natural phenomena for which such a theorem holds is the basis for skepticism about the empirical nature of any general claims for determinism. The simplest concrete models for which the theorem holds are those with a single billiard ball moving on a table on which is placed a convex-shaped obstacle. As the theorem indicates, the motion of the ball as it hits the obstacle from various angles is not predictable in detail, but only in a stochastic fashion. Moreover, there are many reasons for believing what has been proved for certain physical processes is also true of a great many more. A conjecture would be that the isomorphism result of Theorem 4 holds for most physical processes above a certain complexity level.

Deterministic metaphysicians can comfortably hold to their view knowing they cannot be empirically refuted, but so can indeterministic ones as well. Both schools of thought can embrace the presence of randomness and the great variety of quantum phenomena with equanimity. The theorem stated above of course does not cover the case of quantum phenomena, but there is a great deal of current evidence that hidden-variable theories of either a deterministic or indeterministic kind can be consistently introduced, even if they lead to no new experimentally verifiable predictions.

As every philosopher knows, Kant's three great problems of pure reason were the existence of God, freedom of the will, and the immortality of the soul, all problems whose "solutions" necessarily transcend experience. Kant's Third Antimony bears more directly on the analysis given here than the complex of ideas associated with freedom of the will would suggest. The Thesis is:

> Causality, according to the laws of nature, is not the only causality from which all the phenomena of the world can be deduced. In order to account for these phenomena it is necessary also to admit another causality, that of freedom. (A444)

and the Antithesis is:

> There is no freedom, but everything in the world takes place entirely according to the laws of nature. (A445)

The "proof" of the thesis makes clear that Kant has focused the argument on spontaneity of causes, a concept that is exactly right for indeterministic stochastic processes.

> We must therefore admit another causality, through which something takes place, without its cause being further determined according to necessary laws by a preceding cause, that is, an *absolute spontaneity* of causes, by which a series of phenomena, proceeding according to natural laws, begins by itself; we must consequently admit transcendental freedom, without which, even in the course of nature, the series of phenomena on the side of causes, can never be perfect. (A447)

Again, as every philosopher knows, Kant argued at great length to show why the thesis of spontaneous causes should be rejected by any empirical philosopher. Here is a characteristic passage:

> If the empirical philosopher had no other purpose with his antithesis but to put down the rashness and presumption of reason in mistaking her true purpose, while boasting of *insight* and *knowledge*, where insight and knowledge come to an end, nay, while representing, what might have been allowed to pass on account of practical interests, as a real advancement of speculative enquiry, in order, when it is so disposed, either to tear the thread of physical enquiry, or to fasten it, under the pretence of enlarging our knowledge to those transcendental ideas, which really teach us only *that we know nothing*; if, I say, the empiricist were satisfied with this, then his principle would only serve to teach moderation in claims, modesty in assertions, and encourage the greatest possible enlargement of our understanding through the true teacher given to us, namely, experience. (A470)

At the same time Kant warns against the empirical philosopher becoming dogmatic and extending his ideas so as to inflict injury on the practical interests of reason (A471).

The fundamental reinterpretation of the Third Antinomy being proposed in this article is evident. Both Thesis and Antithesis can be supported empirically, not just the Antithesis. The choice must transcend experience. This idea of detailed theories of phenomena that are mathematically inconsistent but equally supported by any possible observational data is not one that is congenial to Kant's philosophy of science, but in other ways can, I think, be a central concept of a revised transcendental philosophy sympathetic to Kant's objective of exposing dogmatic metaphysics for what it is, even when traveling in modern guise.[3] Modern arguments for the universal presence of determinism are of this kind. Arguments like those of Inwagen (1983) that the hypothesis of determinism implies absence of free will are of another sort, which do not belong to dogmatic metaphysics and with which I am quite sympathetic. On the other hand, I am quite unsympathetic with Inwagen's view (p. 209) that the existence of moral responsibility is the chief reason for believing in free will. Arguments against the empirical nature of any universal thesis of determinism are above all necessary to make room for an unlimited range of biological phenomena, much of it intentional, exhibited in animals long before there were even any humans to say "Gavagai," let alone reason about moral problems.

NOTES

Much of this article comes from the draft of the first two of four lectures given at the College de France in April 1988. A further revision was included in the First Ernest Nagel Lecture given at Columbia University in November 1988. Still further changes were made for the Fifth Evert Willem Beth Lecture given in Nijmegen in August 1989 and the Thirteenth Hausser Lecture at Montana State University in June 1991.

I have benefited from many critical comments made at these various lectures, but especially those of Jules Vuillemin, Isaac Levi, Sidney Morgenbesser, and Gordon G. Brittan, Jr., listed in the historical order of their remarks, and also Yair Guttmann who has made several useful criticisms of the final written draft.

1. In introducing various distinctions about determinism, I have, in the interest of avoiding many technical points, neither distinguished between theories and the structures (or models) satisfying them, nor, at the next level, distinguished between set-theoretical structures satisfying a theory and "real" physical phenomena from which the structures were abstracted. Important and useful distinctions are to be made about these matters, but can, I believe, be omitted without too much risk of confusion in an article like this meant to be informal in character. I have also used "system" and "process" in the sense of structure, without really saying so. The various theorems that are stated below are, as mathematical theorems, about set-theoretical structures that satisfy certain theories. However, these structures are not abstract fantasies but ones that approximate quite closely real phenomena that already exist in nature or that can be experimentally produced.

2. I stress that I have in no sense canvassed the many different theories of phenomena that should count as being nondeterministic. In terms of the emphasis I have placed on probability and randomness, it is worth noting the significant generalizations of probability used in various contexts when no proper concept of probability has sufficed. A well-defined concept of nondeterminacy that is of this ilk is discussed in Suppes and Zanotti (1977). The more general setting is the large literature on upper and lower probabilities. Of particular interest conceptually in the present context are upper probabilities that are not supported by any probability measure, because they are nonmonotonic, i.e., there are two events, say A and B, such that the occurrence of A implies the occurrence of B but the upper probability

of A is greater than the upper probability of B. For an application of such upper probabilities to quantum mechanics, see Suppes and Zanotti (1991).

3. My discussion of Kant is too brief to hope to give a proper exegesis of his ideas, but the Third Antinomy is too important historically to be entirely omitted from my analysis. As Christoph Lehner has pointed out to me, I need to go on to distinguish between Kant's concepts of transcendental freedom and of practical freedom, the latter concept being known through experience (B831). It is also clear that my general use of the concept of transcendental is not identical with Kant's.

REFERENCES

Albert, J., and K. Culik. 1987. "A Simple Universal Cellular Automaton and Its One-way and Totalistic Version." *Journal of Complex Systems* 1: 1–16.

Alekseev, V. M. 1968a, 1968b, 1969a, 1969b. "Quasirandom Dynamical systems I, II, III." *Math. USSR Sbornik* 5: 73–128; 6: 505–60; 7: 1–43; *Mat. Zametki* 6: 489–98.

Ayer, A. J. 1954. "Freedom and Necessity." *Philosophical Essays*. London.

Bell, J. S. 1964. "On the Einstein Podolsky Rosen Paradox." *Physics* 1: 195–200.

Inwagen, Peter van. 1983. *An Essay on Free Will*. Oxford.

Kant, I. 1791. *Critique of Pure Reason*. Translated by F. M. Mueller. New York, 1881; second edition, 1949.

Lighthill, J. 1986. "The Recently Recognized Failure of Predictability in Newtonian Dynamics." *Proceedings of the Royal Society of London* A407, 35–50.

Lorenz, E. N. 1963. "Deterministic Nonperiodic Flow." *Journal of Atmospheric Science* 20: 130–141.

Moser, J. 1973. "Stable and Random Motions in Dynamical Systems with Special Emphasis on Celestial Mechanics." *Herman Weyl Lectures, Institute for Advanced Study*. Princeton, NJ: Princeton University Press.

Ornstein, D. S., and B. Weiss. 1991. "Statistical Properties of Chaotic Systems." *Bulletin of the American Mathematical Society* 24: 11–116.

Sitnikov, K. 1960. "Existence of Oscillating Motions for the Three-Body Problem." *Doklady Akademii Nauuk, USSR* 133: 303–306.

Sparrow, C. 1982. *Lorenz Equations: Bifurcations, Chaos, and Strange Attractors*. New York.

Suppes, P. 1987. "Propensity Representations of Probability." *Erkenntnis* 26: 335–58.

Suppes, P. 1990. "Probabilistic Causality in Quantum Mechanics." *Journal of Statistical Planning and Inference* 25: 293–302.

Suppes, P., and M. Zanotti. 1977. "On Using Random Relations to Generate Upper and Lower Probabilities." *Synthese* 36: 397–421.

Suppes, P., and M. Zanotti. 1991. "Existence of Hidden Variables Having Only Upper Probabilities." *Foundations of Physics* 21: 1479–99.

Wolfram, S. 1985. "Undecidability and Intractability in Theoretical Physics." *Physical Review Letters* 54: 735–38.

Wolfram, S. 1986a. *Theory and Applications of Cellular Automata*. Singapore.

Wolfram, S. 1986b. "Random Sequence Generation by Cellular Automata." *Advances in Applied Mathematics* 7: 123–69.

Wolfram, S. 1986c. "Cellular Automaton Fluids I: Basic Theory." *Journal of Statistical Physics* 45:471–528.

MIDWEST STUDIES IN PHILOSOPHY, XVIII (1993)

Idealization and Explanation: A Case Study from Statistical Mechanics

LAWRENCE SKLAR

1

It is a commonplace of the philosophy of science that our description of nature requires that we treat systems in the world as "idealized" if we are to be able to describe them and explain their behavior using our theoretical concepts and laws. In mechanics we talk of point masses, in electromagnetism of perfect conductors, and so on.

We frequently encounter the situation where one and the same physical structure is idealized in more than one way, each idealization being appropriate for some particular descriptive and explanatory purpose. Sometimes the idealizations are "incompatible" with one another. That is, one idealization attributes to the system features quite impossible from the other idealization's standpoint. For an example consider the liquid drop model of the nucleus, which treats the nucleus as a continuum object, as contrasted with the shell model which treats the nucleus as an assemblage of "independent" nucleons all moving in a common general nuclear potential.

Now it is often the case that the idealizations we use are "controllable." I mean by this that we have the theoretical resources to tell us in what ways, and to what degree, the conclusions we reach about the idealized model can be expected to diverge from the features we will find to hold experimentally of the real system in the world. When we are dealing with contrasting idealizations, but where both idealizations are controllable, we will often be in a situation where we can make it quite clear what the circumstances are in which one or another of the models will be the appropriate one to employ in order to generate realistic results. Under these circumstances we will not be dismayed by the incompatibilities of the idealizations. We will know how they each distort reality, and why their distortions go in different directions.

But there are other cases where matters are not so clear. In these cases there is a lack of control over the idealizations, in that we may not be certain just how the idealized model will differ from real world behavior. If this is true for more than one idealization, where the idealizations are incompatible with one another, it may not be at all clear what the appropriate circumstances are in which we ought to use one or the other of the idealizations. We may be left in some ignorance of the relationship the idealizations bear to the real system and the relationship they bear to one another.

It is an example of this kind drawn from statistical mechanics that I intend to focus on here. The particular example I will describe has another aspect that makes it even more important for consideration by anyone concerned with the general issues of idealization in science. Alternative idealizations often result in alternative explanatory accounts of phenomena. Consider, for example, how differently nuclear fission would be treated in the liquid drop and shell models of the nucleus.

But in the case I will be outlining, the difference between the accounts of the real-world phenomena offered by the two idealizations goes a bit further. In the case to be considered each idealized account offers a statistical or probabilistic explanation of a well-known phenomenon. But the idealizations are so different that the very notion of what *kind* of statistical explanation is being offered is quite different from the point of view of one idealization than it is from the point of view of the other. The very structure of how explanation proceeds is contested by the two idealizations.

When one adds to this fact the uncontrollable nature of the idealizations, one arrives at the conclusion that without further understanding one cannot even know, of real world systems, what kind of statistical account of their behavior the theory is to offer. While such a puzzle about idealization is continuous with the ordinary cases, it goes beyond them. And in doing so it emphasizes the risks one runs in relying upon idealized models in one's explanatory accounts.

2

Systems begun in a non-equilibrium state move toward equilibrium, and having obtained equilibrium, they stay in that state. The approach to equilibrium is characterized by a time scale appropriate to the system in question, the so-called relaxation time for the system. And the path followed to equilibrium is characterized by the solution to a hydrodynamical-thermodynamical equation of evolution, such as the Navier-Stokes equation. Why do systems approach equilibrium? Why do they approach it in the time scale that they do? Why do they follow the path described by the hydrodynamical equation?

For the case of the low density gas, L. Boltzmann offered answers to these questions by proposing an equation that would describe the evolution of the function that characterized the distribution of molecular momenta of the molecules making up the system. This is the famous Boltzmann equation. He showed that the equation had a momentum distribution corresponding to equilibrium as a stationary solution. And by means of his famous H-Theorem he showed that any non-equilibrium momentum distribution had to be non-stationary.

But there were immediate objections to Boltzmann's theory. He seemed to show that for any initial state of the molecules that corresponded to non-equilibrium, the system would have to move to states ever closer to equilibrium. But, as J. Loschmidt pointed out, for each such transition there would have to be a possible reverse transition from an equilibrium to a non-equilibrium molecular state, due to the time reversal invariance of the underlying dynamics of molecules. And E. Zermelo pointed out that a theorem of Poincaré guaranteed that a system started in a non-equilibrium state would, after a sufficient time, return arbitrarily close to that state. So Boltzmann's alleged demonstration that for any initial non-equilibrium state the system would get ever closer to the equilibrium state could not be true.

Boltzmann's response was to introduce probabilistic considerations into the argument. The Boltzmann equation was not supposed to describe how systems always behaved. Instead, he said, it was only supposed to characterize the *probable* behavior of a system. So the possible existence of anti-thermodynamic evolutions or of Poincaré recurrence did not prove the equation inconsistent with the underlying dynamical theory.

But the critics were quick to respond. For each molecular transition taking one from a non-equilibrium condition to one of equilibrium, there is a reverse transition. So the "probability" of an anti-thermodynamic evolution ought to be as high as one taking one in the right direction. And the Poincaré result is that "almost all" initial conditions, that is a collection of them of probability one, lead to recurrence. So it cannot be "overwhelmingly probable" that systems go from non-equilibrium monotonically toward equilibrium. Even interpreted probabilistically the Boltzmann equation seems inconsistent with the underlying dynamics.[1]

3

Matters were greatly clarified some time later by P. and T. Ehrenfest in their critical exposition of the Boltzmann theory. Imagine a vast number of systems started in the same non-equilibrium macro-state, with the different systems started in different micro-states compatible with the macro-state. Examine all of these systems at discrete time intervals as time goes on. Depending upon their initial micro-state they will have evolved into macro-states with differing macro-characteristics. But, the claim is, if we put a natural probability distribution over the initial micro-states of the systems, at any later time the overwhelming majority of the systems will be found to share a most common macro-state. Consider the sequence of these "most common" macro-states over time. A curve through those states is called by the Ehrenfests the "concentration curve" of the evolution of the collection of systems. It is this concentration curve that they suggest ought to be identified with the curve that gives the solution to the Boltzmann equation.

The Boltzmann equation does not, then, describe the inevitable evolution of a system. Nor does it describe the most probable evolution of a system. For if we accept the Recurrence Theorem we know that only a set of probability zero of systems could actually monotonically approach equilibrium forever. What the equation does describe is the evolution in time of overwhelmingly most probable states of systems observed at distinct time intervals as the collection of systems evolves due to the dynamic evolution of each of its members.[2]

Now such an interpretation of the Boltzmann equation and its solution does, indeed, eliminate any possibility of showing that the equation is inconsistent with the underlying dynamical description of the evolution of the systems. But is the description of the concentration curve given by the solution of the Boltzmann equation *correct*? It could be consistent with the dynamics but false as a description of how a collection of systems would evolve.

Going beyond the Ehrenfests's consistent interpretation of the import of the Boltzmann equation to some demonstration that it does correctly describe the evolution of a collection of systems, and going even further to explain just why the equation is successful in accomplishing that task, is a complicated project. Attempting to carry it out has resulted in a wealth of insight into the dynamics of systems. But each new insight has been accompanied by additional foundational problems.

How a collection of systems would evolve over time would seem to depend upon how the initial conditions of the individual systems were distributed at the initial time. For the evolution of each of the individual systems in the collection is fixed by its special initial state. Interestingly, however, great progress in rationalizing the correctness of the Boltzmann equation, as the Ehrenfests interpret it, has been obtained by a method that simply ignores the initial conditions of the systems in the collection. As we shall see, though, this approach pays a price for excluding initial conditions from its consideration.

The method focuses only on how the system is constituted out of its micro-components (including the force laws governing the interactions of these micro-components) and on the dynamical laws governing evolution over time at the microscopic level. If one thinks about how the Ehrenfest interpretation works one gets a model that is neatly characterized by J. Gibbs when he describes what comes to be called the "coarse-grained" approach of a collection toward equilibrium. In this account one thinks of an equilibrium state as the state of a collection of systems, and not as the equilibrium state of an individual system. This equilibrium collection of systems consists of systems whose micro-states at a time remain constantly distributed in density in phase-space in the region available to micro-states for the system fixed by its macroscopic constraints. At any time almost all systems in such a collection are in states near that of individual system equilibrium, but some systems are always far from equilibrium in the individual system sense. The collection is an equilibrium collection in the sense that it represents an unchanging distribution of states over time, even as individual systems approach, stay at, or depart from individual system equilibrium conditions.

Next one looks at a collection of systems that represents a system started in some non-equilibrium condition. Will that collection, over time, become a collection that represents an equilibrium system? A simple argument shows that such a non-equilibrium collection cannot evolve to the collection that represents equilibrium in the exact sense. Basically this is because the true equilibrium collection includes systems with every possible micro-state compatible with the equilibrium macro-conditions, but the collection of systems started in a given non-equilibrium condition will consist only of the restricted collection of states that "remember" in their micro-state what non-equilibrium condition they had in the past. Such a collection, then, will always be strictly smaller than the true equilibrium collection.

Instead one claims that the non-equilibrium collection approaches the equilibrium collection "in a coarse-grained sense." Saying in exactly what sense a collection can "coarsely" approach an equilibrium collection would require too much preliminary discussion to be carried out in the confines of this paper. But the reader can get some idea of what goes on by considering a famous analogy given by Gibbs. Imagine a glass filled 90 percent by clear water and 10 percent by ink insoluble in the water. Start with the ink all in the top 10 percent of the glass. Now stir the two fluids together vigorously for some time. The resulting mixed fluid will appear from a distance to be grey all over. Look closely enough, however, and one sees that in any small region each point of the fluid is entirely water or entirely ink. But in any small region that is not too small the same portion of the region, 10 percent, will be occupied by the ink with the remaining 90 percent water. The original separate water-ink fluid has "coarsely" approached a fluid that is one which is uniformly grey to a degree 10 percent of the blackness of the pure ink. But it really is not such uniformly grey fluid. It is actually fluid that is, at every point, perfectly clear or perfectly black. It is in this sense that the original non-equilibrium collection "coarsely" approaches the collection used in the theory to represent the equilibrium state. The original non-equilibrium collection of system points spreads out through the newly available phase-space region in much the way the ink spreads out through the full volume of the glass, even though its size in phase-space remains constant as does the volume of the glass actually occupied by the ink.[3]

Such a coarse-grained approach to equilibrium for a collection neatly captures the Ehrenfests' interpretation of the Boltzmann equation's solution as characterizing the concentration curve of ensemble evolution. Individual systems approach and leave equilibrium. But their approachings and leavings are uncorrelated with one another. The net result is that the overall collection of systems has its distribution become one in which the overwhelming bulk of the systems at a time are found to be in states that get ever closer to the individual equilibrium states.

But can we show that collections of systems started in non-equilibrium really display such coarse-grained approach to the equilibrium collection? The answer is a qualified, very qualified, "Yes." Certain models taken to idealize real physical systems can be shown to have an important "trajectory instability" property. What this means is that systems started with their initial conditions quite near one another will quickly evolve to systems whose micro-states at the later time are quite far from one another. If systems have such a sufficient trajectory instability, then a result can be proven about a collection of such systems that corresponds to an initial non-equilibrium condition. Essentially what the result says is that in the infinite time limit in the future the collection will have approached the equilibrium collection in the coarse-grained sense outlined above. And this will be true no matter what initial non-equilibrium distribution we start with. Such systems are said to be "mixing."[4]

Does this provide the rationale we would like that would justify the claim that the Boltzmann equation really does describe the approach to equilibrium, where that mode of description is understood in the sense made explicit and clear by the Ehrenfests? I said the "Yes" answer was qualified. Let us examine some of the qualifications.

First it must be noted that the conditions under which mixing can be demonstrated for a system are such that they can be shown to hold only for certain idealized systems. If we have a model of a gas as consisting of molecules that do not interact unless colliding, and then suffer infinite forces of repulsion, the so-called model of "hard spheres in a box," then the conditions necessary for mixing can be shown to hold. But for more realistic systems there are theorems that at least hint that the systems cannot, in fact, meet the conditions for being mixing. Under these circumstances much more needs to be said about the role of the idealization that we are making when we use mixing for explanatory purposes.

Next there is the fact that we can only prove mixing to hold in the infinite time limit. Here the uncontrollability of the idealization becomes important. There is nothing in the theory to tell us the extent to which mixing can be assumed for any finite time values, nor the amount of error we are making if we assume that some time is "long enough" for mixing to have taken place. A collection of systems can, for finite time periods of any length, move, in the coarse-grained sense, away from the equilibrium collection, yet still be mixing in the infinite time limit.

In fact the results are limited in other ways. We want not only to explain why systems approach equilibrium, we want also to explain why they take the typical time that they do to reach near equilibrium states (the "relaxation time"). And we want to understand why systems follow the particular path to equilibrium that they do follow, a path, for a rare gas, allegedly described by the solution to the Boltzmann equation. But by giving results only in the infinite time limit, the mixing theory precludes itself from even attempting to offer an explanatory account of the duration of the relaxation time or of the particular path a system will follow in its approach to equilibrium.

4

Is there an approach to the problem that will provide a rationalization for the entire description of the approach to equilibrium given by the Boltzmann equation? Can we go further than the mixing approach does and demonstrate something more than just an infinite time limit approach to equilibrium? One approach to the theory that does just this has been developed, beginning with the work of O. Lanford. This is the so-called "rigorous derivation of the Boltzmann equation." How does it work?

As usual, idealization is crucial. In this case one studies the behavior of a gas system in what is called the Boltzmann-Grad limit. This is the limit obtained where the number of particles in the system, n, is allowed to go to infinity, but where the quantity nd^2 is held constant. Here d stands for the diameter of a gas molecule when measured on a scale of the size of the containing box. So we are working in a limit where the number of molecules has gone to infinity, but where the size of the molecules has become vanishingly small relative to the size of the container. One consequence of this limiting idealization is that we are working in an idealization where the density of the gas has gone to zero.

What is derived in this approach? It is shown that in the limit taken, if one places an appropriate probability distribution over the initial conditions of systems all prepared in an identical non-equilibrium condition, then the systems will "with probability one" evolve in the manner described by the Boltzmann equation. One derives from the probability distribution over the micro-states what is called the one-particle distribution function, f. This, from the collection point of view, is the probability function that corresponds most closely to that distribution function for molecules of a single gas, the function whose evolution Boltzmann's equation was originally designed to describe. What is shown is that if the collection starts with a probability measure that concentrates points of the initial collection sufficiently tightly around those that generate the appropriate f at $t = 0$ (appropriate for the initial non-equilibrium character of the system), then that collection after time evolution to time t' will assign high probability to f for the collection at $t = t'$ being concentrated around the value for f at $t = t'$ that the Boltzmann equation for f would predict. So it is full Boltzmannian behavior that is derived, not just approach to equilibrium in the infinite time limit. Consequently standard relaxation times and the standard path of approach to equilibrium are now being rationalized.[5]

Unlike the mixing results, this approach does require that for any given initial non-equilibrium specification, we adopt a posited probability distribution over initial conditions compatible with that macro-specification of the system. This is hardly a surprise. Any program of rationalization that attempts to derive the short time behavior of the system will have to have something to say about just how likely specified initial conditions are for the systems in question. This is because short time behavior is determined by the underlying dynamics only when that is conjoined with a specification of the initial micro-state of the system.

One version of the Lanford theory uses an initial probability distribution that is "time symmetric." That is, it gives the same probability to a region of micro-states as it does to a region of micro-states that are just the time reverses of the states in the original collection. Paradoxes are avoided by the fact that the initial probability distribution holds only at $t = 0$ and does not propagate into the future. An alternative approach has a characterization of the probability distribution that does continue after $t = 0$, but paradoxes are now avoided by the fact that this probability distribution is intrinsically time asymmetric. Deep issues of the justification for choosing such probability distributions over initial conditions arise, but we shall not delve into these here.

The Lanford proof shows that the evolution of the one-particle distribution function will follow the path predicted by the Boltzmann equation when we are in the Boltzmann-Grad limit idealization and when the initial probability distribution over the micro-states of the systems has its appropriate form. But, alas, it only shows this to be the case for a very short time indeed. In fact one can prove that the evolution is Boltzmannian only for a time equal to one-fifth of the average time taken for the first molecular collision to occur. This is hardly what we would like, a demonstration of Boltzmannian behavior for realistic evolutionary periods. But exponents of this approach argue that even if one can only prove Boltzmannian behavior for very short times, it may still be true that it holds for reasonable times. And there is reason to think that the proof can be improved to demonstrate the high probability of the desired behavior for longer time periods.

Suppose we think of the result as holding for longer time intervals, perhaps for unlimited time intervals. At that point we seemed to be faced with a paradox. The Poincaré Recurrence Theorem tells us that with probability one a system started in a non-equilibrium micro-state will return to a micro-state arbitrarily close to the initial micro-state. In fact, this will happen over and over again without end. But the Lanford result seems to tell us that systems started in non-equilibrium micro-states will, with probability one, show a Boltzmannian monotonic approach to equilibrium as described by the Boltzmann equation. How could both of these claims be true? Remember that the mixing results were compatible with the Recurrence Theorem, for they allowed all or almost all of the systems to show recurrence. It was because these recurrences of individual systems were uncorrelated with one another that the collection as a whole could show the kind of monotonic behavior that the Ehrenfests described. Only the concentration curve was Boltzmannian, not the most probable behavior of an individual system. But the Lanford result is quite different. It tells us that it is highly probable that the individual system will show Boltzmannian behavior. How could this be?

The solution resides, once again, in the idealization chosen in order to derive the Lanford results. We are working in the Boltzmann-Grad limit in which a system has an infinite number of molecules of vanishingly small size and in which the density of the gas is zero. The evolution of a collection of such systems cannot be represented by the normal flow in phase space that is presupposed when results such as the Poincaré Recurrence Theorem are derived. It is this that allows for the consistency between the Lanford rigorous derivation of the Boltzmann equation and the Poincaré Recurrence Theorem.

It is very important to note that the idealization employed in deriving the Lanford results is also one that is uncontrollable. Suppose we are dealing with a system that is not at the Boltzmann-Grad limit of zero density, but that is of very small density. Can we assume that the Lanford type results will hold of this not fully idealized system "to at least some degree" or "within some specifiable range of error?" The answer is that we can make no such assumption. While the results hold rigorously at the Boltzmann-Grad limit, nothing in the proof gives us any reason to suppose that a system near, but not at, this limit will show behavior that is "close" to the behavior that holds at the limit. Once again we are dealing with one of those idealizations that lacks the desirable property of controllability. We cannot show that slightly less idealized systems show slightly less idealized behavior. Nothing we can prove about what happens "in the limit" tells us about what happens when the limit is approached but not reached.

The Lanford results also seem at odds with the mixing results outlined above. Mixing describes a collection in which systems are behaving in a manner that shows radical uncorrelation from system to system. It is the appropriate way for a collection to behave if that collection shows the kind of behavior posited by the Ehrenfests as rationalizing the Boltzmann equation. But a collection of systems obeying the Lanford result shows a uniformity from system to system, the one-particle distribution function of almost all systems monotonically approaching its equilibrium distribution form. Once again it is the nature of the idealization that shows us how inconsistency can be avoided, even if we posit the Lanford results as holding of all times and not just for a very short initial time interval. The systems described by the Lanford result are not systems whose micro-states form the kind of collection for which a mixing result can be proven.

5

We see, then, that the approach to equilibrium of a dilute gas started in a non-equilibrium condition can be modeled within statistical mechanics by two quite distinct probabilistic models.

In one of these models we consider the evolution of a collection of systems. Each system begins with the non-equilibrium macroscopic condition that characterizes our real system. But in the collection the systems have initial microscopic conditions distributed according to a standard probability distribution. The evolution of this collection is then studied, as the individual systems in it evolve as determined by their initial state and the dynamical laws governing the evolution of the micro-states. The approach to equilibrium of a real system is modeled by looking at the dominant micro-states in the collection at a series of discrete times and plotting the concentration curve through those most dominant states at a time. It is this curve that is supposed to characterize an expected evolution of an individual real system. It is assumed that the individual systems in the collection will not show monotonic approach to equilibrium but, instead, will show Poincaré recurrence. It is only the fact that the evolution of the individual systems is so wildly divergent and uncorrelated that the ensemble as a whole shows coarse-grained approach to equilibrium. And this approach can only be proven to hold in the infinite time limit. That the approach will have the time scale and the nature described by the Boltzmann equation is more a posit than a proven result.

In the other model we once more look at a collection of individual systems all possessing the same original macro-condition but with a probability distribution over micro-states. In this approach we deal with systems in the Boltzmann-Grad limit. Within this new account we are able to derive the explicit form of the curve of approach to equilibrium given by the Boltzmann equation. And we derive it as the curve describing the way that "almost all" systems will behave, not as a concentration curve among dominant states. Severe limitations on the amount of time for which such behavior can be proven hold, but there is hope that the results can be extended to longer times. The apparent contradiction to the Poincaré Recurrence Theorem is then eliminated by pointing out that the results hold only for systems at the Boltzmann-Grad limit for which the phase evolution picture needed for the Poincaré result no longer holds.

Now incompatible idealizations are a common thing in physics. But what I want to emphasize here is the way in which these contrasting idealizations demand entirely distinct models of the very kind of statistical explanation that is to be invoked in order to account for one and the same observed physical process.

In one of these accounts, for example, the fact that the system has a large number of components plays only a secondary role. Mixing results hold even for systems with a small number of micro-constituents. The large size of the system is invoked only later in the argument to assure us that the concentration of systems around the typical state at each discrete point, the state that determines the concentration curve, will be a highly compact concentration. But the shape of the concentration curve is independent of the size of the system. In the other account, however, the idealization of large numbers of micro-constituents is central to the derivation. In this latter, Lanford, derivation, in fact the idealization crucially assumes a zero density for the gas and a vanishingly small size for its molecules, idealizations that play no role at all in deriving the mixing results.

In the mixing account the uncontrollable idealization of looking at the system only after an infinite amount of time has elapsed is essential. But the Lanford approach uses no such temporal idealization. In this latter account, however, the results can be proven to hold only for very short times after the beginning of the system's evolution, even if they might be true for larger, but still finite, times.

Most important of all is the way in which we probabilistically rationalize what goes on with an individual system in the two accounts. Experimentally we look at an individual system and we see that it follows a specific path from non-equilibrium to equilibrium. Why does it evolve as it does?

The mixing solution, the solution using coarse-graining and the observation of ensemble evolution at discrete times argues like this: "We are dealing with a specific system prepared in a particular macro-state; but whose micro-state is unknown. Let us look at all systems that have this initial macro-state in common and let us assume that such systems occur with a particular probability for their initial micro-state. Now let us look at this collection of systems as it evolves. At discrete time intervals we will find nearly all the systems in the collection in a specific macro-state, the overwhelmingly most probable macro-state at that time. We may, then, assume that our particular system will display such a sequence of macro-states over time. While we cannot prove that such a sequence of most probable states will follow the course described by the solution to the Boltzmann equation, we can prove that in the infinite time limit the overwhelmingly most probable macro-state will be the equilibrium state. We do not prove, however, that the course of evolution so described will be that followed by 'almost all' systems, for we know this to be incompatible with the Poincaré Recurrence Theorem."

On the other hand, the other rationalization program argues as follows: "We are dealing with a system of low density and small molecular size. So it will be appropriate to idealize the system by moving to the Boltzmann-Grad limit. We know that the system is started in a given macro-state, but with an unknown initial micro-state. We will consider a collection of systems, all of which share the same initial macro-state, but whose initial micro-states vary and which are distributed according to a standard probability distribution. We shall now show that for the idealized system we are considering 'almost every' system in the collection, that is a sub-collection of systems of probability one, will have their evolution of macro-states follow that described by the solution to the Boltzmann equation. So it is overwhelmingly probable that the system with which we are dealing will follow out this course of evolution, moving monotonically toward equilibrium in the time and manner specified by the Boltzmann equation. We can only prove that this will be so for a very short period of time, but there is reason to think that it will be true over the larger time scale needed for the system to get quite close to equilibrium."

What is philosophically most interesting here is that the two idealizations give us two quite distinct posited explanatory structures under which the behavior of the actual system is to be subsumed. Once again we must carefully note that a sequence of overwhelmingly most probable states is not at all the same thing as an overwhelmingly most probable sequence of states.

Which idealization is the right one to use? Which statistical model is the one under which to subsume the observed behavior of individual systems? We really do not know. The second, Lanford style, rationalization is much richer in what can be proven to hold than is the first, mixing, style rationale. The latter gives us results only in the infinite time limit. The actual time of approach to equilibrium and the actual course followed by the system toward equilibrium can only be posited in this approach, not proven to hold. The

Lanford result, on the other hand, gives us, if only for short times, a rigorous proof of Boltzmannian-like behavior in the approach toward equilibrium. On the other hand, the Lanford result only applies to the specific case of the rare gas, the case that can be idealized by going to the Boltzmann-Grad limit, whereas the mixing style derivations are applicable to a wide range of idealized physical systems.

In both cases we get provable results only by relying on idealizations that are uncontrollable. In the mixing case we can prove that a state is reached in the infinite time limit. But we can demonstrate nothing of what we want about what happens before time goes to infinity. We cannot show that physical relaxation times are to be expected, nor that the approach to equilibrium will be monotonic, much less in accordance with the solution of the Boltzmann equation. In the Lanford case we can prove our results only for a system of zero density with vanishingly small molecules. The very nature of the derivation gives us no hope of showing that the results will be "approximately true" for systems near but not at the chosen limit.

It remains an open question in the foundations of statistical mechanics which, if either, of these two idealized models is the right clue as to how to go on and understand the behavior of realistic systems in a more exact and less idealized way. The issue between the two "schools" of non-equilibrium theory is only one of the many that remain matters of contention and debate in the very problematic area of the foundations of statistical mechanics. The details of this debate are, of course, only of interest to those whose special concern is the foundations of statistical mechanics. But, I think, some morals can be drawn of interest to the wider philosophy of science community.

The closer one looks at the role played by idealization in science, the more complex the picture becomes. Controllable idealizations are the ones whose explanatory role is easiest to understand. If we can get a systematic grasp on the kind of error we make in using the idealization, that is, on the degree to which we can expect the behavior of the real system to deviate from that studied in the idealized framework, then we think we have a pretty good idea of the explanatory role the idealized account can play. But, as we have seen, uncontrollable idealizations can play an important role in explanatory contexts as well. But I do not think we yet have a good account within the philosophy of science of just what notion of explanation we are invoking when we think, with some justification, that an idealization has taken us at least part way toward a full explanation of some phenomenon, uncontrollable as that idealization may be.

We also now know that the notion of explanation is one that becomes more complex the more one looks at the details of actual scientific explanatory practice. This is especially true when it is statistical or probabilistic explanations that are in question. In the case study examined above we see that more than one overall explanatory structure can be invoked to account for one and the same physical phenomenon. The difference between the two accounts is not a mere detail of the mechanism invoked in the explanatory account. A wholly different idea of the very *kind* of statistical explanation to be invoked is what is at stake.

And we have seen how the two issues of uncontrollable idealization and alternative explanatory model can interact in a very perplexing way. For the decision to be made about the kind of explanatory structure to be appropriately invoked itself depends upon the choice of uncontrollable idealization. There is much here for methodologists of science to ponder on.

NOTES

1. A detailed exposition of the early debates between Boltzmann and his critics can be found in S. Brush, *The Kind of Motion We Call Heat* (Amsterdam, 1976). A brief survey of these debates is in chapter 2 of L. Sklar, *The Physics of Chance* (Cambridge, 1993).

2. The Ehrenfests' proposals for a consistent reading of Boltzmann are contained in P. Ehrenfest and T. Ehrenfest, *The Conceptual Foundations of the Statistical Approach in Mechanics* (Ithaca, N.Y., 1959).

3. J. Gibbs, *Elementary Principles in Statistical Mechanics* (New York, 1960).

4. A formal discussion of mixing and related concepts is in V. Arnold and A. Avez, *Ergodic Problems of Classical Mechanics* (New York, 1968). Non-technical discussions can be found in J. Earman, *A Primer of Determinism* (Dordrecht, 1986), chapter 8; and in chapter 7 of Sklar, *The Physics of Chance*.

5. For the details of the "rigorous derivation of the Boltzmann equation," see O. Lanford, "On a Derivation of the Boltzmann Equation," in *Nonequilibrium Phenomena I: The Boltzmann Equation*, edited by J. Lebowitz and E. Montroll (Amsterdam, 1983). See also J. Lebowitz, "Microscopic Dynamics and Macroscopic Laws," in *Long-Time Prediction in Dynamics*, edited by C. Horton, L. Reichl, and V. Szebeheley (New York, 1983).

MIDWEST STUDIES IN PHILOSOPHY, XVIII (1993)

The Fabric of Space: Intrinsic *vs.* Extrinsic Distance Relations

PHILLIP BRICKER

S tart with ordinary Euclidean space. Be a realist about points, and about distances between points. I ask: how are the points interwoven to form the fabric of space? Are there direct ties only between "neighboring points," so that points at a distance are connected only indirectly through series of such direct ties? Or are there also direct ties between distant points, so that the fabric is reinforced, as it were, by irreducibly global spatial relations? To fix ideas, roughly, try the following thought experiment. Take a scissors and cut along some plane in space, severing the points just to one side of the plane from the points just to the other. Does space thereby fall into two disconnected pieces, so that points on one side now stand at no spatial distance from points on the other? Or does space, being reinforced, retain its shape?

If space is maximally reinforced by direct ties of distance, then a distance relation, such as *being twenty feet from*, is intrinsic to the points that stand in it. Whether or not the relation holds depends solely upon the intrinsic nature of the two points, and of the composite of the points. On the other hand, if space is not maximally reinforced by direct ties of distance, then a distance relation will not in general be intrinsic to the points that stand in it, and its holding may depend in part upon features of the surrounding space. What features? To fix ideas, roughly, try the following thought experiment. Start with two points, say, twenty feet apart, and remove some of the space directly between the two points. (I do not mean just the matter or energy occupying the space; I mean the space itself.) I ask: now how far apart are the points? Are they *still* twenty feet apart, on the grounds that distance, being intrinsic, is indifferent to changes in the intervening space? Or are they now *less* than twenty feet apart, on the grounds that there is now less space between them? Or are they now *more* than twenty feet apart, on the grounds that the shortest (continuous) path from one to the other is now more than twenty feet long? We have three competing answers, each, I think, with some intuitive appeal. Which is correct? And how can we tell?

The situation is familiar. We start with some notion from ordinary language or thought—in this case, the notion of distance (but compare the notions of person, cause, law, matter). We notice that there are different criteria associated with the notion, depending upon the context of application or of thought. But given the presuppositions under which the ordinary notion has evolved—in this case, presuppositions about the Euclidean nature of space—the different criteria fit together as well as you please. Then, driven perhaps by science, by mathematics, or by analytic philosophy, we consider extraordinary physical or logical possibilities that violate the presuppositions—for example, space with a "hole"—and the different criteria are seen to come apart. We are left with a plurality of competing conceptions, typically none of which captures all that was thought to be essential to the original ordinary notion. The question then arises: which conception should we accept?

It would be wrong in general to expect a univocal answer. Competing conceptions may be evaluated along at least three different dimensions. One can ask: which conception best corresponds to the ordinary notion with which we began? One can ask: which conception is mathematically, or philosophically, more fruitful, say, by leading to more interesting and powerful generalizations? Or one can ask: which conception has application at the actual world according to our best physical, or perhaps philosophical, theories?

In this paper, I evaluate various conceptions of distance. There are clear losers, but no clear winner, no conception that dominates the score on all dimensions of evaluation. I recommend pluralism: different conceptions can peacefully coexist as long as each holds sway over a distinct region of logical space. But when one asks which conception holds sway at the actual world, one conception stands out. It is the conception of distance embodied in differential geometry, the conception that underlies modern treatments of physical space (and spacetime) based upon Einstein's general relativity. On this conception, all facts about distance are analyzed in terms of "local" facts about distances between "neighboring points." Putting quantum mysteries to one side, I would say that this "local" conception gives the best account of distance at the actual world.[1] But there is a problem: the "local" conception, notwithstanding its mathematical and physical credentials, appears metaphysically suspicious. In the final section, I try to give the "local" conception a sound metaphysical footing.

A word of caution. My question whether distance relations are intrinsic to pairs of points should not be confused with the oft-discussed question whether "space has an intrinsic metric." Reichenbach, Grünbaum, and many others held that facts about the congruence of intervals of space are imposed from the outside, as it were, by our conventions for interpreting the behavior of material rods or rays. Without these conventions, they held, there are no facts about congruence; with different conventions, there are different such facts. I simply reject this here. The question here is not whether to be a metrical realist or conventionalist, but rather, assuming realism, what sort of realist to be.[2]

I

I need some preliminary notions and assumptions. I will speak of possible worlds. In this paper, I restrict attention to worlds at which space exists, at which space is composed entirely of points, and at which there are determinate facts about the distance, say in feet, between pairs of points. For these worlds, both intraworld and transworld comparisons of distance are meaningful. For the most part, I also restrict attention to *Newtonian* worlds, worlds at which there are determinate facts about the "identity over time" of points of space, and at which distances between points do not change over time. Newtonian worlds need not have Euclidean space. I assume that a variety of spatial structures will be exhibited at Newtonian worlds: curved and flat, finite and infinite, with and without boundaries, continuous and discrete.[3] Although I speak for simplicity primarily of spatial distance in Newtonian worlds, what I say applies more generally, *mutatis mutandis*, to temporal duration, and to intervals of spacetime in relativistic worlds. In section VI, the focus will switch from space to spacetime.

It is a matter of indifference whether one speaks of a distance *function* assigning non-negative reals to pairs of points, or of a multitude of distance *relations*, one for each non-negative real; I will speak of distance relations. I assume that the distance relations satisfy, at each Newtonian world, the usual constraints. Write '$D_r(\mathbf{p}, \mathbf{q})$' for '$\mathbf{p}$ is r feet from $\mathbf{q}$.' Let 'r', 's', and 't' range over non-negative reals.[4] Then, for any world, for any points $\mathbf{p}$, $\mathbf{q}$, and $\mathbf{r}$ at the world:

(D0) $D_r(\mathbf{p}, \mathbf{q})$, for exactly one r.
(D1) $D_r(\mathbf{p}, \mathbf{q})$, iff $D_r(\mathbf{q}, \mathbf{p})$, for all r.
(D2) $D_0(\mathbf{p}, \mathbf{q})$, iff $\mathbf{p} = \mathbf{q}$.
(D3) If $D_r(\mathbf{p}, \mathbf{q})$, and $D_s(\mathbf{q}, \mathbf{r})$ and $D_t(\mathbf{p}, \mathbf{r})$, then $t \leq r + s$, for all r, s, and t.

I will freely apply mereology to points of space. Whenever there are some points, there is a unique *fusion* of those points. In particular, any pair of points, $\mathbf{p}$ and $\mathbf{q}$, has a unique fusion, $\mathbf{p} + \mathbf{q}$. Whenever X is a proper part of Y, there is a unique *difference*, $Y - X$, which is the fusion of the parts of Y that do not overlap X. Space at a Newtonian world is the fusion of all the points of space existing at the world. I stay neutral as to whether a Newtonian world contains, in addition to the parts of space, entities that occupy those parts. If not, then the properties and relations ordinarily attributed to the "occupants" of parts of space must be attributed directly to the parts of space themselves. I assume that the spaces of distinct worlds do not overlap, that no point inhabits more than one world. Modal assertions that are *de re* points or regions of space must therefore be interpreted with respect to an appropriate counterpart relation.[5]

I assume that there are certain primitive or fundamental properties and relations, the holding or failing to hold of which suffices to determine, at any world, all the qualitative facts at that world. In particular, there are primitive or fundamental spatial properties and relations which suffice to determine, at any world, all the facts about distance. Whether or not the distance relations are themselves among the primitives is a question soon to be addressed. I will call the primitive or fundamental properties and relations *perfectly natural* properties and relations.[6] I assume that the part-whole relation is perfectly natural.

I will need to speak of worlds or parts of worlds being (intrinsic) *duplicates* of one another. I define 'duplicate' in terms of perfectly natural properties and relations:[7] for all X and Y, X and Y are *duplicates* iff there is a one-one correspondence between the parts of X and the parts of Y that preserves all perfectly natural properties and relations. (Remember: everything is a part of itself.) I call any such correspondence establishing that X and Y are duplicates an (X,Y)-counterpart relation; to each part Z of X, it assigns a unique part W of Y to be its (X,Y)-counterpart. (I drop the prefix when context allows.) Note that, for any (X,Y)-counterpart relation, (X,Y)-counterparts are duplicates of one another. However, duplicate parts of X and Y will not be (X,Y)-counterparts, for any (X,Y)-counterpart relation, unless they are similarly related to the other parts of X and Y. Note also that, since there will in general be more than one (X,Y)-counterpart relation, a part Z of X and a part W of Y may be (X,Y)-counterparts relative to some (X,Y)-counterpart relations, but not others. Nonetheless, in presenting examples I will leave the relation unspecified, and say simply that Z and W are (X,Y)-counterparts. That will not lead to trouble because what I say will hold true for an arbitrarily chosen (X,Y)-counterpart relation, assuming, of course, that a single such relation is held fixed throughout the example.

I turn now to the notion of an intrinsic property or relation. Intuitively, a property is intrinsic just in case whether it holds of an object depends only upon the way the object is in itself. Let us take the way an object is in itself—its *intrinsic nature*—to be given by the disposition of perfectly natural properties and relations among the object and its parts. Then we have, in terms of duplicates: A property P is an *intrinsic nature* iff P is had by all and only the duplicates of X, for some X. And, since an intrinsic property of an object is one that depends only upon the object's intrinsic nature, we have: A property P is *intrinsic* iff, for all X and Y, if X and Y are duplicates, then X has P iff Y has P. Note that, on this notion of intrinsic, an object's *haecceity*—the property of being that object—is not one of the object's intrinsic properties, since it is not shared by the object's duplicates.

The notion of intrinsic can be extended to relations in a natural way. Let us say that a (dyadic) relation is *intrinsic* just in case whether or not it holds of a pair $\langle X_1, X_2 \rangle$ depends only upon the intrinsic natures of X_1, X_2, and the fusion $X_1 + X_2$. Then, we have in terms of duplicates: A (dyadic) relation R is intrinsic iff, for all X, X_1, X_2, Y, Y_1, Y_2, if X and Y are duplicates, $X = X_1 + X_2$, $Y = Y_1 + Y_2$, X_1 a counterpart of Y_1, and X_2 a counterpart of Y_2, then R holds of $\langle X_1, X_2 \rangle$ iff R holds of $\langle Y_1, Y_2 \rangle$.[8] (Similarly for relations of three or more places.) If a property or relation is not intrinsic, it is *extrinsic*. Note that it follows immediately from the definitions that the perfectly natural properties and relations are themselves intrinsic. No surprise: the assumption was built into the definitions from the start.[9]

Finally, I will need two modest assumptions about the plenitude of possible worlds. One, a principle of recombination for points, I will introduce when I need it in section VII. The other is this: for any part of the space of a Newtonian world, there is a Newtonian world whose entire space is a duplicate of that part. Actually, I only apply this assumption to rather ordinary parts of a three-dimensional Euclidean space. To illustrate: start with an ordinary world satisfying the laws of Newtonian mechanics. The principle posits a world just like it except for a "hole" in space. At this world, conservation laws fail in the vicinity of the "hole". Objects entering the "hole" simply vanish; objects emerging from the "hole" appear out of nowhere. Bizarre, indeed. But logically impossible? Contemplate that as you drift towards the black hole at the center of the Milky Way!

II

I now present a multiple-choice exam. I invite the reader to try it. The questions involve precisely formulated variations on the thought experiments from the introduction. The original formulations were unsatisfactory. They were naturally understood to involve *de re* counterfactuals—e.g., if some of the space between two points *were* removed, *those* points *would* be closer together. The interpretation of such counterfactuals is not fixed once and for all: different contexts may favor different comparative similarity relations on worlds, and different counterpart relations on points.[10] I had a particular interpretation in mind; only when so interpreted do responses to the thought experiments have the intended metaphysical consequences. Therefore, to rule out unintended interpretations, I bypass the counterfactual formulations of the thought experiments and speak directly in terms of possibilia. (Once the intended interpretation is well established, I will allow myself to slip back into the counterfactual mode.)

DISTANCE EXAM

Part I: Removing Space. Consider a world with a three-dimensional Euclidean space, E. Let $\mathbf{p}$ and $\mathbf{q}$ be points of E twenty feet apart. Let X_i (for i from 1 to 5) be a part of E that includes $\mathbf{p}$ and $\mathbf{q}$. Consider a world whose entire space Y_i is a duplicate of X_i. Let $\mathbf{p}'$ and $\mathbf{q}'$ in Y_i be counterparts of $\mathbf{p}$ and $\mathbf{q}$ in X_i, respectively.

(1) X_1 is $E - A$, where A is an open[11] infinite slab bounded by two parallel planes, ten feet wide, centered on and perpendicular to the line segment connecting $\mathbf{p}$ and $\mathbf{q}$. (See figure 1.) How far apart are $\mathbf{p}'$ and $\mathbf{q}'$ in Y_1?

(a) Twenty feet apart.

(b) They stand in no
distance relation.

(c) Ten feet apart.

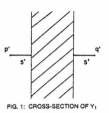

FIG. 1: CROSS-SECTION OF Y₁

(2) X_2 is $E - B$, where B is an open sphere, ten feet in diameter, centered on the point midway between **p** and **q**. (See figure 2.) How far apart are **p**′ and **q**′ in Y_2?

(a) Twenty feet apart.

(b) $10\sqrt{3} + 5\pi/3$ feet apart.
(Greater than twenty feet!)

(c) Ten feet apart.

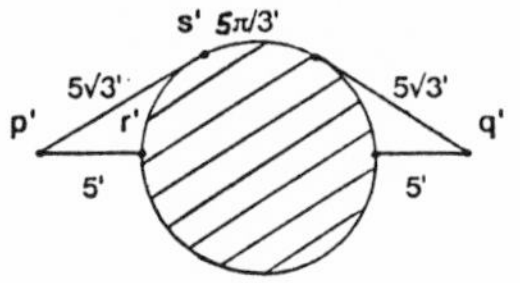

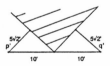

FIG. 2: CROSS-SECTION OF Y₂

(3) X_3 is $E - C$, where C is an open right circular cone with height and radius each ten feet, with vertex on the point midway between **p** and **q**, and with axis perpendicular to the line segment connecting **p** and **q**. (See figure 3.) How far apart are **p**′ and **q**′ in Y_3?

(a) Twenty feet apart.

(b) Twenty feet apart.

(c) $10\sqrt{2}$ feet apart.
(Less than twenty feet!)

FIG. 3: CROSS-SECTION OF Y₃

(4) X_4 is the surface of a sphere with diameter twenty feet and with **p** and **q** at opposite poles. How far apart are **p**′ and **q**′ in Y_4?

(a) Twenty feet apart.

(b) 10π feet apart.
(Half the sphere's
circumference.)

(c) Zero feet apart.

(5) X_5 is an infinite wavy plane whose hills and valleys are an alternating series of infinite half cylinders of diameter ten feet, with **p** and **q** on adjacent summits. (See figure 4.) How far apart are **p**′ and **q**′ in Y_5?

(a) Twenty feet apart.

(b) 10π feet apart.
     (A quarter of the cylinder's
     circumference, quadrupled.)

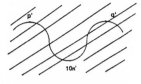

FIG. 4: CROSS-SECTION OF Y₅

(c) Zero feet apart.

Part II: Adding Space. Consider a world with a two-dimensional space, $X_i (i = 6$ or $7)$. Let **p** and **q** be points of X_i twenty feet apart. Consider a second world with a three-dimensional Euclidean space, E, and a part Y_i of E that is a duplicate of X_i. Let $\mathbf{p}'$ and $\mathbf{q}'$ in Y_i be counterparts of **p** and **q** in X_i respectively.

(6) X_6 is the surface of a sphere with circumference eighty feet. How far apart are **p** and $\mathbf{q}'$ in Y_6?

(a) Twenty feet apart.

(b) $40\sqrt{2/\pi}$ feet apart.
     (Length of chord subtending
     a quarter of a great circle.)

(7) X_7 is a Euclidean plane. How far apart are $\mathbf{p}'$ and $\mathbf{q}'$ in Y_7?

(a) Twenty feet apart.

(b) Could be any distance $d, 0 < d < 20$ feet,
     depending upon the nature of E and
     the choice of Y_7.

III

Now for the answers. Unfortunately, no one answer key will do. Different conceptions of distance answer the questions differently. Consider first the conception according to which distance relations are intrinsic to the pairs of points that stand in them. Call this the *intrinsic* conception of distance. (Presumably—though nothing rests on it—distance relations are not only intrinsic on this conception, but perfectly natural; for what other intrinsic features of the two points or their fusion could suffice to determine the distance between them?) On the intrinsic conception of distance, if points **p** and **q** are twenty feet apart, and $\mathbf{p}'$, $\mathbf{q}'$, and $\mathbf{p}'+\mathbf{q}'$ are duplicates of **p**, **q**, and $\mathbf{p} + \mathbf{q}$, respectively, then $\mathbf{p}'$ and $\mathbf{q}'$ are twenty feet apart. Now, for all seven questions, $\mathbf{p}'$, $\mathbf{q}'$ and $\mathbf{p}'+\mathbf{q}'$ are counterparts of **p**, **q**, and $\mathbf{p} + \mathbf{q}$, respectively; and counterparts are duplicates; so, on the intrinsic conception, the answer seven times over is: (a) twenty feet. "Additions to" and "removals from" the space surrounding two points are nowise relevant to the distance between them.[12]

Central to the intrinsic conception is the notion of *congruence*, generalized to apply to parts of space perhaps from different worlds: X is *congruent* to Y iff there is a one-one correspondence between the points of X and the points of Y that preserves all distance relations. On the intrinsic conception, duplicate parts of space are congruent. I assume the intrinsic conception accepts a partial converse as well: congruent parts of space are *spatial* duplicates, that is, they agree with respect to all their intrinsic *spatial* properties.[13] Thus, on the intrinsic conception, congruence serves to delimit the border between the intrinsic and the extrinsic.

The mathematical embodiment of the intrinsic conception is the abstract structure of a metric space. A metric space consists of a non-empty universe of points together with a family of distance relations (or a single distance function—it matters not) satisfying the axioms for distance listed above. The distance relations are taken as primitive, and other features of space—e.g., topological features—are defined in terms of the distance relations. The notion of a metric space is mathematically simple, yet extremely general: it encompasses spaces that are curved, flat, continuous, discrete, and all manner of hybrids thereof.[14] Since the distance axioms all quantify only universally over points, any part of a metric space, and so any duplicate of that part, is itself a metric space. That ensures that one may speak without impropriety of distances between points in the spaces Y_1 through Y_7.

Now consider a second conception of distance: the distance between two points of a space is given by the length of the (or a) shortest continuous path through space from one of the points to the other. (More exactly: the greatest lower bound of the lengths of continuous paths from one to the other, since in general there need be no least length; but I henceforth ignore this complication.) Call this the *Gaussian* conception of distance. The paths through space are themselves parts of space, fusions of points.[15] On the Gaussian conception, the assignment of lengths to paths is prior to the assignment of distances to pairs of points. The Gaussian characterization of distance may have an air of circularity about it; but the air is only apparent. One metrical notion—distance between points—is defined in terms of another metrical notion—length of paths. There is no attempt to analyze away all metrical notions. Later we shall ask how length of path can itself be analyzed; and then, of course, we shall have to be careful not to close the circle.

On the Gaussian conception, I suppose, the length of a path through space is an intrinsic property of that path. However, the distance between two points turns out not to be an intrinsic relation of those points. If some of the space surrounding two points is "removed," some or all of the paths connecting those points may no longer exist, and the length of the shortest remaining path—the new distance between the points—may be greater than it was, or not defined. If the space surrounding two points is embedded in a larger space, new paths connecting the points may come into existence, and the length of the shortest connecting path—the new distance between the points—may be less than it was. In short: the distance between two points does not depend solely upon the intrinsic nature of the fusion of the two points.

To illustrate, turn to the exam. The Gaussian conception answers (b) seven times over. In question (1), none of the paths connecting **p** and **q** in E have counterparts in Y_1. Thus no path in Y_1 connects **p**′ with **q**′, and the distance between them is undefined (equivalently, ∞, given our convention). Y_1 is composed of two "island universes" with **p**′ and **q**′ inhabiting different islands.[16] In question (2), the straight-line path from **p** to **q** in E has no counterpart in Y_2. The shortest path from **p**′ to **q**′ in Y_2 is one that follows a tangent from **p**′ to the edge of the hole, hugs the hole for a sixth of a turn, and then follows a tangent back to **q**′. (See figure 2.) Hence, the answer: $10\sqrt{3}+5\pi/3$ feet apart. In question (3), the straight-line path from **p** to **q** in E has a counterpart in Y_3, so the distance between **p**′ and **q**′ in Y_3 is the same as the distance between **p** and **q** in E: twenty feet. In question (4), the shortest path connecting **p**′ and **q**′ in Y_4 is half a great circle. In question (5), the shortest path connecting **p**′ and **q**′ in Y_5 slides down a quarter circle hill, around a half circle valley, and then up another quarter circle hill. (See figure 4.) In either case, the shortest path between **p**′ and **q**′ is a counterpart, not of the straight-line shortest path between **p** and **q** in E, but of a longer, more circuitous path between **p** and **q** whose length is given by the answer (b). I consider questions (6) and (7) below.

Central to the Gaussian conception is the notion of an *isometry* between parts of space: X and Y are *isometric* iff there is a one-one correspondence between the points of X and the points of Y that (when extended to fusions of points) preserves lengths of paths. (More exactly, the image of a path with endpoints **p** and **q** is a path of the same length with endpoints the image of **p** and the image of **q**.) Since Gaussian distance is defined in terms of lengths of paths, isometries preserve Gaussian distance as well. Duplicate parts of space are isometric, since the length of a path is intrinsic. With that the intrinsic conception can agree. But the Gaussian conception accepts, whereas the intrinsic conception must deny, the partial converse: isometric parts of space are spatial duplicates, and thus agree with respect to all their intrinsic spatial properties. On the Gaussian conception, isometries delimit the border between the intrinsic and the extrinsic: spatial properties are intrinsic just in case they are preserved by isometries, just in case they are *isometric invariants*.

To illustrate the difference between congruence and isometry, consider a "flat" plane F and a "wavy" plane W (such as X_5), each embedded in a three-dimensional Euclidean space E. F and W are not congruent, since no one-one correspondence between F and W can preserve the distances among four points of W not co-planar in E. Therefore, on the intrinsic conception, F and W are not spatial duplicates. However, F and W *are* isometric. Intuitively, this is because something flat, such as a piece of paper, can be made to wave without stretching or tearing, and thus without changing the lengths of any paths confined to its surface. Thus, on the Gaussian conception, F and W *are* spatial duplicates, and agree with respect to all their intrinsic spatial properties.

Let us now see how the Gaussian conception answers questions (6) and (7). For question (6), the Gaussian reasons: X_6 and Y_6 are duplicates; therefore isometric. The only parts of E isometric to X_6 are themselves surfaces of spheres with circumference eighty feet.[17] Therefore Y_6 is one such. Since the shortest path between two points on the surface of a sphere is part of a great circle, and **p** and **q** are twenty feet apart, **p** and **q** are connected by a quarter great circle in X_6. By the isometry, **p**′ and **q**′ are connected by a quarter great circle in Y_6. The distance, then, between **p**′ and **q**′ in Y_6, and so in E, is the length of a "wormhole" through the interior of the sphere; namely, the length of the chord that subtends the quarter great circle connecting **p**′ and **q**′. This length, which is less than twenty feet, is given by answer (b): $40\sqrt{2/\pi}$ feet apart.[18] For question (7), the Gaussian reasons: X_7 and Y_7 are duplicates; therefore isometric. Since E contains both flat planes and wavy planes that are isometric to X_7, Y_7 may be either wavy or flat. If Y_7 is a wavy plane, then **p**′ and **q**′ are closer together in Y_7 (and E) than **p** and **q** are in X_7: **p**′ and **q**′ are connected by a (straight-line) "wormhole" in E. How close together? The distance between **p**′ and **q**′ is just the length of the (straight-line) "wormhole," which may have any value in feet greater than zero and less than twenty, depending upon the "wavelength" of Y_7. Thus, the Gaussian answers (b) to question (7).

The Gaussian conception of distance finds mathematical expression in the development of differential geometry. Here is how the conception is typically motivated.[19] Start with a two-dimensional surface X embedded in a three-dimensional Euclidean space E. Ask: what geometrical features of the surface X could be ascertained by two-dimensional geometers whose measurements were entirely confined to X? These features comprise X's "intrinsic geometry." The length of a path confined to X can be ascertained to any specified degree of accuracy by placing sufficiently small measuring rods end to end along the path; so lengths of paths are intrinsic to X. Moreover, features that depend only upon lengths of paths—the isometric invariants—could all be ascertained by the geometers, and so are intrinsic to X; this includes, most famously, the Gaussian curvature at a point. But the true Euclidean distances between points of X are not intrinsic to X: they cannot be ascertained without "leaving the surface." The geometers do, however, have an intrinsic substitute for distance between points: *distance-within-X*, that is, the length of a shortest path in X between the points. Indeed, if the geometers (wrongly) take X to be all of space, they will (wrongly!) take distance-within-X to be the true distance.[20] The next step is to do away with the embedding space E, to consider a surface Y intrinsically just like X, but not embedded in any larger space. For this surface Y, the intrinsic geometry is all there is to geometry, and distance-within-Y is all there is to distance. The final step generalizes the Gaussian conception to apply to surfaces not embeddable within Euclidean space, and to spaces of dimension greater than two. In particular, since Euclidean distances are identical with distances-within-E, the Gaussian conception applies to E as well.

I will attempt to evaluate the motivation that underlies the Gaussian conception in section V. First I want to consider a third conception of distance. It is not, in my view, a serious contender. But I dare not ignore it: the most popular answer by far to question (1) is neither (a) nor (b), but (c). Call it the *naive* conception. The leading idea is this: the distance between two points should be a measure of the amount of space between the points; but, unlike the Gaussian conception, the amount of space between two points need not be identified with the length of any continuous path. Indeed, although the naive conception agrees with the Gaussian conception that distance relations are not intrinsic, the dependence of distance on the surrounding space is reversed. On the naive conception, if space is "removed" from between two points, the points will then be *closer together*; if space is "added" between two points, the points will then be *farther apart*. How might the amount of space between two points be measured so as to capture these naive intuitions?

Version 1. Many who answer (c) to question (1) have in mind closing up the gap left by the missing slab, and "stitching" the two remaining parts of Y_1 back together. The amount of space between $\mathbf{p}'$ and $\mathbf{q}'$ in Y_1 is then the length of the straight-line path in the stitched-up space. But that won't do. For one thing, on topological grounds the stitching cannot be seamless. One cannot avoid the seam by identifying boundary points on opposite sides of the gap; for that would violate the supposition that X_1 and Y_1 are duplicates (assuming the topological property *being connected* is intrinsic, and so must be shared by duplicates). But a seam composed of distinct, co-present boundary points would violate the distance axiom (D2) requiring that distinct points be some positive distance apart. How bad is that? *Perhaps* violations of (D2), if restricted to boundary points, could be tolerated on the grounds that boundary points may in effect be contiguous, and thus no distance apart.[21] In any case, the stitching idea does not generalize. Depending upon the shape of the part of space removed, the stitching might be done in any number of ways, the choice among which is arbitrary; indeed, this is so for Y_2 through Y_5. Thus, version 1 cannot in general provide determinate answers to questions about distance. We need another idea.

There are, however, ways to make the naive conception more precise; I will consider the two most promising. For each, I assume that the length of a path is an intrinsic property of that path; and I will speak of "disconnected paths," whose lengths (when defined) are the sum of the lengths of their connected parts. *Version 2.* Determine the distance between $\mathbf{p}'$ and $\mathbf{q}'$ in Y_i (i from 1 to 5) as follows: start with the straight-line path in E connecting $\mathbf{p}$ and $\mathbf{q}$; take the part of this path, perhaps disconnected, that overlaps X_i; then take as answer the length of this perhaps disconnected path (equivalently, of its counterpart in Y_i). When applied to questions (1) and (2), the result is answer (c): ten feet apart. But consider question (3). On version 2, the answer would be twenty feet, in agreement with the intrinsic and Gaussian conceptions. But someone who holds the naive conception could say that $\mathbf{p}'$ and $\mathbf{q}'$ are closer together in Y_3 than $\mathbf{p}$ and $\mathbf{q}$ are in E. They could take the amount of space between $\mathbf{p}'$ and $\mathbf{q}'$ to be the amount of space between $\mathbf{p}'$ and the cone-shaped hole plus the amount of space between the cone-shaped hole and $\mathbf{q}'$. (See figure 3.) This illustrates *version 3*. Determine the distance between $\mathbf{p}'$ and $\mathbf{q}'$ in Y_i as follows: start with all the (continuous) paths in E connecting $\mathbf{p}$ and $\mathbf{q}$; for each such path, take the perhaps disconnected part of the path that overlaps X_i; then take as answer the least of the lengths of these perhaps disconnected paths (or, equivalently, of their counterparts in Y_i). (More generally, take the greatest lower bound of the lengths.) When applied to question (3), version 3 gives the answer (c): $10\sqrt{2}$ feet apart.

In the cases considered, either version 2 or 3 may appear to give plausible answers. But trouble looms for both. First, note that on either version distinct boundary points of a space may be zero feet apart in violation of axiom (D2). For example, two points on opposite "shores" in Y_1, or two points on the edge of the "hole" in Y_2. *Perhaps*, as noted above, that is tolerable. But now look more closely at how version 2 applies to Y_2. Since boundary points $\mathbf{r}'$ and $\mathbf{s}'$ in Y_2 are zero feet apart (see figure 2), we have $D_5(\mathbf{p}', \mathbf{r}')$, $D_0(\mathbf{r}', \mathbf{s}')$, and $D_{5\sqrt{3}}(\mathbf{p}', \mathbf{s}')$ in violation of the triangle inequality (D3). Not so tolerable. No family of relations that violates the triangle inequality deserves to be called a family of *distance* relations.

That leaves version 3. Version 3 satisfies the triangle inequality, in much the same way as the Gaussian conception, by defining distance as the length of a shortest "path" (under an expanded notion of path). Since the shortest "path" in Y_2 from $\mathbf{p}'$ to $\mathbf{s}'$ goes by way of $\mathbf{r}'$, $D_5(\mathbf{p}', \mathbf{s}')$ on version 3. But when applied to Y_4 and Y_5, version 3, and version 2 to boot, fail to give plausible answers. Let's focus upon Y_4, the duplicate of the surface of a sphere. On both versions 2 and 3, not only $\mathbf{p}'$ and $\mathbf{q}'$, but any two points of Y_4 are assigned the distance: zero feet apart. That violates axiom (D2) in a big way. It obliterates all distinctions of distance, treating Y_4 in effect as a "space" with but a single point. Moreover, Y_4 is a space without boundaries; no path in Y_4 abruptly comes to an end. Even were one to tolerate violations of (D2) for boundary points, I see no comparable grounds for leniency here. I conclude that neither version 2 nor version 3 gives acceptable answers to questions about distance between points of Y_4.

Could some fourth version of the naive conception answer any differently? The distance between $\mathbf{p}'$ and $\mathbf{q}'$ in Y_4 must be less than or equal to twenty feet, on the naive conception, because space was "removed" from E; yet the distance must be a measure of the amount of space between $\mathbf{p}'$ and $\mathbf{q}'$. How could any such distance, other than zero, be singled out over any other? I reject the naive conception.

IV

That leaves two contenders: the intrinsic and the Gaussian conception. In this section, I question the Gaussian demarcation between intrinsic and extrinsic spatial properties. First, I ask the Gaussian whether the shape of a thing is intrinsic to that thing. Suppose it is. Consider a wavy plane and a flat plane embedded in a three-dimensional Euclidean space. They are isometric; therefore they are spatial duplicates, says the Gaussian; therefore, since shapes are intrinsic, they have the same shape. That is plainly wrong: one is curved and the other is flat! So the Gaussian must deny that shapes are intrinsic. That looks bad. We ordinarily take the shape properties to be the very paradigm of intrinsic properties, of properties that depend only upon the way something is in itself. And we ordinarily would say that two things cannot be duplicates of one another unless they have the same shape. It appears that the Gaussian conception clashes with our ordinary ways of thinking about shape, and so, derivatively, about distance. The intrinsic conception, on the other hand, can uphold the intuition that shapes are intrinsic, since the wavy plane and the flat plane are not congruent with one another.

I suppose a Gaussian might respond as follows. Ordinary intuition, rightly understood, does not conflict with the Gaussian conception. Our intuitions about shape apply only to the objects of our experience, not to mathematical abstractions therefrom; and the objects of our experience are *three*-dimensional and Euclidean. Now, for ordinary three-dimensional parts of a three-dimensional Euclidean space—spheres, cubes, even paper-thin sheets—the parts are isometric only if they have the same shape. That allows the Gaussian to hold that shapes are "intrinsic" in a restricted sense: *for ordinary three-dimensional parts of a three-dimensional Euclidean space*, duplicate parts always agree in shape. And ordinary intuition demands no more. Just as the domain of ordinary intuition is restricted to objects of experience, so should the sense in which ordinary intuition takes shapes to be intrinsic similarly be restricted.

This response lacks conviction. The two-dimensional surfaces of three-dimensional things, no matter how "abstract" or "ideal," seem to be objects of our intuition no less than the three-dimensional things themselves; and intuition pronounces the shapes of the former intrinsic no less than the latter. Nor does it much help to note that many two-dimensional surfaces, such as that of a sphere or a cube, are rigid, and so are isometric only if they have the same shape; for rigidity plays no role in the relevant intuition. The Gaussian should concede the clash with ordinary intuition. It no more condemns the Gaussian conception than, say, the acceptance of continuous paths through space that are nowhere "smooth"—once thought monstrous by intuition—condemns the standard mathematical analysis of continuity. Ordinary intuitions about matters susceptible to mathematical precision have often been found in need of revision. That is a small price to pay for the power and generality conferred by mathematics, or for the explanatory and predictive success of scientific theories mathematically based. Adherence to ordinary intuition is to some extent necessary to keep our bearings; but when mathematics gives us clear vision above and beyond, we should not hesitate to change our course. Agreement with ordinary intuition, *by itself*, favors the intrinsic conception but little.

V

In this section, the Gaussian takes the offensive. Suppose the geometers of a world have access to every part of their space. They have measured the length of every path, the area of every surface, the volume of every region. Then, says the Gaussian, they have the wherewithal to know all there is to know about the structure of their space; there are no spatial facts that are in principle inaccessible to geometric measurement. The intrinsic conception, we have seen, must disagree. Consider again a world with a two-dimensional space isometric to the Euclidean plane. On the intrinsic conception, the space may be "wavy" or "flat," depending upon facts purportedly about "distance"; but these "distance" facts are inaccessible to geometric measurement, even assuming the geometers have access to every part of their space.

Now the Gaussian objects: these facts are mysterious. If the two-dimensional space were embedded in some inaccessible higher-dimensional space, the "distance" facts could be understood in terms of inaccessible facts about the nature of the embedding. But by assumption there is no such embedding space. Rather, the inaccessible "distance" facts reflect the possibilities of embedding, what would have been the case, had the space been embedded in some higher-dimensional space. But to explain such facts in terms of the possibilities of embedding is to reverse the true order of things. What reason could there be, then, for taking these facts which are inaccessible to geometric measurement to be facts about *distance*, to be *spatial* facts? By maintaining that these inaccessible facts are facts about distance, the intrinsic conception posits a *phantom embedding space*, the ghost of a departed embedding.

Note that this argument does not rest upon a positivist premise. The Gaussian need not deny that there could be perfectly natural relations between points of space—even relations satisfying the distance axioms—knowledge of which is inaccessible to geometers who have access to all parts of space. The Gaussian need only deny that such relations could be the *distance* relations.

Before attempting to evaluate the argument, I want to examine the assumption that lies at its heart. It is a supervenience thesis. When stripped of colorful talk of tiny geometers, it comes to this: if the spaces of two worlds are isometric, then the spaces are congruent as well. In short: distances supervene upon lengths of paths.[22] This assumption needs to be qualified in at least two ways. The alleged supervenience is contingent, not logical.

First, consider worlds with discrete space. In a discrete space, there are no continuous paths between points, so no lengths of continuous paths. Therefore, any two discrete spaces with the same number of points agree vacuously on all lengths of continuous paths. But the spaces need not agree on all distances between points, say, by having each point be an island universe all to itself. That is one possibility, I suppose. But I also suppose the points of a discrete space may stand in various distance relations, as long as the relations satisfy the axioms for distance. Indeed, I suppose there could be physical evidence that actual space (or spacetime) is discrete, and thus that the actual distance relations are not Gaussian. Since the Gaussian analysis of distance cannot account for the variety of possible discrete—more generally, disconnected—spaces, the Gaussian assumption must be qualified: distances supervene upon lengths of paths, *for worlds with continuous space.*

Gaussian supervenience also fails, I think, at some worlds with continuous space. Consider worlds with "action at a distance": worlds at which forces act directly from one point to another without being propagated along a continuous path connecting the points. At such worlds, distance relations need not be Gaussian. For example, consider again Y_2, the space with a "hole." Suppose that away from the "hole," Newtonian laws of motion and of universal gravitation have been well confirmed: the force of gravity produces an acceleration that varies inversely with the square of the distance. Then, a measurement of the acceleration of objects located at p' and q' could give evidence against the Gaussian conception. Moreover, the "action at a distance" need not be instantaneous. Suppose that away from the "hole" it is well confirmed that all causal signals travel no greater than the speed of light. Then, a measurement of the time it takes signals to travel from p' to q' could give evidence against the Gaussian conception. The most the Gaussian is entitled to claim is this: there could be no evidence against the Gaussian conception at worlds where it has been established that all action is local, that is, propagated locally along continuous paths. For only in such worlds must evidence for distance relations be, *ipso facto*, evidence for lengths of paths.

The Gaussian supervenience thesis thus applies at most to *local-action* worlds, to worlds with continuous space and no action at a distance. The Gaussian argument against the intrinsic conception must similarly be limited in scope. (The argument erred, in particular, by focusing too narrowly upon what would be accessible to *geometers*, rather than what would be accessible to physicists more generally.) It follows that the intrinsic conception of distance cannot be jettisoned from logical space. The Gaussian must accede to pluralism: at some worlds, distance relations are intrinsic, and presumably perfectly natural; at other worlds, distance relations are extrinsic, and subject to the Gaussian analysis. Under pluralism, worlds with Euclidean space may have either intrinsic or extrinsic distance relations. That points up a flaw in my distance exam. The embedding space E was underspecified. Interpreted one way, the answer is (a) throughout; interpreted the other way the answer is (b). I shall have to give a lot of 'A's.

Pluralism is the best the Gaussian can hope for. Unfortunately, the Gaussian is not yet in a position to demand his share of logical space. Limiting the scope of the Gaussian argument undermines its force altogether. Any world at which Gaussian supervenience fails—be it a world with discrete space or with action at a distance—is a world with a phantom embedding space, no less than a world whose space is a "wavy plane." The argument began as a general indictment of worlds with a phantom embedding space. What remains is a specific indictment—based none too clearly on considerations of physical evidence—of *local-action* worlds with a phantom embedding space. But the intrinsic conception is not committed to such worlds. If one takes a local-action world with intrinsic distance relations, and one "removes," say, a sphere from its middle, one gets a world with a phantom embedding space all right, but the world is no longer a local-action world. I conclude that the Gaussian is left without any argument against the intrinsic conception. For the sake of uniformity, why not take the intrinsic conception to apply to all worlds with distance relations?

VI

Here's why. Our best physical theory of space and time, Einstein's general relativity, is based upon differential geometry, and is Gaussian through and through. I suppose general relativity is logically possible, that is, true at some possible worlds. At these worlds, distance relations are Gaussian, not intrinsic. Moreover, to whatever extent we believe that general relativity is true at the actual world, to that extent we should believe that actual distance relations are Gaussian.

Before turning to the treatment of distance in general relativity, we need to further develop the Gaussian conception. Thus far, length of path has been left unanalyzed. If length of path is taken as primitive, then the Gaussian and the intrinsic conception are both *global* conceptions of distance: both apply primitive metrical notions to pluralities of points, in one case to paths, in the other to pairs. I now want to develop a *local* version of the Gaussian conception according to which the only primitive metrical notions are local properties of points. (I postpone an exact definition of 'local property' until the next section.) Let us again assume that the points of space have a *manifold* structure in terms of which paths can be characterized as continuous and smooth. Now, assign to each point of space a metric tensor g, or ds^2, which supplies information about distances within an "infinitesimal neighborhood" of the point; call the tensor g at a point p the *local metric* at p.[23] Given two points p and q (no matter how "close together"), the distance between them is not determined by the local metric at p and the local metric at q. But given a path from p to q, the length of that path is determined by the local metric at each point along the path: it is the result of *integrating ds* along the path from p to q.[24] (In effect, the local metric at a point provides a set of "infinitesimal measuring rods," one for each direction, to be used for determining lengths of infinitesimal portions of paths passing through the point; integration then corresponds to measuring the length of a path by laying (continuum-many!) appropriately directed measuring rods end to end.) On the local Gaussian conception, the properties of having such-and-such local metric are taken as primitive. These properties then suffice to determine the length of any path through space, and so the Gaussian distance between any two points. There are no primitive global metrical properties or relations.

Now, I claim that general relativity is a local Gaussian theory. (Of course, here the local metric has Lorentz signature, since it provides information about infinitesimal intervals in spacetime, rather than infinitesimal distances in space;

the spacetime interval between two points is given by a longest, rather than a shortest, path.) A key insight behind general relativity is that all physics takes place by local action.[25] When applied to gravitation, this leads to Einstein's field equation, the fundamental law of gravitation in general relativity. Einstein's equation—$G = 8\pi T$—tells how the local mass-energy density, given by the stress-energy tensor T, relates to the local curvature of spacetime, given by the Einstein curvature tensor G. (G is analyzable in terms of the metric tensor g.) More generally: when the fundamental laws of physics are formulated within general relativity, the metrical notions that occur in the laws are all *local* metrical properties, including the metric tensor g. (This contrasts sharply with Newtonian physics: according to the fundamental law of gravitation, the force of gravity varies inversely as the square of the *distance*.) Now, I suppose that all and only the perfectly natural properties and relations instantiated at a law-governed world occur in the fundamental laws of that world. It follows that the properties of having such-and-such local metric are perfectly natural properties instantiated at general relativistic worlds, and that general relativity, as formulated by Einstein, is a local Gaussian theory.

I do not deny that one could give an empirically equivalent reformulation of general relativity in terms of global metrical relations; under specifiable conditions, global and local metrical relations are *inter*definable with aid of the calculus. But that would not show global metrical relations to be perfectly natural, any more than, say, a reformulation of the laws of color (if there were such laws) in terms of grue and bleen would show grue and bleen to be perfectly natural. In general relativity, the local metric at a point is a *dynamic object*, a *prime mover*: it tells objects at that point how to move. It has the same claim to perfect naturalness as the other prime movers, such as the electromagnetic field at a point. It would be arbitrary and absurd to hold that some prime movers are perfectly natural, but not others.

It is well known that Einstein's general relativity eliminates primitive action at a distance. I have been arguing what is perhaps less well known, that general relativity eliminates primitive "distance at a distance." The reduction of global relations to local properties in general relativity applies to metrical relations as well. This has implications for the formulation of philosophically interesting supervenience theses. David Lewis has defended the viability of *Humean supervenience*, according to which all facts, other than facts about spatiotemporal distance, supervene upon local matters of particular fact. At worlds of Humean supervenience: "We have geometry: a system of external relations of spatiotemporal distance between points. . . . And at the points we have local qualities: perfectly natural intrinsic properties which need nothing bigger than a point at which to be instantiated. . . . All else supervenes on that."[26] But have we not just seen that, at least at local Gaussian worlds, even the relations of distance between points supervene on local matters of fact? Do we have, then, a sweeping elimination of all primitive global notions at local Gaussian worlds, a grand supervenience of the global on the local?

That would be too much to ask. On the local Gaussian conception, global distance relations supervene not on local metric alone, but on local metric plus manifold structure. Without manifold structure, no integration. Without integration, no analysis of global distance relations in terms of local metric. Manifold structure is in part topological structure, and topological structure, it is easy to see, is irreducibly global. Consider a two-dimensional Euclidean plane and (the surface of) an infinite cylinder. They are locally indistinguishable: each consists of continuum-many points that are locally Euclidean. But the plane and the cylinder differ topologically. For example, the plane, but not the cylinder, is *simply connected*: all closed paths can be continuously contracted to a point. Just as there are irreducibly global topological features of *space*, so also of *spacetime* at relativistic worlds.

Thus, general relativity suggests no grand supervenience of everything on local matters of particular fact; it suggests something more modest. Call it *Einsteinian supervenience*. At worlds of Einsteinian supervenience: we have a manifold of spacetime points (with topological and differential structure), and a distribution of perfectly natural local properties (including local metrical properties) over those points; all else supervenes on that. Of course, Einsteinian supervenience, like its Humean cousin, is philosophically controversial. I here claim only that it is the right supervenience thesis to consider at general relativistic worlds. Under pluralism, Einsteinian and Humean supervenience are not in conflict. Each holds contingently, and governs its own region of logical space. These regions differ with respect to the instantiation of perfectly natural spatiotemporal relations of distance. Given the success of general relativity, I suspect we are nearer to, if not within, the region of Einsteinian supervenience.

VII

I have argued that modern physics of spacetime is based upon a local Gaussian conception of (spatiotemporal) distance. In this final section, I ask whether the local Gaussian conception is metaphysically suspect. I claim that the local metric at a point, as standardly conceived, is not an intrinsic property of the point.[27] Thus, the local Gaussian appears to be committed to perfectly natural, *extrinsic* properties. That would introduce necessary connections between distinct co-inhabitants of local Gaussian worlds, namely, between points and their surrounding space; it would violate a modal "principle of recombination" that I, for one, would be loath to give up. I take this as a challenge not to the physicist—I am not so bold—but to the metaphysician: provide a coherent metaphysical foundation for modern space-time theories.

First, we need a precise characterization of local properties. The characterization requires topological structure. Say that a part of space, N, is a *neighborhood* of a point $\mathbf{p}$ iff some part of N includes $\mathbf{p}$ and is open in the topology of the space. (For example, in a three-dimensional Euclidean space, N is a neighborhood of $\mathbf{p}$ iff N includes some *open ball* around $\mathbf{p}$, that is, all the points less than some positive distance r from $\mathbf{p}$.) A property of points

P is local iff, for any points $\mathbf{p}$ and $\mathbf{q}$, for any neighborhood N of $\mathbf{p}$ and any neighborhood M of $\mathbf{q}$, if N is a duplicate of M and $\mathbf{p}$ is a (N, M)-counterpart of $\mathbf{q}$, then P holds of $\mathbf{p}$ iff P holds of $\mathbf{q}$. Note that if a property of points is intrinsic, then it is local; for counterparts, being duplicates, share all their intrinsic properties. But in general local properties need not be intrinsic. Call a property of points that is local but not intrinsic *neighborhood-dependent*.[28]

The most familiar examples of neighborhood-dependent properties come from elementary calculus: derivatives of functions. Consider the position of some point-sized object as a function of time; suppose at time $\mathbf{t}$ it is located at point $\mathbf{p}$. The *instantaneous velocity* of the object at $\mathbf{t}$ is the derivative of the position function evaluated at $\mathbf{t}$. This derivative at $\mathbf{t}$ depends upon the object's position not only at $\mathbf{t}$, but also at "neighboring" times. Or, turning this around, the derivative at $\mathbf{t}$ depends upon when the object is located not only at $\mathbf{p}$, but also at "neighboring" points. The object's instantaneous velocity at $\mathbf{t}$ is thus a neighborhood-dependent property of both the time $\mathbf{t}$ and the point $\mathbf{p}$. (In spacetime, of the "event" $\langle \mathbf{p}, \mathbf{t} \rangle$.) In two or more dimensions, position is given by a vector, and so instantaneous velocity is a vector as well, having both a magnitude (speed) and a direction (if non-zero). Both an object's speed and its direction of motion are neighborhood-dependent properties of points and of times.

Now, I claim that the local metric at a point $\mathbf{p}$, as characterized in differential geometry, is a neighborhood-dependent property of $\mathbf{p}$. That is because the local metric at $\mathbf{p}$ is an inner product on the tangent space at $\mathbf{p}$: it takes a pair of tangent vectors as input, and gives a real number as output. (If the inputs are one and the same, the output is the squared length of the tangent vector.) The tangent vectors at $\mathbf{p}$ are defined as the *derivatives* of "smoothly" parametrized paths through $\mathbf{p}$. (A parametrized path is a function from an interval of real numbers to the points of the path. If one thinks of the parameter as "time," then a "smoothly" parametrized path through $\mathbf{p}$ is a trip through $\mathbf{p}$ with no jolts or stops, and the tangent vectors at $\mathbf{p}$ are all the possible "velocities," or "states of motion," when passing through $\mathbf{p}$.) These tangent vectors, being derivatives, give information not just about $\mathbf{p}$, but about the space immediately surrounding $\mathbf{p}$. For example, the dimensionality of the tangent space is the dimensionality, not of $\mathbf{p}$ which is zero, but of the immediately surrounding space. In short: the tangent vectors provide neighborhood-dependent information about $\mathbf{p}$. Since the local metric at $\mathbf{p}$ is an operator on tangent vectors, it inherits neighborhood-dependence from its operands.[29]

Thus, the local metric at a point, as standardly conceived in differential geometry, is neighborhood-dependent; and that is trouble for the local Gaussian conception. For, on the local Gaussian conception, the local metric is also perfectly natural. Apply the definitions from section I. If perfectly natural, then shared by duplicates; if shared by duplicates, then intrinsic. So both neighborhood-dependent and intrinsic. *Contradiction.*

Perhaps we should revise our definitions, not our conception of distance. It was simply built into the definitions that all perfectly natural properties are intrinsic. What was built in can be built out. For a cost. We need to introduce a primitive distinction between the perfectly natural properties that are intrinsic and those that are not. (So we no longer can analyze 'intrinsic' just in terms of perfectly natural properties and relations.) Now the definition of 'duplicate' bifurcates: X and Y are *intrinsic duplicates* iff there is a one-one correspondence between the parts of X and the parts of Y that preserves all *intrinsic* perfectly natural properties and relations; X and Y are *local duplicates* iff there is a one-one correspondence that preserves *all* perfectly natural properties and relations.[30] A *local* property is now defined simply as a property that can never differ between local duplicates. That perfectly natural properties are *local* is now built into the definitions.

Now we face a dilemma. I suppose we will want the revised theory to incorporate the Humean denial of necessary connections between distinct existences, in particular, between a point $\mathbf{p}$ and its surrounding space: any other point $\mathbf{q}$ could have taken $\mathbf{p}$'s place. Of course, at no world is $\mathbf{q}$ itself in the place of $\mathbf{p}$. We need to formulate the principle in terms of duplicates: for any points $\mathbf{p}$ and $\mathbf{q}$, perhaps from spaces of different worlds, there is a world whose space is a duplicate of the space of $\mathbf{p}$ except that it contains a duplicate of $\mathbf{q}$ where the duplicate of $\mathbf{p}$ would be.[31] But, on the revised theory, we must decide: do we mean *local* duplicate or *intrinsic* duplicate? Does the principle require that there be a *local* duplicate of $\mathbf{q}$ where the duplicate of $\mathbf{p}$ would be? It had better not. For example, suppose that $\mathbf{p}$ is surrounded by positively curved space, $\mathbf{q}$ by negatively curved space. Then, a world whose space is a duplicate of the space of $\mathbf{p}$ but with a *local* duplicate of $\mathbf{q}$ in $\mathbf{p}$'s place must be both positively curved and negatively curved in the immediate neighborhood of $\mathbf{q}$. No world is like that. So the principle requires only that there be an *intrinsic* duplicate of $\mathbf{q}$ where the duplicate of $\mathbf{p}$ would be. More generally, the Humean denial of necessary connections is formulated in terms of *intrinsic* duplicates, not *local* duplicates. On the revised theory, however, that will be too weak to capture the spirit of the Humean denial. It rules out necessary connections between the *intrinsic natures* of distinct things. But, on the revised theory, there may be more to a thing than is given by its intrinsic nature. Thus, formulating the Humean denial in terms of intrinsic duplicates fails to rule out necessary connections between the distinct things themselves, in particular, between a point and its surrounding space.[32]

I suggest we drop the revised theory and pursue a different tack. Although the local metric, as standardly conceived, is an extrinsic property of points, and therefore not perfectly natural, perhaps the extrinsic local metric is "grounded" on an intrinsic, perfectly natural property of points. To illustrate the sort of grounding I have in mind, consider mass density. If one assumes that each neighborhood of a point has some determinate (finite) mass and volume, then the *mass density* at a point may be characterized as the limit of the ratio of mass to volume, as volume shrinks to zero. So characterized, mass density is an extrinsic property of points. But it is customary in physics, when considering a continuous matter field, to instead take mass density to be a primitive scalar field: a function that assigns to each point a real number representing (given appropriate units) the *intrinsic mass density* at the point. Given intrinsic mass density, and an assumption about its smooth distribution, mass can be defined by integration. Extrinsic mass density then supervenes upon intrinsic mass density. And, thanks to a fundamental theorem of integral calculus, the values of extrinsic and intrinsic mass density coincide. Note that the smoothness of the intrinsic mass density field is a contingent feature of worlds with continuous matter. Given a principle of recombination for points, there will be worlds whose intrinsic mass densities (perhaps no longer properly so-called) are jumbled up in such a way that no (finite) masses (and no extrinsic mass densities) exist at the world.[33]

The suggestion, then, is to say something analogous about the local metric: the extrinsic local metric supervenes on an intrinsic local metric (plus manifold structure). It is the *intrinsic* local metric properties that are perfectly natural. That is on the right track, I think; but there is a problem. Whereas the mass density at a point is a simple *scalar* quantity, the local metric at a point is a *tensor* quantity. How can a tensor be intrinsic to a *point*? Points are spatially simple. Tensors, being operators on vectors spaces, are spatially complex. It is repugnant to the nature of a point to suppose that a local metric, which is a tensor, could be intrinsic to a point. If we hope to ground the extrinsic local metric on an intrinsic local metric, the latter had better be intrinsic not to a point, but to something spatially complex.[34]

No sooner said than done. If we are willing to posit perfectly natural properties on theoretical grounds, we should be willing to posit appropriate entities to instantiate those properties: in this case, entities that are spatially complex. I propose that we reify talk of the "infinitesimal neighborhood" of a point. The tangent space at a point is now conceived as the infinitesimal neighborhood of the point "blown large," as viewed through a "microscope" with infinite powers of magnification; it no longer depends for its existence upon the manifold structure. Tensor quantities are intrinsic not to points, but to the infinitesimal neighborhoods of points. At local Gaussian worlds, space (or spacetime) has a "non-standard" structure. There are "standard" points, and there are "non-standard" points that lie an infinitesimal distance from standard points. The points along a path in space are ordered like the non-standard continuum of Abraham Robinson's non-standard analysis.[35]

Let us take stock. The local Gaussian conception of distance, if founded upon standard differential geometry, is committed to local metric properties that are both extrinsic and perfectly natural. I propose founding the local Gaussian conception instead upon non-standard differential geometry. That allows the perfectly natural local metric properties to be intrinsic, though not to points, but to their infinitesimal neighborhoods. The intrinsic local metric at a point now comprises a family of infinitesimal distance relations. So there turn out to be perfectly natural distance relations after all; but they are local, not global, because they hold only among points within an infinitesimal neighborhood of a standard point. ('Local' is defined with respect to the topology of the standard points.) The local Gaussian is no longer committed to perfectly natural, extrinsic properties. Metaphysical worries about necessary connections have been resolved.

When non-standard analysis gained mathematical and logical respectability some thirty-odd years ago, the question naturally arose whether the non-standard continuum is instantiated at any possible world, or even at the actual world. Perhaps the mere consistency of non-standard analysis already gives reason to suppose that the non-standard continuum is possibly instantiated. The role that non-standard differential geometry can play in firming up the metaphysical foundations of physical theory gives reason all the more—including reason to suppose the non-standard continuum is actual.

NOTES

1. What if quantum mysteries are taken into account? Should I endorse some conception of space as foamy, or spongy, or stringy, or loopy? I haven't a clue.

2. For the conventionalist line, see Hans Reichenbach, *The Philosophy of Space and Time* (New York, 1957), or Adolf Grünbaum, *Philosophical Problems of Space and Time*, 2nd ed. (New York, 1963), chapter 16. For a realist response, see Graham Nerlich, "Is Curvature Intrinsic to Physical Space?" *Philosophy of Science* 46 (1979): 439–58.

3. For a discussion of which spatial structures are possible, that is, instantiated at some possible world, see my "Plenitude of Possible Structures," *Journal of Philosophy* 88 (Nov. 1991): 607–19.

4. For convenience, I include ∞ among the non-negative reals, and read '$D_\infty(\mathbf{p}, \mathbf{q})$' as $\mathbf{p}$ stands in no distance relation to $\mathbf{q}$. As usual, ∞ is greater than any other non-negative real, and ∞ added to any non-negative real is ∞. This artifice allows the distance axioms to hold at worlds with "island universes," spatially disconnected parts.

5. Those who prefer genuine transworld identity may simply suppose that one of the appropriate counterpart relations is the relation of identity. No trouble arises because, for the counterpart relations introduced below, counterparts are always intrinsic duplicates.

6. Following David Lewis. For discussion, see *On the Plurality of Worlds* (Oxford, 1986), 59–61.

7. Again following David Lewis, *On the Plurality of Worlds*, 61–62. The definitions below of 'intrinsic', 'internal', and 'external' are also adapted from Lewis, although he does not apply the word 'intrinsic' to relations. Note that quantifiers here and below range over all *possibilia*.

8. As Lewis notes (*On the Plurality of Worlds*, 68), on a theory of universals according to which universals are parts of the particulars that instantiate them, we must everywhere replace the fusion $X + Y$ with an *augmented fusion*, which includes among its parts not only X and Y, but the dyadic universals that hold between X and Y (or between parts of X and parts of Y), and the monadic universals that hold of the fusion $X + Y$ (or of its parts). *Mutatis mutandis* for a theory of tropes.

9. The intrinsic relations can be further divided into *internal* and *external*. A (dyadic) relation is internal just in case whether it holds of a pair $\langle X_1, X_2 \rangle$ depends only upon the intrinsic nature of X_1 and of X_2, not upon the intrinsic nature of $X_1 + X_2$. In terms of duplicates: A (dyadic) relation R is *internal* iff, for all X_1, X_2, Y_1, Y_2, if X_1 and Y_1 are duplicates and X_2 and Y_2 are duplicates, then R holds $\langle X_1, X_2 \rangle$ iff R holds of $\langle Y_1, Y_2 \rangle$. R is *external* iff R is intrinsic but not internal. This distinction, however, will not be needed below. It is agreed on all sides that distance relations, if intrinsic, are external.

10. On the interpretation of *de re* counterfactuals, see David Lewis, *Counterfactuals* (Cambridge, Mass., 1973), 36–43.

THE FABRIC OF SPACE

11. That is, the points of A form an open set in the usual topology. A part of space is open iff it excludes all of its boundary points.

12. The intrinsic conception of distance is endorsed by David Lewis in *On the Plurality of Worlds*, 62. Lewis does not consider alternative conceptions.

13. What about a right- and a left-handed glove that are mirror images of one another, and thus congruent? That is no counterexample. The gloves differ in *orientation*, and orientation is not intrinsic, as consideration of a Möbius strip (or its three-dimensional analog) should make clear.

14. Of course, a further generalization is needed to account for the interval relations of Minkowski spacetime, since the interval squared may be positive, negative, or zero.

15. In what follows, unless otherwise noted, I assume that paths are continuous and "smooth"—that is, without corners or cusps. (Technically, I assume that paths can be given a parametrization that is differentiable with non-zero derivative at all points along the path.) In order that the notions of continuity and "smoothness" be applicable to parts of space, the Gaussian must assume that space is a *manifold*, that space has both topological and differential structure.

16. I assume island universes are possible. An upholder of the Gaussian conception who denied this should refuse to answer question (1) on the grounds that there is no world whose entire space is a duplicate of X_1.

17. This is because the surface of a sphere (unlike a flat plane) is *rigid* in E: any part of E that is isometric to the surface of a sphere in E is also congruent to it, and so itself the surface of a sphere of the same size. Intuitively, a surface in E is rigid if it cannot be deformed without stretching or tearing.

18. Note that, on the intrinsic conception, $\mathbf{p}'$ and $\mathbf{q}'$ are separated by more than a quarter great circle of Y_6, since $\mathbf{p}'$ and $\mathbf{q}'$ are twenty feet apart in Y_6, and Y_6 is the surface of a sphere of circumference eighty feet.

19. For an excellent introduction to differential geometry, including the standard motivational asides, see Barrett O'Neill, *Elementary Differential Geometry* (New York, 1966).

20. Unfortunately for our purposes, distance-within-a-surface is often called "intrinsic distance," since it is part of the surface's intrinsic geometry. I will avoid that usage.

21. This notion of a boundary point of space requires differential (but not metrical) structure: a boundary point is one such that some path to the point cannot be "smoothly" extended.

22. *Globally* supervene, that is, since the assumption only applies to entire worlds. On the distinction between various notions of supervenience, see Paul Teller, "A Poor Man's Guide to Supervenience and Determination," *Southern Journal of Philosophy* 22 (1984): 137–62.

23. More exactly, the metric tensor at a point is an inner product on the tangent space of the point; and the metric tensor field is differentiable, it varies "smoothly" from point to point.

24. The information carried by $\mathbf{ds}^2$ is coordinate-independent, though of course calculations of length of path will be done by representing $\mathbf{ds}^2$ and the path in question relative to some chosen coordinates. (Coordinate-free geometric objects are represented in boldface.) In the case of a three-dimensional Euclidean space, there will be x, y, z-coordinates under which $ds^2 = dx^2 + dy^2 + dz^2$. In general, however, ds^2 will be a more complicated quadratic function of dx, dy, and dz, for any x, y, z-coordinates.

25. For an elaboration on this theme, see the introductory chapter of Charles Misner, Kip Thorne, and John Wheeler, *Gravitation* (San Francisco, 1973), 4.

26. David Lewis, Introduction to *Philosophical Papers*, Volume II (Oxford, 1986), 3–4.

27. David Lewis, in the passage just quoted, requires that local properties be intrinsic properties of points (or their point-sized occupants); but I do not think that is how 'local' is standardly used in mathematics or physics.

28. Neighborhood-dependent properties may be *exclusive* or *inclusive*: those are exclusive

that exclude information about the intrinsic nature of points that instantiate them. Thus, a property **P** of points is *exclusively neighborhood-dependent* iff, for any points **p** and **q**, for any neighborhood N of **p** and any neighborhood M of **q**, if $N - $**p** is a duplicate of $M - $**q**, then P holds of **p** iff P holds of **q**.

29. This is a bit fast and loose. Unless the paths through **p** are embedded in some higher-dimensional Euclidean space, the derivatives in question are not defined, and tangent vectors are instead identified with (directional) derivative *operators*. (See O'Neill, *Elementary Differential Geometry*, 182–84.) But the argument is essentially unchanged, since derivative operators, which require manifold structure, are no less neighborhood-dependent than derivatives.

30. One might ask, independently of the question whether perfectly natural properties can be extrinsic, whether 'duplicate' in ordinary usage means 'local duplicate' or 'intrinsic duplicate', or is indeterminate. *Test case.* Consider a cube with sides of two feet and a sphere with a diameter of one foot, each composed of (the same kind of) homogeneous continuous matter. The sphere has continuum-many intrinsic duplicates among the parts of the cube; but the sphere has no local duplicates, since no interior point of the cube is a local duplicate of any boundary point of the sphere. Using our ordinary notion of duplicate, how many duplicates of the sphere are there in the cube? It seems to me one can answer either way.

31. This is an instance of the principle of recombination put forth by David Lewis. See *On the Plurality of Worlds*, 86–92. (Of course, the points in question must be of the same kind, be it Newtonian or spatiotemporal.)

32. The argument is especially compelling if one holds that perfectly natural properties correspond to immanent universals or classes of tropes. For immanent universals or tropes are present in their instances. Now consider a neighborhood-dependent, perfectly natural property of **p**. The corresponding universal, or a corresponding trope, is present at **p**. And, unlike a dyadic universal or trope, it is wholly present at **p**. (Remember: its holding at **p** tells one nothing about any point other than **p**, not even something relational.) Intrinsic or not, how can one deny that it is part of the nature of **p**, and so must be "recombined" along with **p**?

33. Michael Tooley has argued that extrinsic (or "Russellian") velocity should be grounded in this way on primitive velocities that are intrinsic to points. But the theoretical reasons for positing primitive velocities, at least at worlds approximating ours, seem to me much weaker than the theoretical reasons for positing primitive local metrics. See Michael Tooley, "In Defense of the Existence of States of Motion," *Philosophical Topics* 16 (Spring 1988): 225–54.

34. Denis Robinson asks whether vectors could be intrinsic to points, and answers "no," in "Matter, Motion, and Humean Supervenience," *Australasian Journal of Philosophy* 67 (December 1989): 394–409. I concur. Although vectors are spatially less complex than tensors, they have a "tail" and a "tip": too much to fit within a single point.

35. Non-standard analysis is applied to differential geometry, for example, in Abraham Robinson, *Non-Standard Analysis*, rev. ed. (Amsterdam, 1974).

MIDWEST STUDIES IN PHILOSOPHY, XVIII (1993)

Against Experimental Metaphysics

MARTIN R. JONES AND ROBERT K. CLIFTON

The attempt to understand the violation of the Bell inequalities has, for many, reduced to the attempt to understand how physical systems could fail to obey a particular constraint baptized "completeness" by Jon Jarrett (1984). It is fast becoming widely accepted that completeness (COMP), rather than Jarrett's "locality" (LOC) or any other condition, has got to go, and the purported fact that violations of COMP do not, even in principle, threaten to facilitate superluminal signaling is often cited as a basis for this.[1] In Abner Shimony's oft-used words, we are thus supposed to be able to maintain a "peaceful coexistence" between quantum mechanics, which predicts the violations of the Bell inequalities, and special relativity (Shimony 1984, 226). This is important enough in itself, but many are going further. The belief that violations of COMP do not make signaling possible seems to make room for explanations of such violations which do not appeal to direct causal connections between outcomes (a result which is particularly welcome when the events are space-like related). Indeed, it seems that the putative gap between superluminal signaling and violations of COMP is seen as lending support to the claim that correlations which violate COMP are intrinsically noncausal. In the subsequent attempt to uncover alternative explanations of the correlations in question, some are beginning to draw fairly startling metaphysical conclusions from the physical world's supposed violation of COMP: the existence of physical holism of some stripe (Teller 1986, 1989; Howard 1985, 1989; Healey 1989; Jarrett 1989, 79); a new sort of quasi-causal connection between events, dubbed "passion-at-a-distance" (Shimony 1984, 227; Redhead 1986, 1987, 1989); "incompleteness" as a property of nature (Ballentine and Jarrett 1987, 700); the necessity of broadening the classical concept of a localized event (Shimony 1989, 30); even relative identity for physical individuals (Howard 1989, 250–51).

It is said that in drawing such conclusions we are engaged in an activity which might appropriately be called "experimental metaphysics." (The phrase is Shimony's [1981, 572]; Hellman [1983, 601] and Jarrett [1989, 60–63] take up and endorse the label in a vivid and compelling way.) The idea is a simple and attractive one. First we demonstrate that any empirically adequate model of the Bell-type correlations which does not contain superluminal signaling will have a particular formal feature. (According to the influential attempt at experimental metaphysics which we shall be examining, the conclusion is that any adequate model will involve violations of COMP.) Then we adduce an argument which purports to show that the formal feature in question is evidence of a certain metaphysical state of affairs. If this works, we have a powerful argument from weak and general premises (namely, empirical adequacy and a ban on superluminal signaling) to a rich and momentous conclusion about the structure of the physical world.

Even putting aside the intrinsic interest of the metaphysical claims involved, this appears to be a rather striking state of affairs epistemologically. Apparently we are now in a position to learn rich metaphysical lessons on the sole basis of some relatively "bare" experimental data, a little mathematics, and one of the precepts of one of our best-confirmed physical theories—for the advocates of experimental metaphysics usually take the prohibition on superluminal signaling to be at the very heart of special relativity. (In this regard, see n. 7 below.) It is, of course, conceded that our confidence in the conclusions of a piece of experimental metaphysics should fall short of certainty, as the gathering of reliable experimental data is always a tricky business, as is the job of drawing conclusions from them. Nonetheless, the suggestion is that we can have as much confidence in our conclusions as we have in our experimental techniques and in run-of-the-mill scientific reasoning. And the argument is notable for its seeming ability to draw metaphysical conclusions without recourse to metaphysical premises. Our central thesis, however, is that in this case appearances are deceptive. Although quantum mechanics is a theory rich in metaphysical suggestion, the particular version of experimental metaphysics which holds the most sway over current thought about the Bell inequalities does not hold together under closer scrutiny.

We shall prove a simple theorem which demonstrates that if the outcome at one end of an EPR-Bohm experiment is statistically dependent on the hidden microstate of the apparatus at that end, and if one other locality-type condition is satisfied, then violations of COMP can straightforwardly lead to superluminal signaling. (We will also argue that superluminal signaling would almost always be possible even if the additional, locality-type condition were not satisfied.) We do not take this result to be especially surprising at a technical level; nonetheless, we will argue that it has very serious negative implications for the views just outlined under the rubric of "experimental metaphysics." In particular, we will draw the following conclusions. First, considerations arising from our result undercut Jarrett's argument for the claim that any minimally adequate model of the Bell-type correlations must violate COMP rather than LOC. A demand for peaceful coexistence with the basic precepts of special relativity does not, of itself, settle the question of which condition the world violates. And this means that the attempt at experimental metaphysics under consideration cannot get off the starting block. Secondly, the result shows, *contra* Jarrett and Shimony, that there is nothing in the intrinsic nature of failures of COMP which prevents their use in superluminal signaling, and so nothing intrinsically noncausal about such failures;[2] it is not even true that violations of COMP are intrinsically less connected to superluminal signaling than violations of LOC.

We will clarify and defend our claims in Section 3. To forestall certain tempting confusions, however, we should make two small points of clarification. First, we are not claiming here that there are *no* considerations which suggest that violations of the Bell inequalities are due to violations of COMP; nor are we claiming that there are no considerations which might militate in favor of a noncausal interpretation of such violations.[3] Our point is rather that neither of these things can be established purely on the basis of such weak and general considerations as empirical adequacy together with respect for a ban on superluminal signaling (as well, perhaps, as the implausibility or methodological perilousness of recourse to so-called "conspiratorial" explanations). Thus described, the program of experimental metaphysics has failed, at least to date. And it is important to recognize this if we are to have a proper appreciation of the epistemological situation we are in when we attempt to glean the metaphysical implications of the failure of the Bell inequalities. Secondly, we are not proposing, as the correct or most plausible model of the correlations, a model which respects COMP and violates LOC, or one which satisfies the condition we shall call 'MC,' or any other model. We are not even claiming that apparatus hidden variables *are* statistically relevant to outcomes. The simple models we shall discuss are intended to function as counterexamples to unworkable arguments, and hopefully as sources of further insight at a more general level; we are not putting forward a hidden variable theory.

To work, then. After setting out the background to the argument in more detail, we will explicate our technical result in Section 2. Then, in Section 3, we elucidate and defend our central claims about the implications of this result for the debate over the violations of Bell's inequalities. Finally, we provide a critique of a similar result recently provided by F. M. Kronz (1990a, 1990b). We will show that Kronz's arguments miss the mark; an understanding of exactly how they do so highlights the dangers of conflating two different pairs of probabilistic assumptions, a habit which is endemic to the literature. These criticisms notwithstanding, we should emphasize that Kronz's work provided the inspiration for the theorem of Section 2.

1. MOTIVATION AND CONTEXT

The class of experiments which is the cause of so much concern will be represented here by a Bell-type variant of Bohm's spin version of the EPR experiment (Bohm 1951, 611–23). A source emits pairs of electrons, one pair at a time, in the quantum-mechanical singlet state. The electrons fly off in opposite directions towards identical measuring devices. Each measuring device is capable of measuring the spin component of its electron in any direction, and for any particular measurement the result is always either +1 or -1, indicated by the flashing of the relevant light on the measuring device. Notoriously, quantum mechanics (QM) does not tell us which light will flash on a given occasion, although it does require quite striking correlations between the results at the two ends. Experiment vindicates the predictions of QM, striking or not, and to many these results seem to indicate the presence of some sort of nonlocality. To see whether this first impression is correct, we can try to ascertain the general constraints which the correlations in question place upon any theory which is intended to account for them. Our argument in this paper is part of that general project.

We shall need some notation, so consider the following two-dimensional space-time schema for the EPR-Bohm experiment.

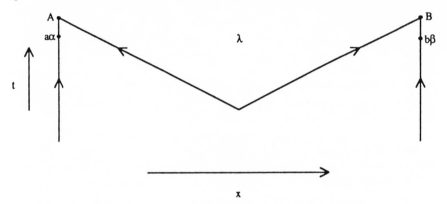

Figure 1

The vertical lines are the world-lines of the left and right-hand measuring devices, and the diagonal lines represent the electrons. (Remember, this is schematic; no express commitment to a particular view of the nature of the electrons or to the idea that they have determinate trajectories is intended.) A and B are outcome events, or just outcomes. λ is the joint state of the two-particle system at the point of measurement (i.e., on a space-like hyperplane which includes A and B, and for convenience, given by a frame of reference in which the source is at rest) and is provided by any theory of these phenomena which falls under our analysis. 'a' and 'b' are the directions in which spin is measured and are macroscopic parameters of the measuring devices. 'α' and 'β' summarize all other relevant (and, if you wish, irrelevant) parameters of the devices; they might be called 'measuring-device hidden states'. a, b, α and β are parameters of the measuring devices at the time of measurement (in the lab frame). They are set immediately before the electrons arrive so that, for example, A and β are space-like related, as in the well-known Aspect experiments (Aspect et al., 1982). We shall also assume that they are set late enough to ensure that there is no significant evolution of the hidden parameters between setting and measurement. Note, finally, that we use 'A', 'B', 'a', 'b', 'α', 'β', 'λ' and similar terms to represent both values of parameters, and the events of the relevant parameters taking those values. It should always be clear from the context which sense is intended.

Now, consider the following two probabilistic constraints one might place upon the outcomes of experiments like these:

$$\forall\ A, B, a, b, a', b', \alpha, \beta, \alpha', \beta', \lambda :$$
$$(A|ab\alpha\beta\lambda) = (A|ab'\alpha\beta'\lambda) \qquad\qquad \text{LOC}$$
$$(B|ab\alpha\beta\lambda) = (B|a'b\alpha'\beta\lambda)$$
$$\forall\ A, B, a, b, \alpha, \beta, \lambda :$$
$$(AB|ab\alpha\beta\lambda) = (A|ab\alpha\beta\lambda)(B|ab\alpha\beta\lambda) \qquad \text{COMP}$$

These are Jarrett's "locality" and "completeness" conditions, first formulated in a slightly different, but logically equivalent manner in his (1984).[4] As Jarrett demonstrates, the conjunction of LOC and COMP is just the well-known "factorizability" condition from which, together with one or two other assumptions, one can derive a Bell-type inequality applicable to both deterministic and stochastic theories.[5] Those other assumptions are usually taken to be relatively weak.[6] Consequently, the repeated violation of Bell-type inequalities by experimental results which are in good agreement with the predictions of QM (Redhead 1987, 108) is taken to sound the death-knell of either LOC or COMP. The question is, which one?

2. VIOLATIONS OF COMPLETENESS AND SUPERLUMINAL SIGNALING

It is standardly claimed that we can and should retain LOC, at the cost of giving up COMP. There are two arguments typically offered for this. The first is simply that QM itself violates COMP while respecting LOC. (Cf., for example, Shimony 1989, 29). This argument has an important limitation, for our faith in metaphysical lessons drawn from the structure of QM can only be as deep as our faith in QM itself. And if we are willing to read such lessons straight off the pages of QM, then why were we considering alternative (or "hidden-variable") theories in the first place?

The argument we are interested in, on the other hand, involves no commitment to the representational accuracy of the internal structure of QM. Instead, it hinges on the claim that, while a violation of LOC would automatically make superluminal signaling possible (at least in principle), no such threat would arise from a rejection of COMP. As the possibility of superluminal signaling is in considerable tension with the dictates of special relativity (SR) on just about any reading of that theory, our commitment to SR leads us to conclude that a good model of the Bell correlations, whatever other features it has, should satisfy LOC and violate COMP.[7] What is more, as we mentioned earlier, the putative fact that violations of COMP cannot be implicated in superluminal signaling is taken to support the attractive possibility that the nonlocality involved in the failure of the Bell inequalities should not be explained by invoking a direct causal connection between distant events—at least, not in any normal sense of the word 'causal'.

The argument for the claim that violations of LOC would make super-luminal signaling possible in principle is quite straightforward (Jarrett 1984, 573–76).[8] Suppose, for example, that LOC fails because for some particular[9] $A, a, b, b', \alpha, \beta, \beta', \lambda$

$$(\underline{A}|ab\alpha\beta\lambda) \neq (\underline{A}|ab'\alpha\beta'\lambda).$$

To allow for the general case in which the terms on either side of this inequality are between 0 and 1, the experimenters will need to employ a large array of EPR-Bohm set-ups. We can imagine perhaps a thousand such experiments lying side-by-side in a large room, and we will suppose that O_R, the right-hand experimenter, is sitting at a large control panel which allows her to set the parameters of all the measuring devices on the right-hand side of the room by the mere manipulation of a few buttons, switches, and dials. O_L, her colleague on the left-hand side of the room, has similar command over the left-hand measuring devices. They have previously agreed to prepare every pair of particles in the state λ.[10] They have also agreed that at some fixed time (in the laboratory rest frame common to them) O_L will perform measurements of the spin-component in direction a of each of the left-hand particles by simultaneously putting each of his devices into the $a\alpha$ state just beforehand. Finally, they have agreed that at a certain time just before the left-hand measurement events, O_R will either set each of her devices into the $b\beta$ state, so as to measure the spin-component of each of her particles in the b direction, or else she will set each of her devices into the $b'\beta'$ state, and thus measure the spin-component of each of her particles in the b' direction. She will decide which course of action to take just beforehand, perhaps by tossing a coin. The experiment is arranged so that there is a space-like interval between each of these latter events (O_R's decision and the setting of her measuring devices) and the measurement events at O_L's end of the experiment. Now, when O_L performs his measurements he will get the result $\underline{A}$ with a certain relative frequency, and this frequency will (usually) correspond, with reasonable accuracy, to either $(\underline{A}|ab\alpha\beta\lambda)$ or $(\underline{A}|ab'\alpha\beta'\lambda)$. In the first case O_L can reasonably infer that O_R decided to set each of her measuring devices to the state $b\beta$, whereas in the second case he can conclude with equal certainty that O_R decided in favor of the $b'\beta'$ state. Thus it is clear that O_L and O_R can use the two-particle system to communicate superluminally if the system violates LOC.[11] Consequently, we have shown that a prohibition against superluminal signaling entails LOC.

What, then, is the argument which shows that violations of COMP are any different? Violations of COMP and violations of LOC are, after all, both just correlations between pairs of distant events. Jarrett suggests an argument to fill this gap in a recent paper (1989, 77). Consider the requirement that a complete specification of the experimental situation should determine the outcome at both ends:

$$\forall\ A, B, a, b, \alpha, \beta, \lambda \qquad (AB|ab\alpha\beta\lambda) \in \{0, 1\}$$

It is easy to show that this condition entails COMP (Jarrett 1984, 580–81). Thus, whenever COMP fails there must be an element of randomness involved in the production of the outcomes. Jarrett draws a strong conclusion from this fact:

> . . . no correlations of the sort associated with violations of completeness can be exploited for superluminal communication because it is a consequence of the failure of determinism that measurement outcomes are not (even in principle) under the control of experimenters. (Jarrett 1989, 77)

There is a certain *prima facie* force to this argument. A failure of COMP involves correlations between distant measurement outcomes, and so the easiest way for an experimenter to signal superluminally in such a situation would indeed be simply for her to bring about the outcome of her choice at her own end of the experiment (or experiments, if the correlations are not perfect). However, as Jarrett points out, because a violation of COMP immediately implies the failure of the generalized "measurement determinism" condition laid out above, the would-be signaler can do no such thing. So far, so good; but the inference from here to the claim that violations of COMP cannot be "exploited for superluminal communication" is too hasty. The possibility remains open that the experimenter might use some controllable feature of the experimental situation as a "trigger" which operates *stochastically* on the outcome at her own end of the experiment. The signaler could then influence, without completely controlling, the result in the individual case, and could thus signal superluminally by employing an array of identically prepared experiments—just as in Jarrett's own argument for the claim that a failure of LOC for stochastic theories makes superluminal signaling possible.

In the rest of this section, we provide a rigorous articulation of this intuitive idea. We begin by introducing a new probabilistic condition, 'Measurement Contextualism':

$$\exists\ B, \beta, \beta', X \qquad (B|\beta X) \neq (B|\beta'X) \qquad\qquad \text{MC}$$

(Here, and throughout the rest of the section, 'X' replaces '$ab\alpha\lambda$' in all formulae, as a, b, α and λ will be held fixed.) MC is not intended as a universal constraint in the same way as, say, LOC. We do not claim that this condition will necessarily be satisfied by the best theory of the phenomena; it is just a condition which is true in some theories. These are theories in which there is a stochastic "trigger" available to the experimenter.[12]

A terminological interjection: The sense of the word 'contextualism' intended here should not be confused with other senses to be found in the literature. The label simply emphasizes that, in a theory which satisfies this new condition, the likelihood of a certain outcome at one end sometimes depends on certain aspects of the state of the measuring device other than its setting angle. In particular, note that there is no reason to think that MC introduces nonlocality of any stripe.

We need to introduce one more condition before we can implicate violations of COMP in superluminal signaling. We call it 'Constrained Locality', or 'CLOC',[13] because it is similar to the first half of LOC, except that the quantifier is existential, and one more parameter—the nearby measurement outcome—is held fixed ("constrained") while measuring device states are varied:

$$\exists\, A, B, \beta, \beta'(\neq \beta), X \qquad (A|B\beta X) = (A|B\beta' X) \qquad \text{CLOC}$$

The motivation for introducing CLOC is as follows: We wish to implicate violations of COMP in superluminal signaling. On one plausible causal reading, violations of COMP indicate the existence of a causal link between outcome events. MC suggests, again on a plausible reading, that β is a partial cause of B. Hence we might well expect to be able to signal from the right side of the left-hand end by manipulating β, *unless* β also has a net causal effect on A by some other route which exactly cancels its effect via B. In that case the relative frequencies of the values of A might remain fixed despite variations in the value of β. However, CLOC provides a guarantee that no such net effect exists, provided only that we hold a, b, α and λ fixed. For CLOC implies that, for some value of β, if we hold a, b, α and λ fixed, varying β has no effect on the relative frequencies with which the two values of A occur. Hence, given that MC obtains and COMP fails, we should expect that if we hold a, b, α and λ at the appropriate values, but allow B to vary, we will be able to use β to vary the relative frequencies of the two possible values of A, as there will be no net β-A effect to cancel out the effect β has on A via B. Note that LOC, on the other hand, could not provide the same sort of guarantee as CLOC, despite the fact that it holds a, b, α and λ fixed, because with LOC we do not conditionalize upon B. Thus LOC may hold because β, say, is acting on A both directly and via B, and in such a way that the two effects cancel out.[14]

So far we have only a plausibility argument for the claim that a combination of MC, CLOC, and failures of COMP will result in the possibility of superluminal signaling. A proof follows. But note first that the various conditions have to be combined in the right way, for implicit in the above analysis is the assumption that some one set of values of the parameters involved satisfies both MC and CLOC, *and* violates COMP. This condition is stronger than just the conjunction of MC, CLOC, and ¬ COMP. Nonetheless, it picks out a possible class of models of the Bell-type correlations, models in which superluminal signaling is possible due to violations of COMP. After proving this, we shall indicate an important way in which the initial assumption can be weakened, thus showing that superluminal signaling can arise from failures of COMP in a much wider class of theories than is at first obvious.

Suppose, then, that in a particular theory there is some one set of values of the parameters which satisfy MC, violate COMP, and satisfy CLOC (for both $B = +1$ and $B = -1$). We can express this more precisely in the following manner:

$$\exists\, A, B, B'(= -B), \beta, \beta', X$$
$$(\{(B|\beta X) \neq (B|\beta' X)\} \wedge \{(A|B\beta X) \neq \qquad \text{ACTION}$$
$$(A|B'\beta X)\} \wedge \{(A|B\beta X) =$$
$$(A|B\beta' X)\} \wedge \{(A|B'\beta X) = (A|B'\beta' X)\}).$$

We shall now show that any theory which satisfies ACTION contains a violation of LOC. It then follows by Jarrett's original argument that superluminal signaling is possible in such theories.

We proceed by *reductio*. If LOC is true, then one consequence is:

$$\forall\, A, \beta, \beta', X \qquad (A|\beta X) = (A|\beta' X) \qquad\qquad \beta\text{LOC}$$

Denote a particular set of values of the parameters which make ACTION true by underlining characters. Then from βLOC we have:

$$(\underline{A}|\underline{\beta}X) = (\underline{A}|\underline{\beta}'X)$$

By the probability calculus:

$$\sum_{B=\pm 1} (\underline{A}|B\underline{\beta}X)(B|\underline{\beta}X) = \sum_{B=\pm 1} (\underline{A}|B\underline{\beta}'X)(B|\underline{\beta}'X)$$

Expanding and rearranging:

$$(\underline{A}|B\underline{\beta}X)(B|\underline{\beta}X) - (\underline{A}|B\underline{\beta}'X)(B|\underline{\beta}'X) =$$
$$(\underline{A}|\underline{B}'\underline{\beta}'X)(\underline{B}'|\underline{\beta}'X) - (\underline{A}|\underline{B}'\underline{\beta}X)(\underline{B}'|\underline{\beta}X)$$

Applying the last two conjuncts in the existential instantiation of ACTION:

$$(\underline{A}|B\underline{\beta}X)((\underline{B}|\underline{\beta}X) - (\underline{B}|\underline{\beta}'X)) =$$
$$(\underline{A}|\underline{B}'\underline{\beta}X)((\underline{B}'|\underline{\beta}'X) - (\underline{B}'|\underline{\beta}X))$$

But by normalization,

$$(\underline{B}|\underline{\beta}X) - (\underline{B}|\underline{\beta}'X) = (\underline{B}'|\underline{\beta}'X) - (\underline{B}'|\underline{\beta}X)$$

and by the first conjunct in the existential instantiation of ACTION (the 'MC' part) this quantity is non-zero. Thus we have:

$$(\underline{A}|B\underline{\beta}X) = (\underline{A}|\underline{B}'\underline{\beta}X),$$

contradicting the second conjunct of the existential instantiation of ACTION. Therefore we conclude that given ACTION, βLOC is false. Hence, as LOC entails βLOC, LOC is false, and we have the sort of correlations which, given in principle control over all measuring-device and particle states, make superluminal signaling possible.

Where there is signaling, it is hard to avoid the conclusion that there is causation. And the most natural causal reading of this situation is one on which the signaling is due, in part, to a direct causal connection between the space-like related outcome events. Although LOC fails in the class of theories we are considering, this is no reason to suppose that it is a *direct* causal link between the β's and the A's which makes the signaling possible; indeed, as we have claimed, the fact that CLOC holds rules out such a link. This means that the probabilistic situation captured by ACTION can quite plausibly be seen as evidencing the following underlying causal structure:

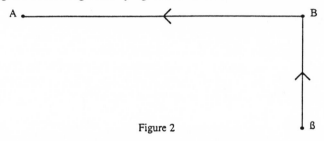

Figure 2

Thus there is nothing intrinsically noncausal about failures of COMP.[15]

We have shown, then, that if on some theory there is a set of values of the relevant parameters which simultaneously satisfies CLOC (for $B = +1$ and $B = -1$) and MC and violates COMP, then that theory allows the use of those states of the measuring devices and particles for superluminal signaling.[16] However, the possibility of superluminal signaling arises in a much wider class of theories than the class of those which satisfy ACTION. Recall that, on a standard causal interpretation of the probabilistic constraints we are discussing, CLOC provides a guarantee that for fixed a, b, α and λ, β's effect on A via B is not cancelled out by any net effect of β on A by other routes (such as by a direct causal link). Thus CLOC is intended as a sufficient condition for signaling given MC and an appropriate violation of COMP. It is not, however, a necessary one. To see this, consider the following diagram:

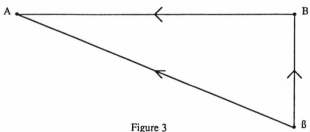

Figure 3

The arrows stand for direct causal influences due to which the "senders" of the arrows make the "receivers" more or less likely when everything else which is causally relevant to the occurrence of the effect ("receiver") is held fixed. The idea here is that, for the particular values of the parameters pictured, MC is satisfied while COMP *and* CLOC are both violated. Provided that the strengths of the causal influences involved add up in the right way, LOC will still be violated, and hence we will still be able to use these causal links to signal. Briefly put (and again underlining characters to denote events involving parameters taking particular values), if β influences the likelihood of $\underline{A}$ more positively via $\underline{B}$ than it influences it negatively directly, then we will be able to make $\underline{A}$ more likely overall by choosing β rather than β'. On the other hand, if $\underline{\beta}$ influences the likelihood of $\underline{A}$ more negatively via the direct route than it influences it positively via $\underline{B}$, then superluminal signaling will again be possible, although this time *despite* the violations of COMP. It clearly requires a careful balancing act to avoid violations of LOC in this sort of situation. In fact, one can construct a measure-theoretic proof which shows that if CLOC is violated, the combination of MC and ¬COMP will still result in violations of LOC in all but a measure-zero subset of the possible cases (Clifton 1991, 28–30). However, if the central point is clear, such elaborate tracery will be irrelevant to our conclusions for the philosophical debate.

3. THE IMPLICATIONS FOR
EXPERIMENTAL METAPHYSICS

The position we are criticizing has it that any adequate model of the Bell-type correlations must satisfy LOC and violate COMP, as this is supposedly the only way to respect the joint constraints of empirical adequacy, on the one hand, and the special relativistic ban on superluminal signaling on the other. The implicit assumption is that violations of COMP do not lead to superluminal signaling. Yet we have just shown that violations of COMP can perfectly well lead to superluminal signaling, if MC is satisfied for the same values of the parameters as those which violate COMP.

Let us now consider a possible line of defense the experimental meta-physician might try to pursue; defusing this strategy will help us to diagnose the exact nature of the flaw in Jarrett's argument. The objection runs as follows: "You have shown us that violations of COMP will facilitate superluminal signaling *if* another condition (MC) is satisfied, and satisfied in the right way. But you have done nothing to make us doubt the claim that violations of LOC lead to superluminal signaling *all by themselves*. Given our commitment to SR, then, you have shown us nothing which should make us revise our original conclusion. Rather, you have simply added to it. For in order to avoid superluminal signaling, we now know that any adequate model of the phenomena will have to respect LOC, violate COMP, *and violate* MC (or at least not satisfy it in the right way). So we are still justified in concluding that the world violates COMP, and in looking for some noncausal way of interpreting that fact."

This objection founders on a forgotten qualification, for it is not true that violations of LOC automatically result in the possibility of superluminal signaling. We also need to assume that in principle the experimenters can control, or at least influence, the values of a, b, α, β and λ. (Cf. Jarrett 1984, 574, n. 13.) Of course, this qualification also applies to our own result about violations of COMP, as we established the connection to superluminal signaling via a violation of LOC. Nonetheless, once we stress this point, it becomes apparent that there is little difference between violations of LOC and violations of COMP with respect to superluminal signaling, the only difference is that to exploit violations of COMP we must *also* be able to influence the value of B. MC is just one way of cashing out the idea that the experimenter can influence B; no doubt there are other ways. The essential point is just that *if the experimenters can control all the factors on which the relevant probabilities are conditional, they will have little difficulty in exploiting violations of COMP for the purpose of superluminal signaling.* And this claim is exactly parallel to the strongest claim we can make about a connection between violations of LOC and superluminal signaling—the same "if" is present in both cases.

This brings home our first main point: The demand for peaceful co-existence does not unequivocally mandate the rejection of COMP and the retention of LOC. There are, instead, various ways to model the phenomena without opening the floodgates to superluminal signaling. One is to satisfy LOC, reject COMP, and reject MC, along with any other way of influencing the value of B. But another is to satisfy COMP, reject LOC, and deny the experimenters sufficient influence over α and β, say, or λ, to turn the violations of LOC to their advantage.[17] One might even be able to reject both LOC and COMP without falling prey to superluminal signaling if one has a theory on which experimenters do not have the right sort of control over the relevant variables. Which of these options is correct is certainly an important matter, but our point is that we are not in a position to draw any general conclusions solely on the grounds of the statistics of the experimental results and a ban on superluminal signaling.[18]

At this point we should note an interesting ambiguity. The argument of the previous paragraph assumes that it is sufficient for peaceful coexistence with SR that superluminal signaling not be *physically* possible according to one's model of the correlations. Given that assumption, our point can be rephrased as follows: There are many sorts of model in which superluminal signaling is physically impossible, and in some of those LOC, rather than (or as well as) COMP, will be violated. What is more, the experimental metaphysician, given the avowed generality of her project, has no way of narrowing the range of acceptable models any further. Now certainly Jarrett takes the crucial question to be whether the failure of a given condition makes superluminal signaling possible "in principle" (1984, 575; cf. also 574, n. 13). But perhaps by this he means that what matters for the question of peaceful coexistence is whether we would be able to signal superluminally with the correlations a given model posits *if* we had the right sort of influence over all relevant (local) parameters, regardless of whether such influence is physically possible. However, even if that is the relevant question, it should by now be clear that in *that* sense violations of either LOC or COMP makes superluminal signaling possible. Consequently, the demand for peaceful coexistence still fails to discriminate between LOC and COMP, with the additional gloomy development that the prospects for peaceful coexistence become very grim indeed.

Hopefully this is enough to establish our central thesis, that this attempt at experimental metaphysics has failed. Considerations of empirical adequacy and peaceful coexistence alone do not settle any interesting questions about the formal structure of the correct model of the Bell-type correlations. We can, of course, turn to QM for clues, but if we do so it must be in the full awareness that we have raised the epistemological stakes considerably.

There is a further lesson to be gleaned from the above discussion, however. For certainly one possibility is that the world violates COMP; indeed, that is just the direction in which QM itself points. And so it is still relevant to the task of understanding the violation of the Bell inequalities to appreciate the interpretive significance of a failure of COMP. Now Jarrett has not been alone in the conviction that failures of COMP are intrinsically incapable of resulting in superluminal signaling. (Cf. Cushing 1989, 11–12; Shimony 1986, 191–92, and 1989, 29; Teller 1989, 219.) And, beginning with Jarrett's seminal paper, there has been a growing tendency to regard failures of COMP as speaking directly of holism, nonseparability, and the like, instead of seeing such correlations as open to several interpretations, one of which invokes direct causal links between the outcomes. (Cf., for example, Howard 1989, esp. 230–32; Jarrett 1984, 581 and 589, and 1989, 79; Teller 1989, 220–21; and, in the context of QM, Redhead 1987, 106–107.) The joint popularity of these two attitudes is somewhat suggestive. It is not that any of the authors mentioned have explicitly avowed a commitment to the idea that signaling capability is a necessary condition on causal connectedness, an idea with links to the notion of the "mark method" in the work of Hans Reichenbach (1956) and Wesley Salmon (1984, chap. 5). Nonetheless, perhaps the view that violations of COMP are definite indications of some sort of noncausal connectedness has seemed attractive to some precisely because they believe that violations of COMP could not, of their very nature, be used for superluminal signaling.

Be that as it may, the discussion of the preceding section should help to clarify matters here. For there we saw not only that violations of COMP *per se* are as likely as the next set of correlations to make signaling possible; we also saw that failures of COMP can lead to signaling in such a way that the most natural account of the signaling mechanism is one in which the correlated outcome events are related as cause and effect (problematic as such an account may be once we throw SR into the equation). So there is, after all, nothing intrinsically noncausal about failures of COMP.[19]

This is not to say that there is no hope for an account of the Bell-type correlations on which the world violates COMP in virtue of its holism, or its nonseparability, or the unusual way in which the events or entities involved should be individuated; far less is it to say that there must be causal connections between space-like related events. Our point is rather that a noncausal account receives no special mandate from the intrinsic features of violations of COMP. If we opt for a noncausal account, it will be because the structure of QM suggests one and we take QM seriously, or because we wish to avoid the perplexities which arise when we try to posit causal connections between space-like related events while maintaining certain metaphysical commitments. It will not be that we are forced into revisionary metaphysics by special relativity and the bare phenomena alone.

To sum up: We have no special reason for thinking that failures of COMP (as opposed to failures of LOC) are the product of a metaphysically exotic relationship between the systems or events involved, unless we turn to the internal structure of QM. We also have no general reason, of the sort which experimental metaphysics supposedly provides, for believing that it is COMP which is violated. The attempt to pull metaphysics out of a hat has failed.

4. COMPLETENESS VERSUS OUTCOME INDEPENDENCE

In closing, let us examine another recent attempt to forge a link between distant outcome correlations and superluminal signaling, one due to F. M. Kronz (1990a, 1990b). Shimony claims that violations of the completeness-type condition he labels "outcome independence" (1986, 188) cannot be used to send signals superluminally. Part of the significance Kronz attributes to his own result is that it is a refutation of that claim, and we concur with Kronz in this respect. However, Kronz also describes his result as a refutation of Jarrett's claim that violations of COMP cannot be used to signal superluminally (Kronz 1990a, 229). Thus, Kronz concludes, his result shows that Jarrett's grounds for rejecting COMP and maintaining LOC in light of the failure of the Bell inequalities are "faulty" (1990a, 235). Here we must take issue with Kronz. Although, as we have seen, violations of COMP could be used to signal superluminally in certain contexts, and although that fact does undermine Jarrett's argument, we will show briefly that Kronz's own technical result does not establish either of those things. Doing so will highlight the dangers of making a certain technical conflation which is, nonetheless, very common in the literature.

To make these points we do not need to examine the details of Kronz's proof, with which we have no quarrel. It is enough to consider his premises and his conclusion. We will call the first premise 'OD', for 'Outcome Dependence', in line with Shimony's terminology:

$$\exists\, A, B, a, b, \lambda \qquad (A|Bab\lambda) < (A|ab\lambda) \qquad \text{OD}$$

Kronz calls this a violation of 'JC' (for 'Jarrett Completeness') in (1990a), and a violation of 'HL$_1$' (from 'Hidden Locality') in (1990b). It expresses a statistical correlation between outcomes for at least one set of values of measuring-device settings plus two-particle system hidden state. But note that calling OD a violation of 'Jarrett Completeness' is not obviously appropriate, as COMP, unlike OD, conditionalizes on measuring-device hidden states.

Kronz's second premise is what he calls 'Hidden Contextualism', and amounts to the demand that the outcome at a given end be entirely determined (in the probabilistic sense) by the parameters of the measuring device at that end, plus the two-particle system hidden state. Applied to the right-hand end of the experiment, this can be formally expressed as:

$$\forall\, B, b, \beta, \lambda \qquad (B|b\beta\lambda) \in \{0, 1\} \qquad\qquad \text{DHC}_\text{R}$$

We append the 'D' for 'Deterministic' to draw attention to this important feature of the condition. We shall see in a moment just how important it is. Finally, the conclusion Kronz derives is:

$$\exists\, A, a, b, \beta, \beta', \lambda \qquad (A|\beta ab\lambda) < (A|\beta' ab\lambda) \qquad\qquad \text{SC}$$

'SC' stands for 'Signaling Conclusion'.[20] Given SC, the right-hand experimenter can signal superluminally to her colleague at the far end by manipulation of the hidden state of her measuring devices,[21] as long as they have a prior agreement to hold a, b and λ fixed. The justification of this is exactly parallel to Jarrett's proof that violations of LOC make superluminal signaling possible.

This is an impressive result, and might seem to do the same sort of work as the result we presented in Section 2: an assumed correlation between distant outcomes for a fixed experimental context, reminiscent of a violation of COMP, is shown to entail the possibility of superluminal signaling, given one other premise (and the right sort of experimental control). However, there is a rather startling feature of this schema lurking in the wings; namely, it is actually *inconsistent with* violations of COMP. This is because DHC$_\text{R}$ entails COMP, and quite trivially so. When a conditional probability is deterministic one can simply conditionalize on anything else without consequence for the value of the probability, provided one does not thereby end up conditionalizing on an event with probability 0. So DHC$_\text{R}$ entails

$$\forall\, B, A, a, b, \alpha, \beta, \lambda \qquad (B|Aab\alpha\beta\lambda) = (B|ab\alpha\beta\lambda),$$

and this is just another way of writing COMP. Consequently, Kronz is clearly not showing that the possibility of superluminal signaling can arise in a theory which violates COMP—or at least, not non-trivially, as the alternative is that he is working with an inconsistent set of premises.

We could stop there. The simple fact is that DHC$_\text{R}$ entails COMP, and so Kronz's result cannot support the claim that violations of Jarrett's completeness condition (COMP) can be used to signal superluminally. Nor can he use the result to undermine Jarrett's grounds for rejecting COMP and maintaining LOC. However, there is something to be learnt by exploring the way in which this confusion has arisen. Consider the following condition:

$$\forall\ A, B, a, b, \lambda \qquad (AB|ab\lambda) = (A|ab\lambda)(B|ab\lambda) \qquad\qquad \text{OI}$$

This is just Shimony's "outcome independence" (1986, 188). Writing it in the equivalent form:

$$\forall\ A, B, a, b, \lambda \qquad (A|Bab\lambda) = (A|ab\lambda)$$

makes it clear that OD does indeed constitute a violation of Shimony's condition. That condition, OI, is generally taken to be an acceptable substitute for, or even equivalent to Jarrett's COMP. So what has gone wrong?

The essential point is a simple one. Rather than drawing out the consequences of a violation of COMP, Kronz has focused on OD, a violation of OI. The difference between COMP and OI, as we have noted, is that the former conditionalizes on measuring-device hidden states, whereas the latter ignores them. OD would thus be a perfectly good stand-in for a violation of COMP if the α's and β's were statistically irrelevant to the outcomes of measurements (for fixed values of a, b and λ). However, Kronz can hardly make that assumption in the context of the other premise of his proof, DHC_R, because as Kronz himself demonstrates, OD and DHC_R entail a very strong correlation between B and β for precisely those values of b and λ which make OD true (1990a, 223, eqn. 15).

In short, Kronz's arguments clearly cannot be used to add anything new to the debate about whether the violation of the Bell inequalities forces us to give up COMP or LOC. Any model which satisfies the premises of his proof either satisfies both, and thus cannot differentiate between them, or satisfies COMP and violates LOC, and thus serves only to remind us of what we already knew— that superluminal signaling is possible in such a situation, at least in principle.

Nonetheless, a look at Kronz's discussion has taught us a valuable lesson. This lesson follows from the fact that Kronz has achieved his aim, most narrowly construed: he has shown that violations of Shimony's "outcome independence" can, in some circumstances, result in the possibility of superluminal signaling. The conjunction of this with the fact that Kronz's scheme satisfies COMP, points to an important logical gap between COMP and outcome independence. These two conditions have been treated as identical, or virtually so, by many authors.[22] Yet Kronz's proof trades precisely on the difference between them. The gap is not only there, it can turn out to be crucial. Consequently, we urge that more care be taken over which of these is meant in future discussion. Presumably we should take the same precaution with regard to LOC and "parameter independence."[23]

5. CONCLUSION

We have tried to articulate, by way of a technical result, some considerations which lead to the following conclusions. First, it is not true that any adequate model of the correlations must violate COMP, rather than LOC. Furthermore, there is nothing intrinsically noncausal about failures of COMP. We may choose to interpret such failures as arising from some sort of holism or nonseparability if doing so enables us to avoid certain vexing problems, but we are not forced into such measures by the intrinsic nature of the correlations. Thus the attempt to draw strong metaphysical conclusions from nature's violation of the Bell inequalities on the sole basis of some weak and rather general considerations—in particular, a demand for empirical adequacy and a prohibition on superluminal signaling—has failed. The relationship between physics and metaphysics is far more complex and delicate than that.

APPENDIX

List of probabilistic conditions (including glossary of acronyms)
ACTION

$$\exists\, A, B, B'(= -B), \beta, \beta', X$$
$$(\{(B|\beta X) \neq (B|\beta'X)\} \wedge \{(A|B\beta X) \neq (A|B'\beta X)\} \wedge$$
$$\{(A|B\beta X) = (A|B\beta'X)\} \wedge \{(A|B'\beta X) = (A|B'\beta'X)\}).$$

βLOC (β-Locality)

$$\forall\, A, \beta, \beta', X(= ab\alpha\lambda) \qquad (A|\beta X) = (A|\beta'X)$$

CLOC (Constrained Locality)

$$\exists\, A, B, \beta, \beta'(\neq \beta), X(= ab\alpha\lambda) \qquad (A|B\beta X) = (A|B\beta'X)$$

COMP (Completeness)

$$\forall\, A, B, a, b, \alpha, \beta, \lambda \qquad (AB|ab\alpha\beta\lambda) = (A|ab\alpha\beta\lambda)(B|ab\alpha\beta\lambda)$$

DHC$_R$ (Deterministic Hidden Contextualism at the Right-hand end)

$$\forall\, B, b, \beta, \lambda \qquad (B|b\beta\lambda) \in \{0, 1\}$$

LOC (Locality)

$$\forall\, A, B, a, b, a', b', \alpha, \beta, \alpha', \beta'\lambda$$
$$(A|ab\alpha\beta\lambda) = (A|ab'\alpha\beta'\lambda)$$
$$(B|ab\alpha\beta\lambda) = (B|a'ba'\beta\lambda)$$

MC (Measurement Contextualism)

$$\exists\, B, \beta, \beta', X (= ab\alpha\lambda) \qquad (B|\beta X) \neq (B|\beta' X)$$

OD (Outcome Dependence)

$$\exists\, A, B, a, b, \lambda \qquad (A|Bab\lambda) < (A|ab\lambda)$$

OI (Outcome Independence)

$$\forall\, A, B, a, b, \lambda \qquad (AB|ab\lambda) = (A|ab\lambda)(B|ab\lambda)$$

QM: Quantum Mechanics

SC (Signaling Conclusion)

$$\exists\, A, a, b, \beta, \beta', \lambda \qquad (A|\beta ab\lambda) < (A|\beta' ab\lambda)$$

SR: Special Relativity

NOTES

The central ideas in this paper were forged while Martin Jones was a visitor to the Department of History and Philosophy of Science at the University of Cambridge in the winter of 1990. Alex Master of Cambridge was especially helpful in bringing out the points made in Section 4. The main themes have since been presented in seminars and colloquia at Cambridge, Stanford, Princeton, and Berkeley, as well as in talks given to the Logic and Methodology Program at Berkeley and the 1991 Pacific Division APA Meetings. We would like to thank all the people who were involved in these discussions, but special thanks should go to Jeremy Butterfield, Nancy Cartwright, Andy Elby, Jim Griesemer, Lisa Lloyd, Michael Redhead, Allen Stairs, and Patrick Suppes.

Martin Jones would like to thank Jeremy Butterfield and Michael Redhead for their hospitality in the Lent term of 1990, and Stanford University for its financial support. Rob Clifton would like to thank Christ's College, Cambridge and the Social Sciences and Humanities Research Council of Canada for their generous support.

1. Jeremy Butterfield is a notable recent dissenter, cf., for example, 1989, 131–35, where he argues that completeness and locality are on a more equal footing with respect to special relativity than Jarrett would have us believe. (Cf. also his 1992.) Many of Butterfield's views can be seen as complementary to the present discussion.

2. By this, we simply mean that when COMP is violated, there is nothing in the structure of the formal situation *alone* which prevents us from interpreting the relevant correlations as resulting from a causal interaction between space-like separated events. Consequently, additional arguments are required if such an interpretation is to be ruled out. And it remains to be seen whether there is such an argument which does not draw, in part, on metaphysical premises.

3. Shimony has argued that our faith in QM should lead us to lay the blame on COMP; in this regard, see the first paragraph of Section 2, below. And Michael Redhead has argued on other grounds that violations of COMP, at least as they appear in QM itself, are intrinsically noncausal. See, for example, Redhead 1987, 102–107, and Redhead 1989. For critiques of that view, see Cartwright and Jones 1991, and Healey 1992.

4. The terms delimited by parenthesis are standard conditional probabilities. Here and throughout we discard the usual prefix, 'p', as this will engender no risk of confusion in the present contexts.

Our only real variation on Jarrett's formulation of LOC and COMP is that we have here notated null measurements (in which one or both devices are switched off) in the same way as any other, with the conventions that in such a case A, for example, ranges only over the 'null outcome', and that its probability (given a null measurement) is one. Cf. Jarrett 1989, 67.

5. The analysis of the factorizability condition into two components of this form was anticipated by several authors. Patrick Suppes and Mario Zanotti first made such a distinction in their 1976. Apart from the fact that they formulate their conditions in terms of conditional expectations rather than probabilities (thus rendering their assumptions weaker), LOC corresponds to their "Axiom 5," or "Locality," and COMP to their "Axiom 6," or "Statistical Independence" (Suppes and Zanotti 1976, 449.) Bas van Fraassen again carved factorizability at these joints in 1982. LOC corresponds to van Fraassen's "Hidden Locality", and COMP to his "Causality" (van Fraassen 1989, 104.) And Geoffrey Hellman not only pointed out that factorizability can be analyzed along these lines, but went on to claim, in effect, that LOC expresses the proscriptions of special relativity, whereas COMP embodies an additional requirement going beyond the demands of that theory (Hellman 1983, 605). As we shall see, this is an essential part of Jarrett's argument. Jarrett's technical innovation was to conditionalize explicitly on measuring-device hidden variables. This can be a crucial difference, as we will show in the final section.

6. Perhaps one of those other assumptions could be denied, comes the objection. This is a legitimate worry, and one we do not have room to address here. For arguments to the effect that any such move could lead to the possibility of superluminal signaling, cf. Kronz 1990b. (Unfortunately his arguments do not explicitly incorporate apparatus hidden variables, but they can be extended quite straightforwardly.)

7. The argument is Jarrett's, but it has since been used or endorsed by many others. Cf., for example, Shimony 1986, 190–93, or 1989, 29. Paul Teller accepts the argument with some reservations (1989, 219). Don Howard reviews the argument and says that its appearance constituted "significant progress" towards understanding the failure of the Bell inequalities (1989, 229–30).

It is arguable that a ban on superluminal signaling is, contrary to a very popular belief, not part of any but the most idiosyncratic formulations of SR. (See Friedman 1983, Chap. 4, sections 5, 6 and 7. For a discussion of this issue in the context of the Bell inequalities, see Jones 1991, 10–13, and, for a far more extensive exploration, Maudlin, forthcoming). If this is correct, then Jarrett's argument might instead proceed with the following supplementary premise: superluminal signaling must be absent from any adequate model of the phenomena because it entails the existence of superluminal causation. Yet it is not clear that even superluminal causation is problematic unless one has certain (perhaps entirely correct) views about the nature of causation. These considerations suggest that Jarrett's argument needs some additional, metaphysical premises. If so, then this is one way in which the argument does not have the advertised features of a piece of 'experimental metaphysics'.

8. We do not wish to endorse Jarrett's claim that the converse entailment also holds; cf. Jones 1991, chap. 2, on this point.

9. Here and throughout, underlined characters represent particular values of parameters.

10. As we shall see, a crucial assumption in Jarrett's argument is that the experimenters could, "in principle," have complete control over the state preparation of both the two-particle system and the measuring-device states. This is why the signaling which would result from a violation of locality is "in principle" signaling.

11. The qualifications 'usually', 'reasonably', and so on, emphasize the fact that we are allowing for cases in which the violation of LOC involves probabilities between 0 and 1. In such cases the signal will not be perfectly reliable, but there is no special problem here.

12. Physical plausibility is, as we have said, not crucial here. However, note that the satisfaction of MC more plausibly evidences a stochastic trigger of the right sort if 'A' and 'B' denote events of the particles deciding which lights to flash, so to speak, rather than the events of the lights flashing. We are not constructing a model in which the lights on the measuring devices are in superluminal communication with one another.

13. We pronounce this 'see-lock', but the reader may prefer to pronounce it 'clock'. Hopefully not.

14. In this situation we would expect CLOC to fail.

15. It is also interesting to note that here we have a failure of LOC without a direct $\beta - A$ causal link, whereas we pointed out earlier that LOC may hold despite the presence of a direct $\beta - A$ link, due to lack of conditionalization on B. This nicely illustrates the fact that we should be circumspect when making inferences from probabilistic to causal facts, or vice versa.

16. That there are such theories is not mere speculation. Clifton has constructed a model which satisfies much stronger premises than the one we have used in the proof, and for a non-zero-measure set of states; yet it reproduces all the quantum-mechanical predictions for spin measurements on a two-particle system in the singlet state, including the strict correlations (Clifton 1991, 25–28).

17. Indeed, as Clifton et al. have pointed out, the so-called "causal interpretation" of QM due to David Bohm arguably provides models of the Bell-type correlations which fall into the latter category, thus providing an apparent counterexample to Jarrett's arguments (Clifton et al. 1992, 120). Maudlin also makes the point in Chapter 4 of his (forthcoming).

18. Kronz (1990a, 1990b) stresses the importance of assumptions about controllability in the arguments concerning superluminal signaling, but rather than drawing skeptical conclusions about the feasibility of experimental metaphysics (and despite his rejection of Jarrett's version of it—see Section 4 below), he seems willing instead to draw new conclusions about the features a "good" model will have. For example, he suggests that a "desirable" model will make control of λ physically possible, thus facilitating "systematic study of [the] system[s]" involved (1990b, 442). Although such considerations may have some methodological force if we are trying to construct a hidden variable theory, they are surely not enough to ground strong claims about metaphysical structure of the situation. (Kronz perhaps recognizes this when he says, in passing, that "there is no evidence that favors one [way of avoiding superluminal signaling] over the others" (1990b, 442). We should also note that, independently, Tim Maudlin seems to have been thinking along lines somewhat in sympathy with our own; cf. Chapter 4 of his forthcoming *Quantum Nonlocality and Relativity*.

19. Nancy Cartwright (1989, chap. 6) draws the same conclusion, but on very different grounds.

20. This is our label—Kronz does not christen the result. It appears in his equation (15) (1990a, 233).

21. In the plural, as the probabilities involved in SC may well be between 0 and 1, and so the single case will not necessarily provide the left-hand experimenter with sufficient information about his friend's choice for successful signaling.

22. Cf., e.g., Butterfield 1992, 69; Cushing 1989, 11; Howard 1989, 229–30; Hughes 1989, 243; Stairs 1988, 319; Teller 1989, 218–19 (although his footnote on that page does suggest a distinction); van Fraassen 1989, 104. Even Shimony occasionally slips into this way of speaking: cf. 1989, 29.

Shimony has argued that his conditions are to be preferred to Jarrett's, as Jarrett's conditions may be "physically empty" (1984, 226 n.; cf. also Shimony et al., 1976). For a discussion of Shimony's claims in this regard, see Jones 1991, 87–97.

23. For examples of a confusion in this regard see, for example, Healey 1989, 148; van Fraassen 1989, 104; and Jones 1989, 79.

REFERENCES

Aspect, A., J. Dalibard and G. Roger. 1982. "Experimental Tests of Bell's Inequalities Using Time-varying Analyzers." *Physical Review Letters* 49: 1804 ff.

Ballentine, L. E., and J. P. Jarrett. 1987. "Bell's Theorem: Does Quantum Mechanics Contradict Relativity?" *American Journal of Physics* 55: 696–701.

Bohm, D. 1951. *Quantum Theory.* Englewood Cliffs.

Butterfield, J. 1989. "A Space-Time Approach to the Bell Inequality." In Cushing and McMullin (1989), 114–144.

Butterfield, J. 1992. "Bell's Theorem: What It Takes." *British Journal for the Philosophy of Science* 43: 41–83.

Cartwright, N. 1989. *Nature's Capacities and their Measurement.* Oxford.

Cartwright, N., and M. R. Jones. 1991. "How to Hunt Quantum Causes." *Erkenntnis* 35: 205–31.

Clifton, R. K. 1991. Nonlocality in Quantum Mechanics: Signalling, Counterfactuals, Probability and Causation. PhD dissertation, Cambridge University.

Clifton, R. K., C. Pagonis and I. Pitowsky. 1992. "Relativity, Quantum Mechanics and EPR." In *PSA 1992*, edited by D. Hull, M. Forbes and K. Ohkrulik. Philosophy of Science Association, v. 1, pp. 114–128. East Lansing.

Cushing, J. T. 1989. "A Background Essay." In Cushing and McMullin (1989), 1–24.

Cushing, J. T., and E. McMullin, Eds. 1989. *Philosophical Consequences of Quantum Theory: Reflections on Bell's Theorem.* Notre Dame, Ind.

Friedman, M. 1983. *Foundations of Space-Time Theories: Relativistic Physics and Philosophy of Science.* Princeton, N.J.

Healey, R. 1989. *The Philosophy of Quantum Mechanics: An interactive interpretation.* Cambridge.

Healey, R. 1992. "Discussion: Causation, Robustness, and EPR," *Philosophy of Science* 59: 282–92.

Hellman, G. 1983. "Stochastic Locality and the Bell Theorems." In *PSA 1982*, edited by P. D. Asquith and T. Nickles. Philosophy of Science Association, v. 2, 601–15. East Lansing.

Howard, D. 1985. "Einstein on Locality and Separability." *Studies in History and Philosophy of Science* 16: 171–201.

Howard, D. 1989. "Holism, Separability, and the Metaphysical Implications of the Bell Experiments." In Cushing and McMullin (1989), 224–53.

Hughes, R. I. G. 1989. *The Structure and Interpretation of Quantum Mechanics.* Cambridge, Mass.

Jarrett, J. P. 1984. "On the Physical Significance of the Locality Conditions in the Bell Arguments." *Noûs* 18: 569–89.

Jarrett, J. P. 1989. "Bell's Theorem: A Guide to the Implications." In Cushing and McMullin (1989), 60–79.

Jones, M. R. 1989. "What Locality Isn't: A Response to Jarrett." In *Bell's Theorem, Quantum Theory and Conceptions of the Universe*, edited by M. Kafatos, 77–79. Dordrecht.

Jones, M. R. 1991. Locality and Holism: The Metaphysics of Quantum Theory. PhD dissertation, Stanford University.

Kronz, F. M. 1990a. "Jarrett Completeness and Superluminal Signals." In *PSA 1990*, edited by A. Fine, M. Forbes and L. Wessels. Philosophy of Science Association, v. 1, pp. 227–239. East Lansing.

Kronz, F. M. 1990b. "Hidden Locality, Conspiracy, and Superluminal Signals." *Philosophy of Science* 57: 420–44.

Maudlin, T. Forthcoming *Quantum Nonlocality and Relativity.*

Redhead, M. L. G. 1986. "Relativity and Quantum Mechanics—Conflict or Peaceful Coexistence?" *Annals of the New York Academy of Sciences* 480: 14–20.

Redhead, M. L. G. 1987. *Incompleteness, Nonlocality and Realism: A Prolegomenon to the Philosophy of Quantum Mechanics.* Oxford.

Redhead, M. L. G. 1989. "Nonfactorizability, Stochastic Causality and Passion-At-A-Distance." In Cushing and McMullin (1989), 145–53.

Reichenbach, H. 1956. *The Direction of Time.* Berkeley.

Salmon, W. C. 1984. *Scientific Explanation and the Causal Structure of the World.* Princeton, N.J.

Shimony, A., M. A. Horne, and J. F. Clauser. 1976. "Comment on 'The Theory of Local Beables'," *Epistemological Letters*, No. 13.

Shimony, A. 1981. "Critique of the Papers of Fine and Suppes." In *PSA 1980*, edited by P. D. Asquith and R. N. Giere. Philosophy of Science Association, v. 2, 572–80. East Lansing.

Shimony, A. 1984. "Controllable and Uncontrollable Nonlocality." In *Proceedings of the International Symposium: Foundations of Quantum Mechanics in the Light of New Technology*, edited by S. Kamef, et al., 225–30. Tokyo.

Shimony, A. 1986. "Events and Processes in the Quantum World." In *Quantum Concepts In Space and Time*, edited by R. Penrose and C. J. Isham, 182–203. Oxford.

Shimony, A. 1989. "Search for a Worldview Which Can Accommodate Our Knowledge of Microphysics." In Cushing and McMullin (1989), 25–37.

Stairs, A. 1988. "Jarrett's Locality Condition and Causal Paradox." In *PSA 1988*, edited by A. Fine and J. Leplin. Philosophy of Science Association, v. 1, 318–25. East Lansing.

Suppes, P., and M. Zanotti. 1976. "On the Determinism of Hidden Variable Theories with Strict Correlation and Conditional Statistical Independence of Observables," in *Logic and Probability in Quantum Mechanics*, edited by P. Suppes, 445–55. Dordrecht.

Teller, P. 1986. "Relational Holism and Quantum Mechanics." *British Journal for the Philosophy of Science* 37: 71–81.

Teller, P. 1989. "Relativity, Relational Holism, and the Bell Inequalities." In Cushing and McMullin (1989), 208–23.

van Fraassen, B. C. 1989. "On Explanation in Physics." In Cushing and McMullin (1989), 109–13. (New appendix to the 1982 paper "The Charybdis of Realism: Epistemological Implications of Bell's Inequality," first printed in *Synthese*, and reprinted in the Cushing and McMullin volume.)

Scientific Realism and Quantum Mechanical Realism

LINDA WESSELS

Since its birth in late 1925, quantum mechanics (hereafter QM) has raised for many, including those who invented it, perplexing questions of interpretation. What does the theory tell us about the nature of the micro-world? Or does it tell us anything at all? In the almost seventy years during which physicists and philosophers have explored the issue of a realistic interpretation for QM, the precise form of the issue has undergone several changes, reflecting the significant progress that has been made toward understanding what kinds of interpretations QM does and does not allow. Philosophers of science have taken a special interest in this situation, because the issue of how to give QM a realistic interpretation seems closely related to the more general issue of scientific realism, an issue philosophers and philosophers of science have wrestled with for quite a while. What is the relation between the philosopher's issue and the physicist's? What can the philosophers concerned with the general issue of scientific realism learn from the work that has been done on the quantum mechanical issue? And what can the physicist learn from the work on the philosophical issue? The aim of this paper is to examine some answers to these questions.

The philosopher's issue of scientific realism has traditionally focused on the following question: When our best physical theories talk about such unobservable objects as electrons, or such unobservable properties as the charm of quarks, should we take this talk literally? Should we understand these theories to be claiming that these unobservable objects and properties exist, and to be giving a description of their unobservable behavior? Or is such talk to be interpreted as only a convenient but elliptical way of saying something else, a convenient or elliptical way of talking about and making predictions about what can be observed? Put even more succinctly, the issue has been: do scientific theories tell us about the unobservable world behind the observable phenomena, or are they just instruments for calculating predictions about observable phenomena and for manipulating these phenomena to our advantage? For the sake of convenience, I will use the acronym PSRI (the philosopher's scientific realism issue) to refer to this issue.

Support for regarding the PSRI as a significant issue has come from the history of science. Many times scientists themselves have raised what seem to be specific versions of exactly the philosopher's question. Should Copernicus's theory be understood literally, as claiming that the sun sits still while the earth moves about it, or is the theory better interpreted as merely a convenient way of calculating the observable properties of planets, moon and sun: their positions and motions relative to the fixed stars? In the eighteenth and nineteenth centuries, chemists worried about whether to take talk of atoms seriously, or to regard atomic models as simply convenient fictions that allow us to categorize and predict the observable chemical phenomena. In the latter part of the nineteenth century, physicists worried about how to interpret Maxwell's theory: is it a compact description of motions and tensions in the unobservable electromagnetic ether, or are Maxwell's equations the full content of Maxwell's theory? Philosophers have pointed to these historical incidents (and others) as examples of science pausing to ask about a specific theory (or type of theory) the very same question that lies at the heart of their more general issue of scientific realism. I will refer to the issue when raised for a specific scientific theory as SRI (the scientist's realism issue).

Of course, the philosopher's reasons for raising PSRI are different from those of the scientist who raises SRI. For philosophers, the realism issue is connected to other philosophical topics that pertain neither directly nor exclusively to scientific theories, topics such as meaning (Are the sentences in a theory that seem to talk about the unobservable meaningful?), reference or truth (Do the sorts of terms/sentences science uses to try to talk about the unobservable refer/have truth value?), or confirmation (Does empirical evidence provide confirmation for claims about the unobservable?). Scientists, on the other hand, have addressed SRI for a specific theory when what that theory seems to say about the unobservable world poses a particular problem. The story the theory tells about the unobservable may conflict with other scientific theories, for example, or with other widely held beliefs, as in the Copernican case. Or there may seem to be no independent way of checking a theory's account of the unobservable, as in the case of early chemical atomism. Or there may be several possible stories about the unobservable that could be attached to a particular theory, as in the case of Maxwell's theory, leaving it unclear which, if any, should be understood realistically.

Thus the way that SRI for a specific scientific theory is resolved will not contribute to an understanding of how to resolve the PSRI. Resolution of the former comes with the dissolution of the conflict; resolution of the latter depends on solving particular philosophical problems. But the particular form that SRI takes for specific scientific theories does help guide the philosopher in understanding how and why the issue of realism is relevant to actual scientific theories, and in developing a formulation of PSRI that more accurately reflects the way genuine questions about realistic interpretation do arise for scientific theories. This sort of guidance could be especially useful right now. As it has traditionally been formulated, PSRI presupposes that a clear distinction can be drawn between what is observable and what is not. Good arguments against the existence of such a clear distinction have left this traditional formulation without a presupposition to stand on, calling into doubt the coherence of PSRI. Nonetheless, something like special cases of PSRI, namely SRI for particular scientific theories, do arise in science, and seem to be both coherent and significant. One would like to reformulate PSRI in a way that captures what does appear to be a genuine realism issue for scientific theories. By looking at the form the realism issue has taken in the case of QM, we will gain some insight into how this reformulation might go. By viewing the quantum mechanical issue in this more general context, we will also gain some perspective on the significance of the current debates over how to interpret QM.

In Part I, we examine the relation between PSRI and the questions more often raised by physicists than philosophers, of how or whether it is possible to give a realistic interpretation for QM. Part II contains a brief description of how progress has been achieved in efforts to address the second issue, and a sketch (also brief) of the current state of research in this area. In Part III, the relation between the philosopher's and physicist's issues is again examined, and some lessons are drawn for both.

I. REALISM AND QUANTUM MECHANICS

In the case of QM, the realism issues that first emerged were not versions of SRI, and they did not surface in the way SRI typically do. The first question that physicists raised concerning a realistic interpretation for QM emerged out of the very process by which that theory was discovered. Werner Heisenberg laid the ground for the theory in mid-1925 in his article, "On the Quantum Theoretic Interpretation of the Kinematic and Mechanical Relations."[1] Out of the reworking of Heisenberg's ideas by Heisenberg, Max Born, and Pascual Jordan, came matrix mechanics—the first form in which QM appeared. Even before Heisenberg's mid-1925 paper, Born[2] and Wolfgang Pauli[3] had argued that the reason Bohr's theory of the atom had failed was because Bohr's theory tried to describe processes that are unobservable. Born and Pauli were thinking in particular about the fact that a description of the orbits and motions of atomic electrons as they travel around the nucleus was fundamental to Bohr's theory, even though experiments related to Bohr's theory do not detect these orbital positions or motions themselves. The primary data for Bohr's theory are the spectral lines that are produced when an atomic electron jumps from one of these orbits to another. Born and Pauli argued further that the key to developing a better theory of atomic structure, one to replace Bohr's theory, would be to abandon any such attempt to describe what is unobservable. Influenced by Born and Pauli, Heisenberg explicitly argued in his mid-1925 paper, that an atomic theory *should not* have the resources to describe what is unobservable, and Heisenberg claimed that he had consciously designed his new theory on that principle.

At the same time, Erwin Schrödinger was developing another theory to replace Bohr's. But Schrödinger's theory was motivated by and explicitly based on assumptions about the unobservable nature and motions of atomic electrons. His idea was that atomic electrons are not really particles, but are wave formations, and he attempted to construct a new quantum theory that would describe the behavior of these wave formations around the nucleus, a theory that came to be called wave mechanics. Heisenberg was appalled at what Schrödinger had done precisely because Schrödinger's theory, at least as he originally presented it, was intended to provide a detailed account of the unobservable, and *mutatis mutandis*, Schrödinger found matrix mechanics intolerable because it did not. Thus in the beginning, physicists addressed a new kind of realism issue in connection with QM: *should* a theory even provide an account of what is unobservable?

In addition to matrix mechanics and wave mechanics, there was a third new quantum theory proposed by P. A. M. Dirac. Like Heisenberg, Born, and Jordan, Dirac took as his starting point Heisenberg's mid-1925 paper and took as his gospel Heisenberg's admonition against trying to describe what is unobservable. But instead of clothing Heisenberg's new ideas in matrices, Dirac used the structure of the mathematical formulae in Heisenberg's paper to develop what he described simply as a quantum mechanical "algebra"—an abstract, symbolic axiomatic system that did its primary calculations using what Dirac called "q-numbers." Q-numbers were introduced simply as the mathematical entities appropriate for quantum mechanical calculations, with certain properties that make them different from ordinary numbers, e.g., they do not commute. Calculations with q-numbers were aimed at determining values for "c-numbers," numbers of the classical type, i.e., real numbers. The c-numbers represented the possible outcomes of measurements on atomic and subatomic objects. Q-number calculations could be used to determine not only these c-numbers, but when supplied with appropriate information about the physical circumstances surrounding a particular atomic or subatomic system, the q-number calculations could also be used to determine a probability on each c-number, representing the probability that that c-number would appear as the result of an appropriate type of measurement. Dirac's q-algebra removed any temptation to attach even the most tentative account of the unobservable to a new quantum theory.

When it was recognized that all three theories, matrix mechanics, wave mechanics, and Dirac's q-algebra, were mathematically equivalent, the question of how to interpret QM came to a head. In the view of many physicists, the issue was settled by the end of 1927, with the formulation of the Copenhagen interpretation by Bohr and Heisenberg. By that time, it had become clear that Schrödinger's original wave-formation interpretation of his theory would not work, and subsequent attempts by Schrödinger and others to attach to the quantum formalism some other account of the nature and behavior of the atomic and subatomic world had failed also. In the hands of the Copenhagen interpretation, the physicists' issue concerning realistic interpretations of QM changed from a question of "should" to one of "can."

As far as I know, Bohr never explicitly addressed the question of whether QM *should* try to describe the unobservable. Bohr argued, rather, that no such description *can* be developed. The crux of Bohr's position was his claim that the conceptual framework previously developed for physical objects and processes out of human experience with ordinary middle-sized objects is simply inappropriate for objects that are far beyond our experience, like the very small atomic and subatomic objects. And we cannot develop new concepts that *would* apply to atomic-sized objects, Bohr went on to argue, because they are unobservable. He was convinced that the only way humans can develop new concepts about physical objects and processes is by abstracting from our experience with these objects and processes, where "experience with" means direct perceptual interaction. Thus, Bohr concluded, even though they are inappropriate, we are stuck having to use classical concepts to describe quantum objects. According to Bohr, QM shows these classical concepts can never provide anything better than "broken-backed" descriptions of the atomic world, descriptions that satisfy what Bohr called a principle of complementarity. This principle says, roughly, that classical concepts like wave and particle, or like position and momentum, can be applied to quantum systems, but only one concept in each such pair can be applied in a given situation. At any given moment, an electron can have either position or momentum, but not both. At some times an electron can be thought of as a particle, at others it can be treated as a wave. Which member of a concept pair applies at a given moment depends on the particular experimental situation in which the electron finds itself at the moment. Hence, it is how we go about measuring or observing an electron that determines which of the classical properties or descriptions might be used for it. There is no further account of the objective (measurement independent) nature and behavior of atomic and subatomic sized objects possible.

Not everyone agreed with this conclusion, of course. Even after 1928, Einstein, Schrödinger (sometimes), and a few others, often as much concerned with reinstating determinism at the atomic level as with retaining some detailed account of an objective reality at that level, argued that Bohr was wrong. While agreeing that QM itself does not explicitly give a "realistic" account of the atomic realm, they argued that such an account was possible, perhaps even implicit in the theory, and urged the search for such a "realistic interpretation" of the theory.

The contrast between SRI, conceived as a special case of PSRI, raised for specific scientific theories, and what by 1928 came to be the physicist's issue of a realistic interpretation for QM is obvious. They are not the same. Recall PSRI: Given the story that a theory does tell about the unobservable, should that story be understood literally, or should it be regarded as simply a useful piece of fiction? The issue for QM after 1928 was: Is there any way at all to give an account of the properties and behavior of atomic and subatomic systems in the absence of measurement that is consistent with QM? For those who denied that QM *can* be given any such "realistic interpretation," SRI for QM cannot even get off the ground: there can be no debate over whether the theory's story about the unobservable should be understood literally if the theory has no story to tell.

Some have taken this as vindication of the anti-realist position with regard to PSRI. The idea seems to be: the fact that QM admits of no realistic interpretation shows that theories do not need to tell stories about the unobservable in order to be good theories. A moment's reflection shows, however, that the situation with QM does not favor either side of PSRI. Even if QM admits of *no* realistic interpretation, this does not decide the question of whether to take literally theories that do have a story to tell about the unobservable. And even if one were able to develop a realistic interpretation for QM, one would still have to decide whether or not the realistic interpretation should be regarded as a description or simply as a useful heuristic that aids in the application of the theory to the observable.

In addition to this obvious difference between the two issues, there is a second, which stems from the way that the notion of a "realistic interpretation" developed historically in the hands of quantum physicists themselves. The philosopher's notion of a realistic interpretation for a theory is very broad. Even those physicists who have argued that QM admits of no realistic interpretation, indeed even those who have claimed that it *should* not admit of such an interpretation have understood the theory to "say something" about what they consider to be the unobservable. For from the beginning, even Heisenberg, Born, and Pauli (and most others) incorporated a 'partial' Bohr model of the atom in their understanding of matrix mechanics. Bohr's old quantum theory of the atom employed a planetary model, with electrons orbiting in elliptical paths about the nucleus, occasionally jumping from one orbit to another when radiation was emitted or absorbed. This planetary model was explicitly rejected in matrix mechanics. But the makers and users of matrix mechanics still assumed that the atomic complex, whatever its nature, has certain stable states, corresponding to Bohr's electron orbits, and that upon emitting or absorbing energy, the atomic complex would change abruptly ("jump") from one of these stable states to another. This partial Bohr model was taken seriously, i.e., it was understood to reflect what actually happens in the atom during emission or absorption of energy. One indication of this is a published exchange between Heisenberg and Schrödinger in 1926. Heisenberg argued that his own quantum mechanical treatment of the phenomenon of resonance between two like atoms could only be interpreted as a description of coordinated sudden jumps in the two atoms, where repeatedly one atom shifts instantaneously to a stable state of higher energy while the other shifts to one of lower energy, and then vice versa.[4] Schrödinger countered with an interpretation of resonance that construed the shifts as gradual and continuous changes in frequency rather than discontinuous jumps in energy.[5] Both took the issue to be about what kind of transition is actually taking place, not simply about what sort of model is most useful. And this partial Bohr model continues to be understood by physicists as a representation of something that QM does tell us about what is "really going on" at the unobservable level, even by those who adhere to the Copenhagen belief that QM admits of no "realistic interpretation." Why is the partial Bohr model nonetheless not considered by physicists to be a "realistic interpretation" of QM? Because in the process of trying to discover what sorts of interpretations can be given to QM, the notion of a "realistic interpretation" for QM has come to mean something fairly specific.

II. OBJECTIVE PROPERTIES FOR QUANTUM SYSTEMS

What do physicists count as a realistic interpretation for QM? Intuitively, it is a descriptive account of the nature and behavior of atomic and subatomic systems that even covers periods when they are not being measured, that in some sense "fits with" or even can be "read off of" the mathematical formalism of QM. This intuition has been refined through attempts to develop rigorous ways to state and then answer the question of whether such an interpretation is possible. Out of early arguments against the possibility of a realistic interpretation, such as those presented in the late 1920s by Bohr, Heisenberg, and then in the early 1930s by von Neumann, and early arguments for the necessity of a realistic interpretation, such as those presented in that same period by Einstein, Schrödinger and others, emerged a more precise but very general notion of what counts as a realistic interpretation for QM. It is the notion of a "hidden variables" interpretation. Roughly, a hidden variables interpretation asserts that micro-systems like electrons or protons exist whether or not they are being measured, that these micro-systems have objective properties independently of what kind of measurement situation they find themselves in, if any, and that these properties cause (or at least are partial causes of) the outcome obtained when the micro-system with these properties is measured.

　　Clearly, this notion of realistic interpretation was developed in opposition to the Copenhagen interpretation. A central part of the Copenhagen interpretation is Bohr's view on property attribution outlined above: the particular type of measurement being done on a micro-entity determines what kinds of properties can be attributed to that entity, in that context, and no properties can be attributed to it at all outside of the measurement context. Hidden variables are characterized as "objective properties" of a micro-system to indicate that their presence or absence does not depend on whether or what kind of measurement an experimenter chooses to perform on the system. Exactly what these "objective properties" are is left unspecified by the general notion of a hidden variables interpretation. Some have taken the objective properties to be classical particle properties such as position and momentum, or classical wave properties like wavelength and frequency, or some, like David Bohm, have proposed a combination of both. On this construal, it is often assumed that these objective properties are the very properties that quantum mechanical measuring devices measure. Others have supposed that the objective properties might be quite non-classical, assuming only that when a system with such properties interacts with a measuring device, these objective properties are the immediate causes for the reading the device exhibits at the end of measurement. On either way of construing them, these objective properties are at best only indirectly and/or incompletely observable, at worst currently unobservable— hence *hidden* variables. The quantum physicist's question of whether there can be a realistic interpretation for QM, has become, then, the question of whether the attribution of such objective properties to quantum systems is consistent with QM itself.

This notion of an "objective properties" interpretation is so elementary and apparently unrestrictive that it may be hard to imagine how it could ever be shown to conflict with QM. The basic strategy for proving the presence or absence of a conflict is to compare the empirical consequences of assuming such objective properties with the empirical predictions of QM. QM is capable of generating only certain sorts of empirical predictions: for each of the physical quantities QM recognizes as experimentally measurable, i.e., for each quantum mechanical "observable" (qm-observable), the theory gives the set of all the values that could possibly appear as outcomes in a measurement for that qm-observable, and gives the probability for each value in the set. To compare these predictions with the implications of an objective properties interpretation, some explicit assumption must be made about how the objective properties of a micro-system relate to the results that will be obtained when that system is measured for a qm-observable. Thus, while the general notion of an objective properties interpretation for QM allows for a broad range of assumptions about the connection between these properties and the results obtained in a measurement for a qm-observable, any particular attempt to show that QM does or does not admit of an objective properties interpretation must start with a particular assumption about the connection.

Since the early 1930s there have been several such attempts. Each involved a slightly weaker assumption about the connection between the objective properties and measurement results than its predecessor, and almost all resulted in a rigorous proof that the assumed connection cannot yield empirical consequences in agreement with the empirical predictions of QM.[6] The complexity of some of these "no-hidden-variables" proofs reveals the subtle difficulties that QM puts in the way of an objective properties interpretation. There are also some proofs that show QM does, nonetheless, admit of an objective properties interpretation—if the right sort of connection between the properties and measurement results is posited.[7] To even sketch what that "right sort" of connection must be would plunge us into more quantum mechanical detail than is appropriate here. But it is not a particularly odd sort of connection. Beginning in the early 1950s, David Bohm has led the way in constructing objective property interpretations that incorporate versions of this "right sort" of connection that have some intuitive appeal.[8] A disturbing feature of Bohm's interpretations (and others like it) is that they require causal connections among events that seem to violate the special theory of relativity. The problem occurs in certain sorts of experiments involving two quantum systems A and B that had previously interacted and are then separated by such a distance that even according to QM itself, they are no longer interacting. QM predicts that if A and B are measured for certain qm-observables, the pair of results will be correlated in a particular way. In order to match these predictions of QM, the Bohm-style interpretations allow the outcome of the measurement on A to be influenced by what goes on during the measurement of B. Since the distance between the two measurement sites can be made arbitrarily large, the influence from A to B must be able to cross between the sites at a speed greater than that of light.

Does QM allow an objective properties interpretation that has the "right sort" of connection but does not require the assumption of superluminal influence? This question was first posed by John Bell in the early 1960s.[9] His answer is known as Bell's theorem: if it is assumed that the outcomes of measurements on A are determined simply by conditions that are local to A (i.e., they are fully determined by the objective properties of A and the type of measurement done on A), and likewise for B, then the particular sort of correlations predicted by QM cannot be reproduced—QM cannot be given a local objective properties interpretation. In the almost thirty years since then, Bell's inquiry into the possibility of local objective properties interpretations has been refined and extended, by him and others. These efforts have inspired a variety of new experiments that have checked hitherto untested empirical predictions of QM. A variety of new proofs have been produced, some positing an indeterministic rather than a deterministic connection between the objective properties and measurement outcomes, with appropriately refashioned notions of "locality," all showing that QM does not admit of local objective properties interpretations.[10]

What do these results—the "yes-hidden-variables proofs" as well as the "no-local-hidden-variables proofs"—tell us about whether QM can be given a realistic interpretation? On the one hand, the "yes-proofs" show that there are objective property interpretations for the theory. On the other, the "no-proofs" of the last thirty years show that the objective properties cannot be local. But what exactly is involved when objective properties are not "local"? The locality constraints imposed on objective properties in these recent "no-proofs" prohibit the outcome of a measurement on A (or B) from depending on what happens during the measurement of the distant B (or A), i.e., the locality constraints are designed to make the causal behavior of objective properties consistent with relativity theory in obvious ways. But is causal behavior that lacks locality necessarily in conflict with relativity theory? Or is there a weaker constraint which can be placed on the causal behavior of these objective properties that both allows their causal behavior to satisfy relativity theory and also allows empirical predictions consistent with QM? How crucial is "locality" to a realistic interpretation of QM, anyway? Does the absence of locality in an objective properties interpretation automatically make it unacceptable? These are exactly the questions that are the focus of the work being done today (by philosophers and physicists) on the issue of how to interpret QM.

The final verdict is not in. The issues at the center of current work in this area range broadly over the philosophical landscape: What criteria of individuation apply in the quantum case—are A and B (above) really two separate entities, or can they be construed as parts of a single, indivisible individual?[11] Do there exist objective relations between two entities that are not supervenient on their individual properties, in particular, objective relations that might account for an "uncaused" correlation between distant events?[12] What criteria distinguish the presence or absence of the causal relation: sufficiency and/or necessity, probability relations, robustness, connection by a causal process, energy or particle exchange?[13] Does the appropriate explication of the causal relation allow the measurement processes on A and B to be related causally without violating relativity theory?[14] Can the order of the causal relation be reference frame dependent (A causes B in one frame, but B causes A in another)?[15] Do law-like correlations always require a causal explanation, or might the correlations on which Bell has focused be such a basic feature of the quantum world that they are simply unexplained explainers?[16] This brief list gives a suggestive glimpse of some of the avenues currently being explored in order to develop acceptable objective properties interpretations for QM. It also gives a glimpse of the kinds of radical revisions in our physical worldview that are being contemplated in order to fashion such an interpretation—and an insight into why one might not be too optimistic about reaching widespread agreement on what even counts as *acceptable* in this case.

III. SOME LESSONS FOR SCIENTIFIC REALISM AND FOR UNDERSTANDING QM

It is tempting to conclude that the issue of a realistic interpretation for QM has finally been transformed from a question of whether there *is* such an interpretation for the theory into a traditional SRI, i.e., into the issue of whether any of the non-local objective properties interpretations allowed by QM should be taken as literal descriptions of the unobservable nature and behavior of the atomic and subatomic world. And surely this is roughly right. With the proofs that there are hidden variables interpretations for QM, but that they cannot be local, the physicist's issue for QM has moved closer to becoming a special case of PSRI, i.e., closer to becoming an SRI. The "yes-proofs" show that there are objective properties interpretations for QM, but the "no-local-proofs" show that all of them will conflict in some way or other with our previously unquestioned assumptions about the physical world or will conflict with one of our currently best theories about the physical world—relativity theory. This is typically the way an SRI arises for the scientist—when what a scientific theory seems to tell us conflicts with what for one reason or another we already believe to be true.

But a more careful assessment shows that this tempting conclusion is not quite right. Seeing why it is not quite right will shed some light on two ways that the traditional conception of the philosopher's issue of scientific realism can be fruitfully revised.

IIIa. What Does a Theory "Say?"

Traditionally the philosopher has taken the controversy over how to interpret Copernicus's theory as a paradigm SRI. But the realism issue for QM is significantly different from this paradigm. An explicit part of the Copernican theory is the supposition that the sun sits at the center of our planetary system, while the earth (and other planets) orbit about this center. Historically, the issue was whether this supposition of the theory should be regarded as one of the claims the theory makes about our planetary system, or just a counterfactual assumption adopted for the sake of calculating planetary orbits. QM, on the other hand, makes no explicit suppositions whatsoever about whether or not micro-systems have objective properties. Thus the realism issue for QM has not been simply a matter of looking at what the theory says about some unobservable part of the world, and then asking whether this should be taken seriously. It has concerned, rather, whether and/or how QM can be supplemented with an account of the micro-world that is more detailed than the meagre one given by the theory itself, an account aimed at explaining what the micro-world must be like in order that the predictions of QM could be true.

The fact that QM yields only probabilities for measurement outcomes has provided a special motivation to search for such a supplement: perhaps micro-systems have some deterministically evolving objective properties that underlie the statistical phenomena QM talks about. But QM is not alone among recent physical theories in seeming to call for some supplementary account. With the move to mathematization in the nineteenth and twentieth centuries, scientific theories in general, and physics theories in particular, have been formulated in terms of abstract mathematical formulae. QM is only one of several cases in which the original intuitive picture of the reality that inspired these formulae turned out to be inadequate; the picture was rejected but the formulae remain. The formulae give accurate empirical predictions because they express relations among certain sorts of measurable physical quantities. But they no longer say anything explicit about what kinds of objects and processes are causing these measurable phenomena. Maxwell's theory of electromagnetism is another example. Maxwell developed his equations for electricity and magnetism using a relatively detailed model of the nature and behavior of the electromagnetic ether. The model turned out to be incoherent, but the equations correctly related measurable electric and magnetic quantities such as charge, current, and electric and magnetic field strengths. During the same period in which empirical evidence for the correctness of these equations was accumulating, the search continued for an adequate "interpretation" of them: what must the ether be like in order to give rise to the phenomena described by Maxwell's theory? The history of attempts to understand the significance of Einstein's theory of special relativity displays a similar pattern.

Whether such issues of "interpretation" have historically been classed as "philosophical" or "scientific" seems to have depended on the historical circumstances. In the case of Maxwell's theory, the question of what kind of ether gives rise to the phenomena covered by the theory was not considered a "philosophical" one, but a scientific one. There was little doubt in the late nineteenth century that the phenomena covered by Maxwell's equations were caused by motions and tensions in the underlying electromagnetic ether, so the job of giving a detailed account of those motions and tensions was pursued by many physicists as the next logical step in developing the implications of the theory. There was no question that some such account existed and that when constructed should be taken as literally as Maxwell's theory itself. In the case of QM, on the other hand, the interpretation issue has not been considered scientific, but "philosophical." The early and rapid acceptance of the Copenhagen interpretation carried with it the conviction that at worst, an objective properties interpretation for QM is not even possible, and that at best, the issue of whether such properties exist or not can never be put to an empirical test—so that whether or not to believe such an interpretation, if one were to be found, would be simply a matter of "philosophical taste."

Whether or not interpretation issues have historically been thought of as "realism" issues has also depended on the historical circumstances. Since those working with Maxwell's theory in the late nineteenth century had no doubts about the existence of the electromagnetic ether, there was no reason for them to even raise the realism issue with respect to any account of the ether proposed as a supplement for Maxwell's theory. But the generally accepted Copenhagen interpretation fostered the belief that micro-systems have no objective properties, or at least none we can ever know about. Thus an objective properties interpretation of QM cuts against the prevailing belief about the micro-world. And the now established fact that only non-local objective properties interpretations are possible leaves the assumption of objective properties for micro-systems in conflict with either relativity theory or other basic beliefs. As already noted, such conflicts have frequently raised the question of whether to adopt a realistic attitude toward the account generating the conflict.

But whether or not the historical circumstances conspire to raise questions for the scientist about how to regard one of these interpretational supplements to a theory, the philosophical question of how to regard them is basically the one the PSRI is meant to capture—how to regard what a theory says about unobservable things. For while QM and Maxwell's theory do not say anything explicitly about what the objective properties of micro-systems are like or what the electromagnetic ether is like, they do say something implicitly about these things. For they place constraints on what the objective properties or the electromagnetic ether can be like. And it is attempts to learn which kinds of objective properties interpretations are allowed by QM that have allowed us to uncover these constraints and that have demonstrated just how complex and subtle such constraints can be. Traditionally, the philosopher has assumed that a theory wears its story about the unobservable on its sleeve, and the only question is whether to take this story seriously. On this assumption, it is easy to suppose that the realism issue then has to do with issues of reference or truth—issues tied to questions of how the language of the theory works, or with questions about how far confirmation reaches—whether empirical evidence can be brought to bear not only on those claims of a theory that are about observable things, but also on those claims the theory makes about unobservable things. In the case of QM, however, the realism issue does not concern whether certain of the explicit claims of the theory refer or have truth value or receive confirmation. Rather, the question is how to construe the constraints placed implicitly by the theory on what kinds of objects and processes are causally responsible for the phenomena that are covered explicitly by the theory. A more general conception of the PSRI is needed, one that includes this question, in order to capture the realism issue in its full generality.[17]

IIIb. What Part of What a Theory Says Is at Issue?

What counts as a "realistic interpretation" for QM? This can be stated fairly specifically: an account (more or less detailed) of the way that objective properties or hidden variables possessed by quantum systems are related to the "observables" of QM and to the results of measurements for those observables. But this (not atypical) way of characterizing it is misleading. It suggests that the objective properties attributed to quantum systems by such an interpretation are unobservable while the physical quantities for which QM can give predictions are. But the high technology needed to measure for qm "observables" hardly makes them what most traditional scientific realists have called "observable." And the fact that one of the most thoroughly worked out non-local interpretations of QM takes measurements for most qm-observables to be in fact (more or less disturbing) measurements of the objective properties it attributes to quantum systems shows that the "hidden variables" of QM need not be "unobservable" in the traditional scientific realist's sense. Thus, while the current issue of realism for QM is usually characterized as an issue about whether to take literally an account about what is "hidden" behind the "observable," the terms in quotations are understood by those who use them in a very special way. Shifting from the specialized physicist's language to the philosopher's, the issue is not accurately described as being about whether to believe an account about the "unobservable" world lying behind the "observable." It is, rather, about whether to accept an account of the nature and behavior of quantum systems that would explain what causes measurements to give the results that QM says they will.

Thus the SRI for QM does not depend on supposing there is an observable-unobservable distinction. Nor does the SRI for Maxwell's theory. In both cases, the theories provide successful predictions about the outcomes of measurements of certain physical quantities, where these "measurements" are in general not simply direct observations of the quantities, but involve complex and carefully constructed measuring devices. The question is how to construe accounts of what causes these measurement results (accounts of objective properties for QM, of the ether for Maxwell's theory). This suggests that there is a genuine "realism" issue for scientific theories that does not depend on assuming this distinction either, an issue that centers on the question of whether to construe as descriptive the account provided by a theory (either explicitly or implicitly) of the underlying causes of the measurement outcomes it predicts. Of course, a reformulation of PSRI and SRI along these lines will have little advantage over the traditional formulations if the distinction between "what is measured" and "the underlying cause of measurement outcomes" is as elusive as the distinction between what is "observable" and what is "unobservable." But it does not appear to be that elusive. In the cases of QM and Maxwell's theory, for example, the theories themselves, augmented by auxiliary accounts of how the measuring devices work, determine quite clearly what physical quantities are the objects of the various type of measurements on which the empirical success of the theories rest. A quick glance suggests that this is also true for recent theories of astronomy and thermodynamics. If a more thorough examination of these and other cases bears out the suggestion, then we might conclude that at least on a theory by theory basis, the difference between "what is measured" and "the underlying cause of the measurement outcomes" is well defined and marks a difference that can ground versions of both SRI and PSRI that are of genuine interest to both science and philosophy.

IIIc. On Interpreting QM

I suggested above that attempts to assess the acceptability of various non-local objective properties interpretations of QM show how difficult it may be to come to a clean and generally acceptable resolution for the problem of how to interpret QM. Viewing the interpretation issue as an SRI puts it into historical perspective, however, and suggests that under certain circumstances, general agreement on how to resolve the interpretation issue for QM might not be so hard to reach. The controversy over how to interpret Maxwell's equations is another instance of SRI that has much in common with the SRI for QM. And recall how the controversy over Maxwell's equations was resolved. It was resolved when someone took a completely different approach not only to Maxwell's equations, but also to kinematics and dynamics. In the hands of Einstein, Maxwell's equations were transformed and given an interpretation that no one had even thought of before 1905. This provided a resolution of the question of how to interpret Maxwell's equations that was clean and generally accepted. This resolution also revealed that the old interpretation question, which has been conceived as a question about the electromagnetic ether, was thoroughly misleading. The question of how to interpret QM may well be resolved in a similar way.[18]

NOTES

1. Werner Heisenberg, "Uber quantentheoretische Umdeutung kinematischer und mechanischer Beziehungen," *Zeitschrift für Physik* 33 (1925): 879–93; English translation in *Sources of Quantum Mechanics*, edited by B. L. van der Waerden (New York, 1967).

2. Max Born and Pascual Jordan, "Uber Quantenmechanik," *Zeitschrift für Physik* 26 (1924): 379–96; English translation in *Sources of Quantum Mechanics*, van der Waerden.

3. See chapter 10 of David Cassidy, *Uncertainty: The Life and Science of Werner Heisenberg* (New York, 1992), and references for that chapter.

4. Werner Heisenberg, "Schwankungsverscheinungen und Quantenmechanik," *Zeitschrift für Physik* 40 (1926–27): 501–506.

5. Erwin Schrödinger, "Energieaustausch nach der Wellenmechanik," *Annalen der Physik* 83 (1927): 955–68.

6. See Max Jammer, chapter 6 of *The Philosophy of Quantum Mechanics* (New York, 1974).

7. Stanley Gudder, "On Hidden-variable Theories," *Journal of Mathematical Physics* 11 (1970):431–36; John S. Bell, "On the problem of hidden variables in quantum mechanics," *Review of Modern Physics* 38 (1966):447–552; Allen Stairs, "Value-definiteness and Contextualism: Cut and Paste with Hilbert Space," in *PSA 1992*, Volume I, edited by David Hull, Mickey Forbes, and Kathleen Okruhlik (East Lansing, 1992), 91–103.

8. Bohm's first proposal of an objective properties interpretation with the "right connection" was in his "A Suggested Interpretation of the Quantum Theory in Terms of 'Hidden' Variables," Parts I and II, in *Physical Review* 85 (1952): 166–93; other versions have since been developed by Bohm and various co-workers.

9. John S. Bell, "On the Einstein-Podolsky-Rosen Paradox," *Physics* I (1964): 195–200.

10. For a survey of the proofs and experiments up to late 1970, see J. F. Clauser and Abner Shimony, " 'Bell's Theorem' Experimental Tests and Implications," *Reports on Progress in Physics* 41 (1978): 1881–927; for important recent developments, see A. Aspect, J. Dalibard, and G. Roger, "Experimental Tests of Bell's Inequalities Using Time-varying Analyzers," *Physical Review Letters* 49 (1982): 1804-807, and Daniel M. Greenberger, Michael A. Horne, Abner Shimony and Anton Zeilinger, "Bell's Theorem without Inequalities," *American Journal of Physics* 58 (1990): 1131–43.

11. See Don Howard, "Holism, Separability and the Metaphysical Implications of the Bell Experiments," in *Philosophical Consequences of Quantum Theory*, edited by James T. Cushing and Ernan McMullin (Notre Dame, Ind., 1989), 224–53, and Don Robinson, "Dissecting the Holist Picture of Quantum Systems," forthcoming.

12. Paul Teller, "Relativity, Relational Holism and the Bell Inequalities," in *Philosophical Consequences of Quantum Theory*, 208–23; Richard Healey, "Holism and Nonseparability," *Journal of Philosophy* 88 (1991): 393–421; and Don Robinson, "On Healey's Holistic Interpretation of Quantum Mechanics," *International Studies in the Philosophy of Science* 6 (1992): 227–40.

13. Nancy Cartwright, chapter 6 of *Nature's Capacities and their Measurement* (Oxford, 1989); Michael Redhead, "Nonfactorizability, Stochastic Causality and Passion-at-a Distance," in *Philosophical Consequences of Quantum Theory*, 145–53; Richard Healey, "Chasing Quantum Causes: How Wild is the Goose?" *Philosophical Topics*, forthcoming.

14. Jon P. Jarret, "On the Physical Significance of the Locality Conditions in the Bell Arguments," *Nous* 18 (1984): 569–89; Jeremy Butterfield, "A Space-Time Approach to the Bell Inequality," in *Philosophical Consequences of Quantum Theory*, 114–44.

15. Richard Healey, "Chasing Quantum Causes: How Wild is the Goose?"

16. See Bas van Fraassen, "When is a Correlation not a Mystery?" in *Symposium on the Foundations of Modern Physics*, edited by Peter Lahti and Peter Mittelstaedt (Singapore, 1985) 113-28, and Arthur Fine "Do Correlations Need to be Explained?" in *Philosophical Consequences of Quantum Theory*, 175–94.

17. The semantic view of theories allows for a very natural formulation of such a generalized PSRI, since the constraints placed by a theory on its extensions are displayed quite explicitly by the range and variety of the theory's models—a topic for another paper.

18. Earlier versions of this paper were read for the History and Philosophy of Science Colloquium at Johns Hopkins University, and the Philosophy and Physics Colloquium at DePauw University. Special thanks to Fred Suppe for raising the question I attempt to answer in part IIIb.

Vacuum Concepts, Potentia, and the Quantum Field Theoretic Vacuum Explained for All

PAUL TELLER

1. THE PROBLEM

The vacuum is supposed to be a state in which there is nothing at all. How, then, could observation in the vacuum state reveal a material like structure or issue in results like those arising from objects having properties? The vacuum of quantum field theory appears, to some, to offer just such food for interpretive indigestion. In the following pages I will try to set out these surprising ideas and show how they can nonetheless be comfortably accommodated within a certain interpretive framework for quantum theories.

WAIT! Don't quit just because you think you know nothing about this somewhat forbidding subject. The problems and their solutions will all be laid out without presuposing any knowledge of quantum theory. As a bonus, if you will read on, I offer a little pseudo-historical sketch of the development of conceptions of the vacuum, a kind of conceptual reconstruction of how we have arrived at a perfectly coherent but surprisingly complex idea of the "state of nothingness." Actually, this sketch, I hope of independent interest, will take more than half the paper.

For technicians, let me make clear what sort of a case I hope to be making: I hope to show that on certain commonplace, though certainly not unique, interpretive views of quantum theories, facts about the vacuum state will fit in coherently, even comfortably. I will not here defend the background interpretive stance, nor will I argue that this approach to the quantum vacuum is the only or the best one possible. I will count it a good day's work if I can show you that there is at least one coherent way of thinking about quantum vacuum phenomena.

Here is a preview of the vacuum phenomena I will consider. By definition, the quantum field theoretic vacuum is the state in which there are definitely no particles. Yet the theory tells us that in such states one can observe non-null values for certain quantities, some of which even have a positive expectation value. A second kind of problem involves something called Rindler particles. In the vacuum state any inertially moving particle detector will detect no particles. But in this very state an accelerating detector will react as if it were moving through a thermal bath of what are called Rindler particles! This would seem to make no sense: Presumedly, particles are either there or not there. Particles, one would have thought, are not entities which appear or disappear depending on whether or not one is accelerating.

Some readers will have noted a strange feature of the very statement of these problems: the problems are supposed to be about the *vacuum*, by which one is supposed to mean a state in which there is nothing at all. A problem is then said to arise because of what one can *observe* in such a vacuum state. But observation requires the presence of some physical apparatus, and if an apparatus is present the state in question is no longer, strictly speaking, a vacuum!

I presume that this oddity arises, and goes without comment, because of the long standing practice in physics of idealizing observation: One talks about observation as if the process would not in any way affect the state observed. This is a plausible enough idealization in the context of classical physics where one can coherently describe measurement interactions which affect the observed system as little as one likes. But this idealization is challenged, and generally rejected, in the context of quantum phenomena. The problem is exceptionally difficult because the whole question of the nature of measurement in quantum theories presents their most pressing interpretive problem. I cannot hope to resolve these issues here. So I will simply swallow the idealization in the statement of the problem and help myself to the same kind of idealization in my statement of a solution. If such idealization be thought to undermine my solution, it ought by the same token be thought to undermine the very problem that was to be solved.

2. THE GROWING VACUUM

The puzzle which quantum field theory presents is that the theory countenances a state which, apparently, is said to be at once utterly empty, yet admits of non-null observable consequences. The ancients had a prior worry: Does it make sense even to talk about a vacuum state, an absolute void, free of physical objects, of any sort of material substance, of structure of any kind? Let's think about this worry in terms of two ancient conundrums: First, how could absolute non-being be? And second, what sense is there in an idea of location independent of any located objects?

I do not know if anyone has advocated the extreme conception of the void as pure capacity to "receive" or to "make room" for physical objects, a void not only empty but also free of all structure. I take it that advocates of an idea of empty space have at least taken it to be three dimensional, and, until recently, have supposed it to be governed, in some ways, by the laws of three-dimensional Euclidean geometry. Those who see these laws as necessary truths might be fooled into seeing them as no structure at all. For example, anyone who thought that all necessary truths are analytic would see such necessary truths as involving no factual structure. Even on more robust conceptions of necessity one might wonder: if emptiness is unavoidably Euclidean, what are we thereby saying about the emptiness? This is now a non-issue because we now know that space may be characterized by geometries other than Euclidean geometry, which fact makes it quite plain that attribution of Euclidean geometry even to empty space is to attribute to it structure of some kind.

But what kind of structure is this? We can get at one way of answering this question by considering the Newtonian picture. Newton gave laws of motion, that is to say, a characterization of which trajectories for physical objects are physically possible. Simply saying this much already presupposes that there is a larger collection of trajectories which are, in some sense, possible. To make this out a Newtonian postulates a collection of things—the spatial points (or spatial volumes) at which things can be, but do not have to be located. Continuous sequences of these spatial points, thought of as occupiable at successive times, constitute the required logically or metaphysically possible trajectories; and Newton's laws then tell us which of these are physically possible. In this background, Euclid's laws apply to structure the collection of spatial points.

On such a Newtonian conception the spatial points are thought of as things of some kind—hence the view is often called *substantivalism*. We should take note of two considerations about such a substantivalist conception. On it a vacuum state is a world composed of spatial points at which no material objects are located. But this is a pretty meaty sort of vacuum! It is stuffed full of the spatial points themselves, and the points are collectively highly structured by the laws of Euclidean geometry. Secondly, presentation of Newtonian space as a collection of spatial points, substantivally conceived, obscures the point that this metaphysics is deep not only in some kind of substance, but also in possibilia, as I was at pains to bring out in the previous paragraph. Newtonian spatial points are things at which physical objects *can* be located.

This conception responds to the first ancient conundrum, of how absolute non-being could be. The response works by backing down: Emptiness is still stuffed full of the spatial points, and it still involves a network of possibilities for location, highly structured by Euclid's laws. When put this way, the Newtonian response to the first conundrum seems heavy handed. Why postulate two kinds of things, the spatial points and the network of possibilities, as a characterization of a vacuum state, now best thought of as a state of minimal being? We cannot postulate the spatial points without the possibilities, for the points are understood, perhaps in their entirety, in terms of possibilities for location. But why not just postulate the network of possibilities?

This suggestion becomes less obscure if we draw it out a bit in the context of a response to the second ancient conundrum and in terms of the Leibnizian attack on the Newtonian view. Leibniz noted a prima facie absurd consequence of the Newtonian postulation of substantival space. Postulate such a space, and consider the way things are actually located in the space, throughout time. Now consider a Leibnizian alternative to the actual situation, namely one exactly like it except that everything has been, throughout the course of history, moved uniformly over three feet. With all spatial relations held constant, the actual case and the Leibnizian alternative are utterly indistinguishable from each other— they differ in no way except for a difference in location in the postulated space. The suggestion is that it is just silly to count these two cases as really different.[1] Since such a silly conclusion arises from postulating substantival space, we should wash our hands of it.

Leibnizians conclude that all there is to space are the spatial relations between physical objects, a doctrine widely known as *relationalism*. On the face of it, this would seem to rule out any sort of void or vacuum. Indeed, one might see Leibnizian relationalism as an expression of what I called the second ancient conundrum: what sense is there in an idea of location independent of located objects?[2] I, for one, am not satisfied. Consider two space voyagers' surveyors' flags placed ten feet apart from each other somewhere in outer space, with no physical objects on the straight line between them. The fact is that there is nothing exactly midway between these flags, but there could be. A space voyager *could* have stuck a third flag midway between the first two. We can understand talk of a local, or circumscribed, vacuum between actual physical objects in terms of unactualized possible spatial relations, in terms of relations which do not but could have held. Although I have no analysis of such counterfactuals, I submit that they are straightforwardly intelligible and true.

Acknowledging these counterfactuals requires extending Leibnizian rationalism to the doctrine I call *liberalized relationalism*.[3] The spatial facts are comprised not only by the actually exemplified spatial relations, but also by those merely potentially exemplified. Every bit as much as Newtonian spatial points, one can take the network of actual and potential spatial relations to be governed by the laws of Euclidean geometry, resulting in a structure as rich and theoretically powerful as Newtonian substantivalism, but free of the excess of the substantival points.[4]

Liberalized relationalism will support the idea of local or circumscribed vacua, that is unoccupied—that is to say potentially occupied—locations bearing only potential relations to actual objects. But what about the idea of a total vacuum—a world with absolutely no physical objects in it? Will failure to make sense of the idea of a total vacuum, in the framework of liberalized relationalism, vindicate the ancient worry about the intelligibility of the idea of location independent of *any* located objects? I do not think so. Although only a philosopher trying to make a philosophical point would say the following, it makes perfectly good sense to say: "there might have been no physical objects at all. Still, relative to that possibility, there could have been two surveyor flags ten feet apart and. . . ." In other words, even on the supposition of no physical objects, it makes perfectly good sense to talk about what spatial relations could have been realized by physical objects if they were to have existed. And this provides exactly the required mock-up of talk about completely empty space.

While I prefer liberalized relationalism to substantivalism as an account of the nature of space, nothing in what follows will turn on a choice between these two points of view. Henceforth I will use the language of substantivalists because it is the more familiar in talking about space and spatial (soon space-time) points. Please understand that in so doing I mean to be neutral between interpreting such talk in terms of spatial points substantively conceived or in terms of potentia for relative location.

Whether seen through the substantivalist or relationalist lens, the ideas current in the time of Newton and Leibniz support a conception of the vacuum, albeit already a rich and highly structured thing. Such a vacuum is no absolute non-being, but is complete with a rich structure of unactualized possibilities, or with these and substantival spatial points, on both views structured with the laws of Euclidean geometry. And as the study of the nature of space continued, ideas of the vacuum—of space as empty as it could be—continued to take on more baggage.

In the seventeenth and eighteenth century, natural philosophers worried about the action at a distance which Newton's theory of gravity seemed to require. Eventually these worries were resolved, or at least sidestepped, by the development of field theories, which attribute values of relevant physical quantities to the spatial points.[5] In a simple example, a distribution of matter in a region of space would be characterized by giving a mass density at each point. For another example in a characterization of the action of gravity, one attributes a gravitational potential to the spatial points. Field theories proved particularly useful in describing the interaction of electric and magnetic fields occurring in the phenomenon of light and other electromagnetic radiation. In general, field theories avoid action at a distance by describing point by point propagation of influences, roughly speaking in terms of a change in the value of a quantity at one point causing correlative changes in neighboring points.

Field theories continued to broaden the question of what counts as a vacuum state, for on such theories space could be completely devoid of physical objects and still be filled with light. Probably one should take a field theoretical vacuum, as described by classical theories, to be the state in which all field quantities take on their null value at all spatial points.

Relativistic theories added further substance to the idea of the vacuum. Special relativity showed that space and time must be treated together, so that we will henceforth talk about space-time points instead of spatial points. In addition, special relativity attributes a non-Euclidean geometry to the collection of space-time points. This move does not add substance or structure to the vacuum state, but it rubs our noses in the fact that even when no physical objects are present and all field values take on a null value, the vacuum state is highly structured. In Newtonian theories the structure was Euclidean. In relativistic theories it is now a radically different non-Euclidean geometry. As I mentioned, when Euclidean geometry was the only alternative, and thought necessary, it was easy to ignore the fact that it involved a lot of structure. The radical change in structure attributed to space-time reminds us of how very much structure even the vacuum state involves.

General relativity further grinds our noses in this very same fact when the geometry itself becomes a variable of the theory! In Newtonian theories the distance between spatial points (in special relativity the distance between space-time points) is fixed. In general relativity the distance between space-time points is a variable of the theory. Of course, since we are talking about distance between space-*time* points, these separations are not variable in the sense that they can change over time. But they are contingent on the overall distribution of energy and matter. If energy and matter had been distributed differently then the space-time separations between the same pairs of points would have been different. To see that these differences make a difference in geometry, think of the spatial separations on a flat rubber sheet. Systematically change the separations between points on the sheet, say by poking your finger into it. The resulting curved surface now has a new geometry of a curved surface. The angles in a triangle no longer add up to 180°, Euclid's parallels postulate fails, etc.

It is of particular relevance to conceptions of the vacuum that in a world described by general relativity, many possible structures—geometries—are consistent with the presence of no matter in the usual sense. In a way somewhat analogous to light waves in empty classical space or special relativistic space-time, a general relativistic space-time devoid of concrete objects can still have "gravity waves." We must here also take into account the relativistic equivalence of mass and energy. Such gravity waves involve energy, and so, in an unconventional sense, some kind of matter. The sense is highly unconventional, or unaccustomed, because the mass-energy must be attributed to space-time regions as a whole but cannot be attributed to individual space-time points. Such a space-time is devoid of concrete material objects and classical-style fields but is still not in every way devoid of "stuff." And again, it is highly structured. Still, in general relativity we have vacuum states of minimal being in completely flat-space times, states devoid of mass-energy but still highly structured in both actual and potential geometry.

Let me try to summarize what we have so far learned by reviewing from a somewhat different perspective. Think of the development of vacuum concepts as one facet of the historical search for a more general theory of the behavior of physical objects. Part of this more general theory is an account of the "backdrop" or "stage," the arena in which all the action takes place. We might start with minimalist assumptions: the vacuum as utterly empty and unstructured susceptibility for being located—possibly the conception which some of the ancients found unintelligible. Intelligible or not, developing physical theory required a richer structure for this backdrop or vacuum. Newtonian theory required the structure of Euclidean geometry and the spatial points, or at least the network of possibilities for relative location. Field theories added the potential for values of fields, such as the electromagnetic field.[6]

General relativity further enriches the vacuum concept at least by enriching the structure which must go along with the vacuum state, by making the geometry a variable and by including the potential for matter and energy in a sense radically different from physical objects and field quantities in the conventional sense.

The quantum field theoretic vacuum fits right into this progression, with the addition of yet more structured potentia to the "backdrop" or vacuum.

3. THE QUANTUM FIELD THEORETIC VACUUM

With this development, I can express the additional feature of the quantum field theoretic vacuum as a one-liner. Fix in your mind the idea that classical and relativistic field theories attribute values of field quantities to the space-time points. In these theories one most sensibly takes the vacuum, the state of minimal being, as the one in which no quantity actually takes on a non-null value. The one-liner is the fact, which non-technicians will simply take on faith, that in all quantum theories, *not all values can simultaneously take on their null value.* This fact subsumes all peculiarities of the quantum field theoretic vacuum. While there can be a state in which no quantities actually take on a non-null value, in any state at least some quantities must involve a potential for a non-null value, in a sense which I will discuss. Hence the quantum field theoretic vacuum adds to prior vacuum conceptions additional structure in terms of what is potential or possible. It is because I have wanted to highlight this fact that I have leaned in such a heavy-handed way on the role of possibility in characterizing vacuum structure in prior conceptions.

This summary statement needs spelling out, which I think one could do on a wide variety of interpretive approaches to quantum theories. I will spell out the interpretive approach I like best. The approach is controversial, of course— all interpretive approaches to quantum theories are controversial. I will not here try to defend my assumptions about background interpretation—I will simply set them out, in a form which should be accessible to anyone who has read this far.

A quantum theory specifies a range of possible quantum states, from any one of which one can calculate a probability for, as it is usually put, a range of possible outcomes of measurements. Let's present the general ideas in terms of a concrete and familiar example. In conventional quantum mechanics, position, conventionally indicated with the letter 'Q', is a measurable physical quantity, which can take on specific values, q.[7] One quantum state, ψ, will specify a probability of 1 for one specific value of Q, and a probability of 0 for all others: $\text{Prob}_\psi(Q = q) = 1$ and $\text{Prob}_\psi(Q = q') = 0$, for all $q' \neq q$. In state ψ I take the quantity Q, position in our example, actually to have the value q. But in other states no value for position gets a probability of 1. When the probability for a given value of position is also not 0 I take that value to be potential but not actual, and I will say that in such a state there is an (actual) *propensity* for the value in question. I understand propensities as objective single case probabilities. Think of these propensities as similar to dispositions, but differing in that their display properties, here a specific value of position, will occur under the right activating conditions, not with certainty, but with the probability calculatable from the description of the quantum state.

What are the right "activating conditions"? Usually these are said to be "measurements." It is a cardinal interpretive problem for quantum theories to explain what counts as a "measurement" and how, or why, outcomes occur when one makes a measurement. On some interpretations, activating conditions are constituted more broadly than what one would ordinarily call a measurement. I will persist in using the term 'measurement' for the appropriate activating conditions, but I intend thereby to stay neutral as to details of how these conditions are more carefully specified. In particular I do not want to rule out some ways of construing "activating conditions" which may be broader than ones which we would ordinarily think of as measurements.

Conventional quantum mechanics also recognizes a quantity called momentum, conventionally indicated with the letter, 'P', which can take on values, p. (Sorry. You would think that 'P' would stand for position. I am just following standard physicists' conventions here!) Exactly the same kind of comments about actual and potential values hold for momentum (and, indeed, for all quantum mechanical quantities) as I described for position.

One of the peculiarities of quantum mechanics is that *no quantum state assigns an exact, actual value to both Q and P*. More specifically, if some state, ψ, makes some value, q, of Q actual, $\text{Prob}_\psi(Q = q) = 1$, then ψ makes *no* value of P actual: $\text{Prob}_\psi(P = p) < 1$, for all values, p, of P. As long as $\text{Prob}_\psi(P = p) > 0$, the value of p is still potential, in this example true for all values, p, of P.

When $0 < \text{Prob}_\psi(P = p) < 1$ I also want to be wary of saying that P does *not* have the value p. I am committed to such statements as long as one is clear that by "P does not have the value, p" one means that the value p is not actual in state ψ. But this expression must not be misinterpreted as saying that in state ψP certainly will not manifest the value p. On the interpretation I espouse, there are three cases: When $\text{Prob}_\psi(P = p) = 1$, the value p is actual. When $\text{Prob}_\psi(P = p) = 0$, the value p is *certainly absent*. And when $0 < \text{Prob}_\psi(P = p) < 1$ the value p is neither actual (-ly present) nor certainly absent. Rather it is potential, more specifically the state has a propensity, an objective probability, for the value p to arise "on measurement."

Q and P, and any pair of quantum mechanical quantities which behave in the way I have just described, are said to be *complementary*.[8] Conventional quantum field theory happens not to use a position quantity. We will be interested in other pairs of complementary quantities. I have used position and momentum for illustrative purposes only. The pivotal idea is that one can say three different things about a value of a quantum mechanical quantity. It may be actual; it may be actually, that is certainly, absent; or it may be neither of these but potentially present, with a propensity calculatable from the description of the quantum mechanical state.

We are now ready to turn to quantum field theory and its vacuum state.

Classically, the electromagnetic field constituted a continuous quantity. But it is the mark of quantum theories to recharacterize such quantities in discrete units. Thus the quantized electromagnetic field recharacterizes the field in terms of "particles,"[9] called photons. Thus there may be exactly 0, or 1, or 2, or. . . . photons in a given state. All quantum field theories follow this pattern, and I will talk generally about quantum field states with 0, 1, or 2 or other definite numbers of particles. The number of particles constitutes a quantity, N, which may take on values $n = 0, 1, 2, \ldots$, or any other non-negative integer. States, ψ_n in which N has an actual value, n, $\text{Prob}_{\psi_n}(N = n) = 1$, are called exact number states.

I can now quickly explain what seems so incongruous about the quantum field theoretic vacuum, and how it can be coherently understood in the interpretive framework I have sketched above. The vacuum will be the state in which there are *no* particles, that is the $N = 0$ state, ψ_0, with $\text{Prob}_{\psi_0}(N = 0) = 1$. But in this empty state, a state with *nothing* in it, one would expect that measurement of just any quantity would have to come up with a null value. Thus one is shocked when in this vacuum state one measures for certain characteristics of the field and often comes up with non-null values! Many are even more shocked when the theory tells us that measurement of the field strength, corresponding to the field energy, has a *positive* expectation value in any small volume, that is a positive value for what one can expect to measure on the average in a small volume.

Here is another way to bring out the prima facie shocking aspects of this state of affairs. The quantum field is the quantum theoretical mock-up of a classical field. One would expect that the $N = 0$ state would be a state in which the field was utterly absent. The facts described above seem to contradict this expectation.

These surprising facts are well-known, really trivial consequences of the facts of quantum mechanical complementary quantities, as I will explain in a moment. What will shortly be seen also to be obvious, although perhaps not sufficiently noted, is that on my way of construing quantum states and quantities, these facts about the quantum field theoretic vacuum fit in comfortably with the ideas I have described for pre-quantum vacuum conceptions, in which a vacuum state is one with nothing actual, yet highly structured in terms of what is possible or potential.

To understand this you need to know just a little about superposition in quantum theories. Consider two quantum field theoretic states, ψ_n and $\psi_{n'}$, the exact number states, respectively with $N = n$ and $N = n'$. The theory also admits a new kind of state called a superposition of the first two, written $\psi_s = \psi_n + \psi_{n'}$. Furthermore, in state ψ_s, neither value n nor n' are actually present or certainly absent. Both are potential: $0 < \text{Prob}_{\psi_s}(N = n) < 1$ and $0 < \text{Prob}_{\psi_s}(N = n') < 1$. In this particular example all other values of n are actually absent, with $\text{Prob} = 0$. In more interesting cases ψ_s will, in the same kind of way, be a superposition, or sum, of all exact number states so that in ψ_s all values, n of N, will be neither actually present nor certainly absent. All will be potential. There are in fact a great number of different such states, $\psi_s, \psi_{s'}, \psi_{s''}$, differing by taking different weightings on the terms in the sum. The indexes, $s, s', s'' \ldots$ can be interpreted as values of a new quantity, S, so that in state ψ_s, S actually has the value s: $\text{Prob}_{\psi_s}(S = s) = 1$, and similarly for $S = s', s'', \ldots$. Finally, the quantities, N and S, are complementary. In any state which gives one an actual value, the other takes on values only potentially.

The consternation about the quantum field theoretic vacuum, the $N = 0$ state, arises by forgetting these facts about complementary quantities in all quantum theories. Setting the value $N = 0$ does completely specify a state of a quantized field. But the field is also described by the complementary quantities, S. The $N = 0$ vacuum state is an honest vacuum in that it actually has no quanta, and it also does *not actually* have any specific value for any complementary quantity. But, every bit as much as in pre-quantum vacuum states, the quantized vacuum is structured with possibilities or potentia. Quantities, S, have no actual values in the vacuum. But the vacuum does include propensities for outcomes on measurement of these quantities. Newtonian and relativistic vacua already included structure, in terms of networks of possibilities. The quantum field theoretic vacuum, on the interpretation I have recommended, is likewise structured with possibilities, differing at most in that the structure of possibilities is of a different, and one might fairly judge, richer kind.

I will close with another illustration of puzzles about the quantum field theoretic vacuum, one which submits to the same treatment, as will all puzzles in this family. We again consider the $N = 0$ vacuum state and now characterize it with an operational style test by considering a class of particle detectors designed so that they will, when operating in the vacuum state, certainly register no particles when these detectors are unaccelerated. Of course, by calling them particle detectors I imply that they will register one particle in the $N = 1$ state, two particles in the $N = 2$ state, and so on. Now, the theory tells us that if we take such a detector and *accelerate* it in the $N = 0$ state the detector will respond as if it were moving through a thermal bath of so-called Rindler particles! This seems crazy: Particles, which we want to think of as hard little objects, are supposed either to be there or not there. It should not matter whether or not a well-designed particle detector is accelerating or not. This line of thinking has induced physicists as well as philosophers to tear their interpretive hair.[10]

The apparent contradiction depends on the tacit assumption that if one finds particles on measurement it can only be because the particles were actually there before measurement. The interpretive stance I have advocated denies that this has to be so. While surprising, those familiar with the interpretive problems of quantum theories will immediately recognize my stance as a familiar one, well motivated by consideration of the oldest problems in understanding quantum mechanics, such as the double slit gedanken experiment, problems which arise quite independently of questions about the quantum field theoretic vacuum. In the present case, we need only note the fact about the formalism, that ordinary particles are described by one quantity, N; and the Rindler particles are described by a *complementary* quantity, R. In the $N = 0$ state, no value of R is actual. Thus in the $N = 0$ vacuum state one can correctly say that there are *no* actual particles of any kind, making the $N = 0$ state as honest a vacuum as pre-quantum vacua. But even more than pre-quantum vacua, the $N = 0$ vacuum state bulges with potentia, in the current example in the form of propensities for Rindler particles to arise should one go and hunt for them with an accelerating detector.

NOTES

1. I won't here discuss the Newtonian counter argument, the so-called bucket argument. In fact I believe both arguments are flawed. I have examined these arguments in Teller (1991).

2. I have never seen this analogy mentioned in the extensive literature on relationalism.

3. I discuss liberalized relationalism in greater detail in Teller (1991).

4. Again, this is spelled out in Teller (1991). The key is that liberalized relationalism will support use of coordinate systems which do all the theoretical work allegedly done by the substantival points allegedly described by exactly the same sort of coordinate systems. In addition, contemporary versions of substantivalism involving space-*time* points are subject to a souped-up version of Leibniz's indiscernibility argument, the Einstein-Earman-Norton hole argument, while liberalized relationalism is free of this difficulty.

5. Where substantivalism attributes a value of a quantity to a space-time point relationalism takes advantage of mass-energy equivalence and the fact that values of physical quantities involve energy to say that a bit of substance exemplifying the quantity in question occurs with the requisite space-time relations. Note that in this context, depending on what fields are in question, one may be able to make a case for a plenum, in which case the distinction between Leibnizian and liberalized relationalism dissolves.

6. Or, on a relationalist way of construing things, the susceptibility for relative location of matter in a broader sense, namely, the mass/energy which general relativity associates with the field values.

7. I am going to talk as if Q were a discrete quantity for the purpose of setting out ideas to those hearing them for the first time.

8. I have actually drastically underdescribed complementarity of quantum mechanical quantities. But the interpreted facts I have presented will suffice for the ideas which follow.

9. Better called 'quanta', for they differ in important respects from particles as conceived of in classical theories. These differences are discussed in Redhead and Teller (1990) and Teller (forthcoming), chapters 2 and 5.

10. Davies concludes that "Particles do not exist" (1984). See Ruger (1989) for a good summary of the problem and interpretive stances in the literature.

REFERENCES

Davies, P. C. W. 1984. "Particles Do Not Exist." In *Quantum Theory of Gravity*, edited by S. M. Christensen, pp. 66–77. Bristol.

Redhead, Michael, and Paul Teller. 1990. "Particles, Particles Labels, and Quanta: The Toll of Unacknowledged Metaphysics." *Foundations of Physics* 21, no. 1.

Ruger, Alexander. 1989. "Complementarity Meets General Relativity: A Study in Ontological Commitments and Theory Unification." *Synthese* 79: 559–80.

Teller, Paul. 1991. "Substance, Relations, and Arguments About the Nature of Space-Time," *Philosophical Review* 50, no. 3: 363–97.

Teller, Paul. Forthcoming. *An Interpretive Introduction to Quantum Field Theory*.

Genic Selection, Molecular Biology and Biological Instrumentalism

ALEX ROSENBERG

Genic selectionism is the thesis that the forces of natural selection operate to select properties of the genes, and not properties of individual organisms, as adaptive. In recent years the biological controversy surrounding genic selectionism has excited a lively debate among philosophers, several of whom have taken sides on what appears to be a contingent issue to be settled by factual considerations.[1] It is a reflection of the degree to which factual considerations involve conceptual issues that philosophers may have shed a little light on this debate. In this paper I want to shed a bit of biological light on the dispute among philosophers. In particular I want to examine the ramifications of molecular biology for this dispute about whether selective forces operate exclusively at the level of the individual gene or not. But I am also interested in a larger thesis about the nature of biological theorizing and its differences from theories in the physical sciences.

In a couple of recent papers (Rosenberg [1988; 1989]) I have argued that biological science is more strongly contingent on the cognitive and calculational powers of *Homo sapiens* than are the physical sciences. Indeed, the dependence is so much greater as to suggest that biology is an instrumental discipline. By 'an instrumental science' I mean a discipline whose theories reflect the needs, interests, and limits of human scientists, instead of reflecting the way the world is. Of course, if some approaches to science are correct then all theories are like this. But the claim that biology is instrumental does not rest on philosophically tendentious theses, but on a distinct difference from physics and chemistry. If we, *Homo sapiens*, were different, biological theory would be different. By contrast, the claims of physics and chemistry seem constant under actual or possible changes in the cognitive powers of physicists. Thus, the shape of biology is more strongly governed by instrumental considerations than is the shape of the physical sciences. Recent work by Sober (1984) on the role of statistics in the theory of natural selection and by Kitcher (1984) on the relations between classical and molecular genetics seem to some to substantiate this view.[2]

The dispute over genic selectionism provides another context in which the same conclusion seems inescapably to be emerging: the kinds of biological theory reflect the interests and capacities of biologists, not the way nature's joints are cut. In fact, the philosophical problem of genic selectionism recapitulates the instrumentalist conclusions to be drawn from the statistical character of the theory of natural selection and the practical impossibility of biological reductionism. Or so I shall argue.

1. GENIC SELECTION AND CAUSAL REALISM

The Nature of Selection (Sober 1984, chaps. 7–9) argues against the thesis, advanced vigorously by Richard Dawkins (1976; 1982), that all adaptations exist solely because they benefit individual genes. Refuting this thesis of "genic selectionism" has generally been taken to require showing that selection operates at other levels, and that selection at these levels cannot be explained smoothly by selection at the level of genes. Thus, refutations of genic selectionism have either proceeded by or presupposed arguments in favor of the existence of natural kinds automomous from and not reducible to the macromolecules that constitute living things. Opponents of genic selectionism make common cause, whether their favored level of selection is the genotype, the organic individual, the population, deme, or species. Much of the debate about genic selectionism has focused on exegetical issues about how its proponents characterize the environment in which traits are selected for: is the relevant environment in which selection takes place the extra- and intra-cellular milieu of the polynucleotides that make up the gene, or the more usual niche of the organism bearing the genes. These exegetical matters have been conveniently sorted out by Waters (1991).

But the entire controversy surrounding genic selectionism has an air of unreality about it. Though all parties to the dispute agree that the issue is one about alleles—Mendelian genes that come in pairs and code for phenotypes— almost all parties also agree that there are no Mendelian genes of the sort required to motivate the debate.[3] The debate is about whether the sole locus of selection is the Mendelian allele, but if advances in breeding, physiology, cytology, and molecular biology from 1900 to the present have shown anything, they show that any phenotype in which we are interested is the result of the interaction of the environment and a vast number of diverse and distributed sections of the genetic material, the polynucleotides that make up the body's DNA. The view that we can connect anything like a Mendelian allele or a pair of them with a phenotype of interest has long been stigmatized as "bean bag" genetics.

Mendelian genetic theory still has an instrumental role to play, and a crucial one, in population genetics. As an instrument for enabling us to predict the distribution from generation to generation of certain phenotypic properties, the theory is unrivalled. But as a useful fiction, a heuristic calculating device, the theory will not support a serious debate for or against genic selectionism.

If there is to be a serious debate about genic selectionism, it must be a debate not about whether the Mendelian allele is the only locus of selection; it must be a debate about whether the genetic material, the DNA polynucleotide, however subdivided into functional parts, is the sole level of selection. So understood, however, genic selectionism has much to recommend it. And the only reason to adopt any other view reflects the instrumental character of biology, its character as a discipline whose kinds are contingent on our powers and interests, and are therefore instrumental and not "natural." Keeping in mind that only on an interpretation of 'gene' as genetic material can this debate get off the ground, let us assess Sober's arguments against and Sterelny and Kitcher's argument for genic selectionism.

Sober's argument against genic selectionism has been challenged by Waters (1985; 1990) and by Sterelny and Kitcher (1988). All three defend a thesis the latter two label "Dawkinspeak," or *pluralist genic selectionism*: "there are alternative, equally adequate representations of selection processes and that, for any selection process, there is a maximal adequate representation which attributes causal efficacy to genic properties" (358).[4] They contrast this thesis, which they endorse, with another stronger one sometimes embraced by Dawkins: *monist genic selectionism*: "for any selection process, there is a uniquely correct representation of that process, a representation which captures the causal structure of the process, and this representation attributes causal efficacy to genic properties" (358). The difference between these two theses turns on the difference between 'maximal adequacy' and 'uniquely correct'. A representation may be maximally adequate compatibly with other representations being maximally adequate, while a uniquely correct representation cannot share the limelight, whence the contrast between the monism they reject and the pluralism they embrace. (This paper in effect argues for monistic genic selectionism.)

In order to understand Sterelny and Kitcher's claim, let's take it that representations are adequate if they are true, so that a maximally adequate representation is one which is completely correct.[5] Accordingly, a representation can be compatible with other equally adequate representations only if it is incomplete in its description of phenomena, and it will be maximally adequate only if it cannot be made more complete in its description of the phenomenon. A representation of some phenomenon X will be uniquely correct only if it is true and complete in its description of X, so that there is no room for alternative representations of equal adequacy.[6]

Sterelny and Kitcher claim that there are alternative ways of representing selection processes: either as effecting genes or effecting individuals that bear them. Employing their example, we may talk of selection for "genes" for spider-web building and their selective advantages (understanding such expressions as elliptical for claims about large numbers of varying quantities of the genetic material which we cannot as yet identify, and which may not even be units in the sense of bearing a single molecular structure or even a small number of molecular structrues in common). Alternatively we may consider the selective advantages for spiders disposed to web-building. They conclude, as between spiders and their genes, "[t]here is no privileged way to segment the causal chain and isolate the (really) real causal story." To focus exclusively on the webs—properties directly connected with survival and reproduction of the individual organisms that bear the genes—is fallacious, they insist. Doing so begs the question as to whether the environment within which selection operates is the individual's environment, or the allele's—the individual gene's. "Equally, it is fallacious to insist that the causal story must be told by focusing on traits of [the genes] which contribute to the survival and reproduction of those individuals. We are left with a general thesis of pluralism: there are alternative maximally adequate representations of the causal structure of the selection process" (358).[7]

Leave aside for the moment the fact that the genetic material cannot really be individuated into allelic units of function. Unless there is some reason in principle why a complete representation of the causal structure of the selection process is not possible, pluralism cannot be correct. Sterelny and Kitcher tell us "[t]here is no privileged way to segment the causal chain and isolate the (really) real causal story" of the selection process (359). These are two separate claims. Apparently the reason for the first of them is that it is equally true today that phenotypes manifested by individual organisms (or other units larger than the individual allele) are selected for as it is to say that properties of genes are selected for.[8] So, the individual organism's having a phenotype is part of the causal story of its survival, and of the survival of genes it contains. And the individual allele's having a certain property is causally necessary for the survival of its lineage, and perhaps also for the survival of the individual's lineage. We cannot deny the causal property of being selected for to either of them, and so there is no privileged way to segment the causal chain. But even assuming there is no privileged way to segment the causal chain, this is no reason to infer that we cannot "isolate the (really) real causal story" of what gets selected for.

Why suppose that there is no such *unique* story, no theory which gives *all* the details correctly, and so excludes alternative representations? One possible reason is to hold that in science there is no such thing, that data always underdetermine theory. Accordingly we should embrace an instrumental conception of theories on which their adequacy consists not in truth, but in some other epistemic or practical virtue. Then of course, it will not be surprising that for different purposes different theories are more suited. In other words, pluralism might recommend itself in biology because, in general, pluralism about scientific theories is in order.

Now, an antecedent commitment to pluralism for scientific theory in general will sustain pluralism about the maximally adequate account of selection, and will sustain pluralism about every other phenomenon of scientific interest as well. But in the present connection such a rationale would be impertinent.

For Sterelny and Kitcher's thesis is not a claim about the nature of scientific theory as such. It is a claim about one particular theory, not every scientific theory. Sterelny and Kitcher need some fact specifically about selectionist phenomena that renders them inaccessible to representations that reconcile truth and completeness. They need an impossibility proof.[9]

Without such a proof, there is no reason to suppose that what they call "the (really) real" story cannot be isolated, at least in principle. Would it suffice for their conclusion that the (really) real causal story is complicated beyond the cognitive powers of mere mortals? What if the best we can do is provide several different stories all of which are maximally adequate in the sense that we cannot do any better, even though there is a (really) real story beyond our poor powers to express? This view would of course fail to undermine *monist genic selectionism* as a theory about the world, and it would substantiate *pluralist genic selectionism* at most as a thesis about us and our powers of discernment and expression. It would underwrite pluralism as a sort of subjectivist second best. Given our feeble powers, the evidence makes pluralist genic selectionism look right to us, even though to a more powerful intellect, monist genic selectionism might suggest itself.

Though I think this sort of subjectivism is near the truth of the matter, it is not an interpretation of Sterelny and Kitcher's argument that they intended. For their view is that there is no determinate fact of the matter about the level of selection, not just that whether there is or not, it is epistemically inaccesssible to us. They liken *pluralist genic selectionism* to "instrumentalism" in the philosophy of science:

> Pluralism of the kind we espouse has affinities with some traditional views in the philosophy of science. Specifically, our approach is instrumentalist, not of course in denying the existence of entities like genes, but in *opposing the idea that natural selection is a force that acts on some determinate target, such as the genotype or the phenotype.* (359, emphasis added)

It is worth asking what Sterelny and Kitcher mean by the phrase I have emphasized: how can selection be a causal force at all if it does not act on some determinate target or other?

One way a causal force might fail to act on a determinate target is when its individual targets are not instances of a single determinate type of target. As a claim about type-causation, Sterelny and Kitcher's thesis may simply be that tokens of the type 'selective force' do not act on tokens of any single homogenous type. Sometimes tokens of the type 'selective force' act on 'genotokens' and sometimes on 'phenotokens', sometimes on traits-tokens of particular kin-groups, demes, populations, species, etc. Then there is no single determinate type of target for the forces of selection to aim at. In this case there will be no interesting generalizations about the selection of traits. For at most selection will target an unmanageable disjunction of traits of different types of biological systems. This conclusion would be a defense of a default version of pluralism: there will be no uniquely accurate representation because all representations will be equally uninteresting; there will be no single systematic representation of the effects of selection at all—or at least no such generalizations of interest to cognitive agents of our powers.

Such an argument for pluralism bears close connections to Sober's controversial one that disjunctive types are causally inert (Sober 1984, 92–96). If a type is causally inert, it is not a natural kind. We understand what it means to say that there are no entities of a certain kind, while accepting the existence of tokens identified by the use of the kind-predicate in question. Thus, there will be no determinate type of target for selection, because the disjunction of types realized by all the particular tokens selected for does not itself constitute some complex disjunctive natural kind. This argument too needs a justification for the claim that disjunctive types cannot figure in scientifically interesting generalizations just because these generalizations would be too complicated for us to express and exploit.

Instead of acting on differing targets on differing occasions, another way a causal force might *fail* to act on any single determinate target is by acting on tokens of several different types at the same time—a bit of genetic material, its organism, its kin-group, etc. But this way of not acting on determinate targets provides no reasons to suppose that "the (really) real" causal story cannot be told. It only suggests that when the uniquely adequate story is told, it will be a complicated one with a lot of interconnections between different levels of organization. Again this is not a conclusion which will sustain pluralism.

Sterelny and Kitcher's appeal to conventionalist theories of space and time as paralleling their own arguments reinforces the suggestion that there is no *biological* basis for their claim that no uniquely right causal story is possible:

> Another way to understand our pluralism is to connect it with conventionalist approaches to space-time theories. Just as conventionalists have insisted that there are alternative accounts of the phenomena which meet all our methodological desiderata, so too we maintain that selection processes can usually be treated, equally adequately, from more than one point of view. The virtue of the genic point of view, on the pluralist account, is not that it alone gets the causal structure right but that it is always available. [359]

This passage in fact suggests that the basis for their pluralism is a general thesis of underdetermination of theory by evidence. But what Sterelny and Kitcher need is at least evidence for the less general, more specific thesis of the underdetermination of *biological* or *evolutionary* theory by evidence.

In the absence of such an argument, pluralism about selection has little to recommend it over and above the virtues of pluralism everywhere and always.

2. TOKEN GENIC SELECTIONISM

If we accept that there is in principle a unique maximally adequate representation, a (really) real causal story, the attractions of monistic genic selectionism increase. For if we take seriously the fact that talk about genes is talk about the genetic material in all its complexity, a deeper understanding of biological processes reveals that selection for genic properties *is* privileged above selection for other properties.

The only traits that are selected for are inheritable ones, traits that can be passed on to subsequent generations hereditarily. And this will be true whether the traits are predicated of species, populations, demes, kin-groups, individuals, gametes, gene cluster, homologous paired alleles, individual genes, or sequences of nucleotides, for that matter. This means that at every "level" except the last—the hereditary material—a trait is at best an *indirect* cause of survival and reproduction of the item that instantiates it.

By *indirect* I mean this: item *a* is the indirect cause of *c*, if, *a* brings about *c* only as a result of the transitivity of causation, via some further event *b*. By a causal process being *more indirect* than another I mean that the more indirect one requires more intervening events transitively carrying the causal "signal" from the cause to the eventual effect. It is difficult to count the links in a causal chain, especially when the whole story is still unknown, as it is in the molecular biology of gene-expression. But it is clear that properties of individuals, organisms, kin-groups, demes, at a generation are more indirect causes of survival-and-reproduction of the next generation than are properties of the genes of organisms in the earlier generation. If a phenotype is selected for, then this causes a genotype in the germ-line to be selected for—the one whose somatic twin carried the genetic information that resulted in the appearance of the phenotype. And selection for the genotype causes appearance of the phenotype in the next generation. Without selection for the germ-line genotype, the adaptiveness of the phenotype would have no effects on subsequent generations and would not be selected for. The immediate or direct cause of reproductive fitness is thus the genotypic properties of the germ-line.[10]

Thus, taking Sterelny and Kitcher's example, web-spinning enhances the evolutionary fitness of spiders because of its connection to the genetic material in the reproductive cells of spiders. That is, when the type of adaptive property, 'web-spinning' is selected for, this is just because (a) its tokens are the *indirect* somatic cell products of interaction between properties of germ-line gene-tokens—individual nucleotide sequences—interacting with the sequences' intra- and extra-cellular environments, (b) web-spinning *indirectly* enhances the frequency with which the properties of these gene-tokens are represented in the germ-line and thus in subsequent generations. So, web-spinning, or any other somatic trait of individuals, for instance, is always only indirectly selected for, while traits of the genetic material in the germ-line are always *more directly* selected for. The same goes for traits of supra-individual items: these traits are indirectly selected for because of the direct selection of portions of germ-line genetic material that, given an environment, control their appearance.

This means that *pluralist genic selectionism* misleads when it suggests merely that a representation of selection in terms of genic properties is always merely "available" and at least as good a representation as any other. It is not just "available": from the perspective of *token*-causation, genic selection is indispensable to higher level selection. Without it, no properties at higher levels of organization could ever be subject to selection in successive generations. So, the (really) real causal story for selection, *wherever it occurs*, will advert to selection for properties of the genetic material of the germ-line, and selection for these properties will provide the ultimate biological explanation for the selection of somatic traits instantiated at higher levels of organization.

The philosopher of biology's favorite case, the sickle-cell trait, provides an excellent illustration of indirect selection of phenotypes and direct selection of properties of the genetic material. In malarial environments, the sickling of red blood cells is indirectly selected for because it confers a slight immunity to malaria. What are directly selected for are germ-line hemoglobin genes with a codon that codes for valine instead of glutamate at the sixth position of the beta chain of the hemoglobin molecule. In the presence of malaria, the sequence with this property has higher reproductive fitness than the sequence which codes for glutamate, because it produces somatic genes that code for hemoglobin molecules that stick together in the red blood corpuscle, reducing the potassium concentration, which prevents the malarial parasite from surviving in the blood cell. So, the portion of the genetic material in the germ-line that codes for the beta subunit of the hemoglobin molecule with a valine instead of a glutamate has the complex adaptive property of producing somatic hemoglobin molecules that link up together in a red blood corpuscle to prevent the survival of malaria parasites. This complex adaptive property of the germ-line genetic material is selected for every time the person's property of resistance to the parasite is selected for. And the latter property will not be selected for, except through its connection to the property of the genetic material. The property of malarial resistance is indirectly selected for because it causes differential reproductive success only via its connection to the germ-line genetic material. The property of producing somatic cell DNA that results in resistant hemoglobin molecules is more directly selected, because it is a property of the germ-line genetic material itself.

To the extent that properties at higher levels have their selective significance only indirectly—through a chain of processes that communicates their effects transitively to the polynucleotide—it follows that for any garden variety phenotypic property-exemplification we identify as selected for, like web-spinning, for example, there is some other complicated relational property of one or more segments of the genetic material that is also selected for. In the case of web-spinning, this genic property is the property of coding for the expression of all those enzymes and structural proteins individually necessary and jointly sufficent (within the range of environmental variables to be specified) for web-spinning behavior under conditions where it is appropriate. This complex relational property of the gene is selected for whenever spider web-spinning is selected for.

But this genic property is selected for more *directly*, while the garden variety property attributable to an organism, or some larger unit, is selected for *more indirectly* because the garden variety property enhances the survival of the sequence of germ-line polynucleotides which are indispensable to its appearance in the next generation and thus whose properties are more directly selected for. But this certainly sounds like a conclusion that will support the claim of monist genetic selectionism, at least as a thesis about token-selection.

3. TOKEN MONISM AND TYPE-INSTRUMENTALISM

The complicated relational property of one or more segments of the genetic material that is selected for whenever some trait of the individual coded for by that gene is selected for may give monist genic selectionism cold comfort. After all, the property selected for will be too complex to be a property of recognizable biological interest. And it will either be so restricted in its instantiation, or so disjunctive as to be without systematic significance. So, this vindication of monist genic selectionism as a thesis about *tokens* will be of limited methodological interest if there is a viable alternative to it that can be defended at the level of *types*.

In fact, it may be argued against what has been claimed above, that the entire dispute about genic selectionism is one about types: no one doubts the thesis of token-genic selectionism. It is genic selection as a thesis about types of genes and individuals that is at stake. But this cannot be correct, for token genic selectionism suggests that there is no defensible general theory about autonomous selection above the gene-token—the polynucleic acids.

To begin with, we require a criterion for type-selection, a test that will tell us whether a trait-type is selected for. That is we need a test we can impose on biologically interesting traits that are repeatedly instantiated by genes, organisms, kin-groups, etc., to determine whether having the trait is selected for. A biologically interesting trait is, roughly, a trait we can recognize as the same from instantiation to instantiation, whose incidence is of actual or possible practical, therapeutic, agricultural, industrial, or theoretical interest, as opposed to some gerrymandered artificial trait cooked up to confirm or disconfirm a claim about selection.

Sober (1984) proposes such a criterion for selection, which he holds to be part of a strong argument against monist genic selection. This principle is important for Sober because it is part of his argument against genic selectionism. The argument proceeds roughly like this:

1. Principle A: a property is selected for if the presence of that property raises fitness in at least one causally relevant background context, and does not lower it in any.

2. If a property is selected for, then the items which exemplify that property selected for together constitute a category, kind, level or type of unit at which selection operates.

3. Any unit on which selection operates, and which is composed of constituent units not selected for (because condition 1 does not obtain), is a natural kind of the theory of natural selection. As such these

units must be noted in any theory of natural selection which does not overlook significant explanatory generalizations.

It is evident that the claim that a property is selected for should be understood as having counterfactual force. For to say that a trait is selected for is to say, among other things, that its incidence is a cause of survival and reproductive success in the disjunction of causally relevant background conditions within which it is instantiated, and presumably its persistence over time is not a merely accidental fact unexplained by its adaptive significance.[11] Moreover, if the items whose properties pass the test of principle A are to be deemed causally homogeneous natural kinds, kinds that figure in the generalizations of evolutionary biology, then it cannot be on the basis of some merely accidental property they share.

So, if a trait passes the test of principle A, and is a trait properly predicated of items at a level of organization above the gene, genic selectionism must be wrong.

But, claim Waters and Sterelny and Kitcher, principle A can only be defended at the cost of misrepresenting the theory of natural selection. The problem is that there seem to be many properties which Sober, among others, wishes to recognize as selected for, but which fail the test of principle A. Ironically enough Kitcher and Sterelny instance as a counterexample frequency-dependent selection, which has always been a favorite example of Sober himself in other contexts (Sober 1982). In frequency dependent selection a property is selected for until it becomes heavily represented in its population, after which it is selected against. Thus, in the same causally relevant background context, a trait is sometimes selected for, and sometimes selected against. The obvious way to defend principle A in a case like this is to split a population up into two different subgroups—one in which the frequency of the trait is low, and its selective effect is uniformly favorable, and one in which the frequency is high and its effects are not uniformly favorable. But, Sterelny and Kitcher write,

. . . a defence of this kind fails for two connected reasons. First the process of splitting populations may have to continue much further—perhaps to the extent that we ultimately conceive of individual organisms as making up populations in which a particular type of selection occurs. . . .

The fact that a process of splitting populations up to find subpopulations that are uniform in their causally relevant background conditions *may have to continue* until the subpopulations are singleton sets—individual organisms, or particular genes, for that matter—is on the one hand no guarantee that it will do so, and on the other by itself no objection to principle A. The biosphere is sufficiently complex that we should not be surprised if no very interesting nomic generalizations about the selection of any particular trait in a large class of cases can be uncovered.[12] Indeed, as I shall argue, we should be surprised if any of the traits in which biologists interest themselves really are selected for (in accordance with principle A) in populations much above the size of the individual organism.

Keep in mind that selection for types proceeds by selection for tokens: types themselves are abstract. There may be disagreement about what the tokens are, and what types they instantiate, but there is agreement on this point: abstract objects do not have causal relations. The direction of selection—for or against—is always determined by the interaction of the token with its particular environment. Over the course of an individual's lifetime, a given trait may be selected for, selected against, or be neutral in its selective role, sequentially, as its local environment changes. For a given trait and a given population, the background conditions causally relevant for selection will include all the differing environments in which members of the population find themselves. The probability that in every one of these environments a trait of biological interest will not be selected against is vanishingly small. It is for this reason that there are almost no interesting exceptionless generalizations about the traits to be found in evolutionary biology. At most one finds 'tendency' statements, like 'polar species tend to have higher volume to surface area ratios than non-polar species'. In fact principle A underwrites and helps explain this paucity of generalizations. Therefore, we have an additional reason to endorse it as reflecting the nature of selection.

Sterelny and Kitcher offer a second reason for rejecting principle A:

Second, as many writers have emphasized, evolutionary theory is a statistical theory, not only in its recognition of drift as a factor in evolution but also in its use of fitness coefficients to represent the expected survivorship and reproductive success of organisms. The envisaged splitting of populations to discover some partition in which principle A can be maintained is at odds with the strategy of abstracting from the thousand natural shocks that organisms in natural populations are heir to. In principle we could relate the biography of each organism in the population, explaining in full detail how it developed, reproduced, and survived, just as we could track the motion of each molecule of a sample of a gas. But evolutionary theory, like statistical mechanics, has no use for such a fine grain of description: the aim is to make clear the central tendencies in the history of evolving populations, and to this end, the strategy of averaging, which Sober decries, is entirely appropriate. We conclude there is no basis for any revision that would eliminate those descriptions which run counter to principle A. (Sterelny and Kitcher, 1988, 345)

This is a highly controversial passage, which attributes aims and strategies to a theory instead of attributing them to theorists who make use of it. Once we see this, Sterelny and Kitcher's objection to principle A is seen to have little force, except on an instrumentalist conception of evolutionary theory.

Evolutionary biology is indeed a statistical enterprise, and is so in part because of the role it accords drift—the effect on variation in hereditary traits caused by the small size of interbreeding populations. But drift is something to be contrasted with selection—the great twentieth-century debates about evolution, between Fisher and Wright among others, are debates about how much of a role is played by each of these differing forces, drift and selection. Natural selection's effects in principle can be separated from those of drift (see Rosenberg 1988). What is more, these drift-independent effects are not indeterministic.

Where then does statistics enter into the theory? Sterelny and Kitcher say it enters because the theory's aim "is to make clear central tendencies in the histories of evolving populations" and its strategy is one of "abstracting from the thousand natural shocks that organisms are heir to." This tendentious claim at most reflects one of the aims of evolutionary theorists; it does not reflect the aim of the theory itself. For theories do not have aims. Theorists have aims, and these aims, along with theorists' means—their capacities for dealing with data—jointly dictate their strategies. Sewall Wright made this point effectively long ago:

> Natural selection is an exceedingly complex affair. Selection may occur at various biological levels—between members of the same brood, between individuals of the same local population, between such populations (as through differential increase and migration) and finally between different species. . . . Selection among individuals may relate to . . . mating activities, . . . differences in attainment of maturity, to differential fecundity and to differential mortality. Selection may act steadily or may vary both in intensity and direction. . . .
>
> In such a complex situation, verbal discussion tends towards a championing of one or another factor. We need a means of considering all factors at once in a quantitative fashion. For this we need a common measure for such diverse factors as mutation, crossbreeding, natural selection and isolation. At first sight these seem to be incommensurable but if we fix our attention on their effects on populations, rather than on their own natures, the situation is simplified. Such a measure can be found in the effects on *gene frequencies*. (Wright, "Statistical Genetics and Evolution" Gibbs Lecture, 1941, reprinted in Wright [1984] 468)

As the passage makes clear, it is not the aim of the theory of natural selection to explain central tendencies, it is *our* aim, to which we apply the theory. Were our aims different, or our means of attaining these aims different, our employment of the theory might not be statistical, or might involve the employment of different statistical measures. Indeed, were our capacities different, our identification of "central tendencies"—statistical or otherwise—would be different too.

Wright's point is that statistical description is introduced so that we can measure the combined effects of diverse evolutionary forces by providing a quantitative description of their *effects* on the distribution of traits. The theory does not "abstract from the thousand natural shocks that organisms are heir to." *We* abstract from the description of the effect of these shocks on organisms. The theory of evolution is statistical because averages are employed to describe its *explanandum phenomena.* Natural selection in itself is not an indeterministic process, or a stochastic one. And it operates on individual items *one at a time.*[13] As such it produces effects which can be summed and from which average values, mean values, variances, and other statistical properties of whole populations can be constructed. But these statistical properties and changes in them describe the explananda of evolutionary biology; they are causally inert, epiphenomena, which we track because they provide a convenient description of the course of evolution, a description convenient *for us.*

If, as Sober has written, without statistics we would miss interesting evolutionary generalizations, then these are generalizations that are of interest *to us,* because of our cognitive and computational capacities. Cognitive agents with greater powers to "crunch" data, to track the fate of larger numbers of smaller lineages over more finely grained periods of time, would compute averages and statistical measures that more closely reflected the processes that underlie and explain the "tendencies" that we identify as "central" because of our cognitive limitations. And a cognitive agent of unlimited powers, a Laplacian demon, would have no need of statistical information at all to explain evolutionary processes by appeal to natural selection over blind variation.

Comparisons between evolutionary theory and statistical mechanics have been made before, and Sterelny and Kitcher are correct to say that thermodynamics has no use for fine grains of description like those involved in tracking the motion of a molecule. But the parallel breaks down just because an evolutionary biologist of even our limited cognitive powers must be interested in the fate of individual organisms, even as the thermodynamic theorist cannot have any interest in the individual molecule. For the individual molecule has no thermodynamic properties. By contrast, every organism has a property that figures in evolutionary theory: fitness. And a single individual variation, like the white eye of Morgan's fruitfly, can amplify over generations until it is the property whose representation in the population becomes the statistical fact to be explained. The property that does the explaining—fitness—is a property of individual items, whose arithmetic average value for the whole population has itself no explanatory role beyond its summing of individual values. By contrast, entropy is a thermodynamic concept that has withstood straightforward reduction to mechanical properties of constituent molecules and their aggregations, from Boltzman's time to our own. If the theory of evolution is statistical, it is for different reasons than thermodynamics is statistical.[14]

So, the strongest conclusion about principle A that the statistical character of the theory of natural selection will substantiate is this: agents beyond our cognitive capacities who employ it may identify and explain tendencies of evolution that we are not equipped to identify. But there is no reason to conclude that they will miss any facts about evolution that we can explain. And the only reason to surrender principle A appears to be that it is an inconvenient one for us to employ, given our finite powers and the coarseness of the evolutionary detail in which we are interested.

4. PRINCIPLE A'S SUPPORT FOR GENIC SELECTIONISM

In fact the only level of biological organization at which principle A has any chance of working is the level of the genetic material. That is, only at this level are selective effects actually uniform enough to pass principle A's test for being selected for.

At every other level of organization, counterexamples to principle A are the stock in trade of contemporary biomedical science. For example, consider a line of laboratory animals raised because, say, they lack a functioning immune response and so make suitable model systems for the study of immunity. Either this is a group within which having a working immune system is selected against (specimens that have one are removed from the breeding population), so that having an immune response cannot be said to be adaptive and so cannot be said to be selected for (which seems false), or we must try to exclude this case as outside the range of causally relevant background conditions envisioned by principle A.

But what argument can be given for excluding these laboratory cases? Surely, the argument cannot be merely that these are cases of artificial selection. For artificial selection is just a subcategory of natural selection in which the selective force happens to be *Homo sapiens*. It is well known that nature frequently reveals cases in which one species has had selective effects on other species. Consider predator-prey relations, for example. Artificial selection is just another example of a case in which the behavior of one species provides a strong environmental effect on the adaptation and fitness of another. What can happen in the laboratory can happen outside of it.

Can we exclude such counterexamples by invoking the qualification in principle A that implicitly restricts it to the actual environment in which a trait is manifested and not all physically possible environments? A defender of principle A will argue that we should do so, and that we must distinguish the environment from the causally relevant background conditions which work with a trait to effect selection within the given environment. Presumably then, the molecular biologist's lab is a different environment from the wild and not a legitimate souce of counterexamples. Such an approach is ultimately unavailing. If we can distinguish environment from causally relevant background conditions, the resourceful experimentalist can replicate the environment and select against any trait we like. And if we cannot distinguish the environment from the causally relevant conditions in a non-arbitrary manner, then every trait selected for against some causally relevant background conditions will pass principle A. To avoid this trivial conclusion, we will have to draw a distinction that returns the force to the counterexamples.

It does not take a science fiction writer to imagine a nomologically permissible background environment that will make any particular maladaptation adaptive, and vice versa. For almost any biologically interesting trait that has been selected for, it takes only a little imagination to dream up perfectly possible "Rube Goldberg" arrangements that will turn the trait into a maladaptation in at least one causally relevant background condition, even while holding the environment constant. What are clearly excellent adaptations for almost all environments will be turned into serious maladaptations.

But this means that for any interesting biological trait, that trait either actually fails principle A, or there is no nomological obstacle to that trait's failing principle A. Accordingly, the claim that some property passes the test of principle A, and is selected for, may be true, and yet lack any nomological force. For there will always be some physically possible background condition in which it would maladaptive, and so would not be selected. But given Sober's argument for the autonomy of units of selection and its reliance on a counterfactually supportive version of principle A, it follows that there are no units of selection.

Or rather there are none above the level of the gene. For no biologically interesting property satisfies principle A counterfactually construed . . . except, that is, just possibly for genic properties involved in gene-expression and gene replication! The biochemical properties of the genetic material— the DNA polynucleotide, constitute the sole class of biologically interesting properties that might pass Sober's test—that are selected for in at least some actual causally relevant background conditions, and are selected against in none, including actual and counterfactual ones. The reason is simple, these properties of the DNA polynucleotides are the only ones that are indispensable *in all causally relevant background conditions* to the mechanism of selection on this planet. For they are the ubiquitous mechanism of hereditary transmission, without which selection cannot obtain.

Though we can design an experiment that will make *almost* any adaptation into a maladaptation, we cannot do this for significant molecular properties of genes. For example, the DNA has the property of copying itself in all causally relevant background conditions where selection operates. We cannot eliminate this property in the laboratory, while preserving the hereditary role of DNA. For to eliminate it is to eliminate hereditary transmission. Much the same can be said for the effects of the genetic material in metabolism. Given the ubiquity of the basic processes like oxidative phosphorylation, eliminating the traits of the genetic material in the germ-line cells that allow for the production of the relevant enzyme types in the somatic cells of off-spring simply eliminates the hereditary lineage within which selection is supposed to operate. To the extent that correct protein synthesis in the somatic cells is necessary for survival and reproduction via the germ-line, the genetic material's ability to direct protein synthesis will be selected for in all contexts in which there is survival, reproduction, and evolution.

These properties of the genetic material required for gene expression and replication stand a chance of satisfying principle A, while no other properties do so. This important difference between biochemical properties of the genetic material and other interesting biological properties is in part a consequence of the fact that the latter are supervenient on the former. In general, where supervenience obtains, the supervening predicates do not figure in discoverable exceptionless nomological generalizations (whence, for example, the anomalousness of the mental: the absence of laws of intentional psychology is a consequence of its supervenience on the neurological). But the properties in the supervenience base can figure in such laws.

Compare apparently smooth reductions: in these cases, the regularities in the reduced theory are shown to relate *indirectly* connected properties as a consequence of more fundamental relations between *more directly* connected properties in the reducing theory. Thus, heating up a balloon is an indirect cause of the balloon's bursting because of more direct causal links between the kinetic energy of the constituent molecules and the elasticity of the balloon. The reductive explanation of why the balloon bursts upon heating is thus simple and straightforwardly given by appeal to just a small number of variables that come into play every time the sequence obtains.

In biology matters are never so simple. Above the macromolecule, every biologically interesting property is realized in virtue of the instantiation of a disjunction of more fundamental properties, which are themselves supervenient on still other properties down to the level of the macromolecule. At each level there are relations of redundancy and divergence among the properties that make for the realization of "higher level" ones. For example, a given coat color, which has a uniform effect on adaptation in at least some causally relevant background conditions, can be constituted by a disjunction of different pigments, which are themselves the products of a disjunction of enzymes, substrates, and products, most of which have more than one functional role, any one of which is dispensable in producing the relevant coat color, but all of them will turn out to be at least in part indirect products of genes indispensable for survival.

Moreover, like any other biologically interesting trait, any particular coat color is not the only "strategy for survival" available to an evolving lineage. And in some causally relevant background conditions, it is maladaptive. By contrast, many of the genes whose products are involved in its availability, are indispensable in all causally relevant background conditions for all of the available strategies for survival.

Properties of the genetic material responsible for gene expression and gene replication are the only ones that we can have much confidence will not fail the test of principle A.[15] Every non-genetic property of biological interest fails principle A. So, if we are to identify biologically interesting properties above the level of the gene as selected for, we shall need another test, a weaker one than principle A provides. For example, the substitute principle might hold that a trait is selected for if it is adaptive in most causally relevant contexts, or a majority, or some otherwise interesting causally relevant contexts. But installing a weaker test of selection for a trait is tantamount to the admission that claims about selection lack the universality indicative of nomic force. And such a weaker test would be no basis on which to identify natural kinds, units or levels of selection. Yet biologists do identify traits—genic, individual, kingroup, and even more holistic traits, as selected for. And they recognize that such claims about higher level traits are not fully explained by the selection for types or traits of the genetic material that codes for these individual and other higher level traits. At most the individual instances of such traits are so explained.

5. BIOLOGICAL INSTRUMENTALISM

What biologists aim at is not the explanatory derivation of their rough and ready generalizations from laws of a complete theory of the macromolecules. Rather they seek the potential completion of token-explanations, the completion of explanations of how any particular model-system works, ultimately in terms of selection operating on the macromolecules. Accordingly, claims about the selection for any given trait of the individual organism, kin-group, deme, etc., should be understood as singular causal judgments, or sets of them. Such explanations are advanced in full recognition of the fact that there are always some causally relevant conditions in which the trait will not be adaptive. And the explanation of selection for a particular trait in one or more cases does not proceed through the claim that the trait satisfies principle A, and is therefore a natural kind, in a causally homogeneous class. The explanation proceeds in an altogether different direction, identifying the particular factors of the local environment that make the trait conducive to survival of the organism and its reproduction. It is only well below these levels that nomic generalizations set in, generalizations about traits of the genetic material that together with descriptions of the local environment—intra- and extra-cellular—that nominally link initial conditions with the trait-instances whose selection we seek to explain.

Why then do we individuate kinds—and seek generalizations—at levels above those of the genetic material? Not because nature carves at these joints, but because *we need to do so* and because it is the best we can do. As inhabitants of this biocosm, our interests in creatures extend beyond those of scientific curiosity. And fulfilling these interests requires information different from that which is provided by a theory about the right natural kind categories expressing the true laws of nature. We need to know a good deal about the details of a particular set of organisms in a particular niche for reasons of health, or nutrition, etc. This is information the search for nomological generalizations will not provide. And even the search for such generalizations is conditioned by our own limitations as organisms. Because we have finite calculational and computation powers, we find it useful to pursue the kind of statistical information that Sterelny and Kitcher think is the main subject-matter of evolutionary biology—the "central tendencies." This latter fact about us leads us to weaken principle A, to accept that a trait we identify as interesting is selected for just in case, on average, it increases the fitness of lineages we recognize. Because our cognitive limitations determine which tendencies we will recognize, combine, and seek explanations for, these traits and lineages will be ones dictated by practical agricultural, ornamental, nutritional, medical agendas, as well as our observational capacities. If all we were interested in were nomic generalizations, then monist genic selection would be vindicated from an instrumental point of view. For the other generalizations about traits of interest to us are either false or reports of local singular causal matters of fact.

To the extent our interests in our fellow creatures—speaking broadly enough now to include all the eukaryotes and all the prokaryotes—are practical and non-theoretical, a focus on genic selection will not be convenient, no matter how "adequate" in principle. Representations of selection at other levels will be preferred, and the description of local matters of fact will suffice. But this preference has to do not with cognitive virtues, but instrumental ones. If we want to preserve a species, it is enough to know in what environments on average their traits will be selected against. If we want to improve crop yields, information about averages is enough. If we want to control a pest, it will suffice to know which of its traits that directly annoy us are on average selected for. Knowing more, knowing the finer details underneath what are to us the "central tendencies," the full story for every token is not only beyond our information storage powers, it is beyond our computational powers to factor together. Knowing more is also unnecessary given our interests. These practical interests make the subjective, observer-relative individuation of levels of selection interesting and important. But they do not underwrite the existence of these levels as objective, observer-independent natural kinds above that of the molecular gene and its properties.

Thus, we may opt for pluralism about selection, but only as a relational thesi a claim about the world and our biological theories, the ones suitable to the needs and capacities of creatures like us. Our biological science will turn out to be an observer relative science, an instrument for getting around in our world, and not something the scientific realist will embrace as the best guess as to what the world is really like. This best guess will embody no laws above the level of the genetic material, and perhaps even none at that level. For if the nucleic acids prove to be just one among a large number of local solutions to nature's problem of providing the means for hereditary transmission, then genic selection will be local matter of fact, and not a nomological truth.

NOTES

I wish to thank Robert Brandon, Barbara Horan, Elliot Sober, and Kenneth Waters for discussion of these issues and detailed comments on previous drafts. No agreement with my views should be attributed to them however.

1. See Sober and Lewontin (1982), Sober (1984), Brandon (1982), Waters (1985), Sterelny and Kitcher (1988), Waters (1991), and rejoinders by Sober (1990), and Sterelny, Kitcher, and Waters (1990).

2. Though neither draw the conclusion that biology is an instrumental science, both inadvertently provide powerful arguments for the claim that the structure of biological science reflects constraints on biologists' cognitive powers. Thus, Sober argues that evolutionary biology is a statistical enterprise because without the appeal to probabilities we would miss important explanatory generalizations. Doubtless this is correct. But Sober does not show that these are generalizations which reflect quantum indeterminism. Nor does he show that a Laplacian demon would need these generalizations to explain any particular matter of

biological fact. The explanation of why we require them must be that we are not Laplacian calculators. Thus, the shape of evolutionary biology reflects our cognitive and calculational limitations. I discuss Sober's arguments further, and try to substantiate these claims in Rosenberg (1988).

Kitcher has argued that molecular biology does not explain classical genetics, it extends it. Like Sober he holds that there are phenomena that can only be explained for us at the level of functional physiology, for instance, and cannot be explained further and more deeply by adverting to the molecular phenomena that underlie such physiological processes. Again, though Kitcher does not explicitly say so, the reason for this limitation appears to be our own powers to absorb and organize the complex relations between the supervenient theory and the supervening theory. This makes "higher level" theory autonomous from more fundamental theory just because we cannot express the disjunctive biconditionals required to forge deductive links between them. For further discussion of the instrumental upshot of Kitcher's view see Rosenberg (1989).

3. Even the rare dissenter recognizes there is something approaching consensus on this view. Cf. Waters (1990), "Why the Anti-reductionist Consensus Won't Survive: The Case of Mendelian Genetics."

4. We may contrast this sort of pluralism with Sober's, according to which in different populations for different traits, selection may operate at different levels. Sterelny and Kitcher's pluralism allows that with respect to a single population and a given trait there are multiple equally true descriptions of how selection operates.

5. According to several philosophical accounts of causation, such a complete description of the causal structure of a process may be unmanagably complex and lengthy. Depending on the theory of causation adopted, the completely correct description might even have to include mention of the whole state of the universe at each instant of the causal process in question. But this possibility is no reason to doubt the coherence of the notion.

6. A serious but largely philosophical problem lurks here. How do we establish the completeness of a description of any particular phenomenon that includes less than a whole history of the universe? If relations are properties which a complete account must include, this problem cannot be solved. On the other hand, scientists have little difficulty with this issue.

7. Waters (1991) advances the same view:

We can no longer maintain that a true description of a selection process provides a unique correct identification of the operative selective forces and the levels at which each impinges. Instead we must accept the idea that the causes of one and the same selection process can be correctly described by accounts which model selection at different levels.

Waters pleads for a pluralism only tenable under a thoroughly non-realist point of view of the forces of selection.

8. This is Waters's (1990) reason for pluralism. On his view genic and individual selection theories do not conflict once we understand them fully. But then they should be combinable into a more correct or perhaps even a uniquely correct account, thus undermining Waters's pluralism.

9. What they need is something like Gödel's proof, which effects the required sort of result in mathematics, or von Neumann's proof that hidden variables are impossible in quantum mechanics as it is currently formulated. But there is no reason to think that anything like this prevails in evolutionary biology. Instead what Sterelny and Kitcher offer us is something much weaker, something like the Copenhagen interpretation of quantum mechanics—an epistemological thesis instead of an account of the way the world is.

10. Compare this direct-indirect distinction with Mayr and Gould's. On their view, the phenotypic property of the individual organism is directly selected for, since it is what "faces" the environment directly, and confers advantage or disadvantage on its bearer as a

result. As Waters has shown, this argument hinges on an equivocation about the environment. If we include the cellular milieu as part of the environment, the genetic material faces the environment just as directly as any of its most distant products. My distinction is meant in part to capture this insight.

11. This is the upshot of the distinction for which Sober has been made famous, between selection for—a causal notion, and selection of—an epiphenomenal notion. See Sober (1984).

12. I try to provide an argument for this expectation and an explanation for the lack of any such generalizations in biology in Rosenberg (1984, chap. 7).

13. This is a point made forcefully, though in opposition to genic selection, in B. Horan (1990a).

14. For an illuminating discussion of the statistical character of the theory of natural selection, see Horan (1990b). My discussion reflects the influence of this paper, though Horan's conclusions about the comparison between thermodynamics and evolutionary theory are quite different from my own.

15. And even they may not be invulnerable. Indeed, the existence of the RNA viruses is a constant challenge to the nomological force of the claim that the properties of DNA are everywhere selected for, or at least not anywhere selected against. That it is not yet an actual challenge reflects the fact that the RNA virus still employs the DNA of a host to survive and reproduce, and that many of the indispensable properties of DNA are also properties of RNA.

REFERENCES

Brandon, Robert. 1982. "The Levels of Selection." In *PSA 1982*, v. 1, edited by Asquith and Nickles. East Lansing, Mich.

Dawkins, R. 1976. *The Selfish Gene*. New York.

Dawkins, R. 1982. *The Extended Phenotype*. San Francisco.

Horan, B. 1990a. "In Defence of the Individual." Manuscript.

Horan, B. 1990b. "The Statistical Character of Evolutionary Theory." Manuscript.

Kitcher, P. 1984. "1953 and All That: A Tale of Two Sciences." *Philosophical Review* 93: 335–73.

Rosenberg, A. 1984. *Structure of Biological Science*. Cambridge.

Rosenberg, A. 1988. "Is the Theory of Natural Selection Really Statistical?" *Canadian Journal of Philosophy*, supp. vol. 14: 187–207.

Rosenberg, A. 1989. "From Reductionism to Instrumentalism." In *What The Philosophy of Biology Is*, edited by M. Ruse, 245–62. Dordrecht.

Sober, E. 1982. "Frequency-dependent Causation." *Journal of Philosophy* 79: 247–53.

Sober, E. 1984. *The Nature of Selection*. Cambridge, Mass.

Sober, E. 1990. "The Poverty of Pluralism." *Journal of Philosophy* 97: 151–58.

Sober, E., and R. Lewontin. 1982. "Artifact, Cause and Genic Selection." *Philosophy of Science* 49: 157–80.

Sterelny, K. and P. Kitcher. 1988. "The Return of the Gene." *Journal of Philosophy* 85: 339–62.

Sterelny, K., P. Kitcher, and C. K. Waters. 1990. "The Illusory Riches of Sober's Monism." *Journal of Philosophy* 97: 158–61.

Waters, C. K. 1985. Models of Natural Selection from Darwin to Dawkins. dissertation, Indiana University.

Waters, C. K. 1990. "Why the Anti-reductionist Consensus Won't Survive: The Case of Classical Mendelian Genetics." 125–39. *PSA 1990* East Lansing, Mich.

Waters, C. K. 1991. "Tempered Realism about the Forces of Selection." *Philosophy of Science*, forthcoming.

Wright, S. 1984. *Evolution: Selected Papers*, edited by W. Province. Chicago.

Could There Be a Science of Economics?

JOHN DUPRÉ

Much scientific thinking and thinking about science involves the assumption that there is a deep and pervasive order to the world that it is the business of science to disclose. A paradigmatic statement of such a view can be found in a widely discussed paper by a prominent economist, Milton Friedman (a paper which will be discussed in more detail shortly):

> A fundamental hypothesis of science is that appearances are deceptive and that there is a way of looking at or interpreting or organizing the evidence that will reveal superficially disconnected and diverse phenomena to be manifestations of a more fundamental and relatively simple structure. (1953/1984, 231)

On the other hand, the person sometimes described as the father of modern science, Francis Bacon, wrote:

> The human understanding is of its own nature prone to suppose the existence of more order and regularity in the world than it finds. And though there be many things in nature which are singular and unmatched, yet it devises for them conjugates and relatives which do not exist. (1620/1960, 50)

I myself find the empiricism of Bacon's position much more attractive than Friedman's rationalism. Assumptions of the kind Friedman's remark illustrates lead naturally, almost inevitably, to philosophical doctrines such as reductionism (of various kinds) and determinism. In a recent book (Dupré 1993) I have argued at length against various doctrines of these kinds and, more generally, against the assumption of underlying metaphysical order with which they are inextricably connected. In the concluding chapters of that book I suggested that more modest and defensible views of the prevalence of order in the world would have important consequences for our general understanding of science, and even for the practice of science. More specifically, I proposed that certain scientific projects might be seen to be quite misguided when deterministic or reductionistic assumptions about the domains under investigation were abandoned. And finally, the rejection of such assumptions suggests a much greater role for antecedent value judgments in scientific investigation than is generally allowed.

In the previous work just mentioned, the arguments against reductionism and determinism were developed mainly in relation to a discussion of biological science. Although biological ideas, through their relevance or alleged relevance to conceptions of human nature, have important implications for normative questions, further investigation of the relation of values to science may more readily be pursued in relation to some part of the social or human sciences. A particularly suitable candidate is the science of economics. The complexity of the phenomena investigated by economics is such as to make claims of fundamental lack of order at least superficially plausible; and how, if at all, social or political values relate to economic investigation is a question of obvious importance. This paper provides a sketch of the argument that economics does indeed provide an illustration of the need for normative input in the investigation of a domain with quite limited preexisting order. In the course of developing this thesis I shall touch on various more specific objections to aspects of economic theory. None of these objections are new: they have existed since the beginnings of neoclassical economics in the late nineteenth century, in the writings of Weber, Veblen, Commons, and numerous successors. Where the skepticism developed in the present paper goes beyond this critical tradition is in the suggestion that the failures of economics derive not merely from excessively simplistic assumptions, crude theories of human nature, and so on, but rather from a fundamental mismatch between the kinds of phenomena with which economics is concerned and widely held conceptions of what it is for an investigation of any realm of phenomena to be genuinely scientific. It is this mismatch, I contend, that raises the question whether a *science* of economics is possible.

I should emphasize that I use the word "sketch" advisedly. One part of the picture that is in particular need of filling in is the question of the extent to which there is empirical support, particularly predictive success, for the broad theoretical claims of economics. Though the skeptical position I adopt in the paper does not seem to me unreasonable, clearly it is in need of more detailed investigation. At any rate, one of the aims of the present paper will be to assess the relevance of such further investigation to the general thesis I am developing. Another point that needs to be stressed at the outset is that throughout the paper I am addressing the foundational, abstract, and typically mathematical theories of economics rather than local, generally qualitative knowledge of specific economic systems. I do not know how the actual labor of professional economists is divided between efforts that fall (roughly) into each of these categories. Arguably much of the work of, for example, labor economists, development economists, and even econometricians, has little connection with this supposedly foundational core. But it is surely the articulation of the formal and abstract core of economic theory that is the most conspicuous and prestigious aspect of the activity of economists.

In the first section of this paper I shall be concerned with economic methodology. In particular I shall criticize a conception of economic methodology that might serve to defuse the worries about the lack of empirical success in economics that have been so widely emphasized by critics of economics. The second section addresses the relevance of questions of underlying order to the prospects for the development of economics, and the third section explores the consequences of skepticism about underlying order for the relation of economic fact to value. I shall argue that the most plausible escape from pressing doubts about the possibility of a scientific study of economic phenomena is to recognize a much greater interpenetration of science and value than is generally recognized.

1. THE METHODOLOGY OF ECONOMICS

One obvious motivation for the question in the title of this paper is the remarkably widespread belief that the study of economics has failed almost totally in at least one of the traditionally central marks of science, prediction. (As Thomas Love Peacock put it: "Premises assumed without evidence, or in spite of it; and conclusions drawn from them so logically, that they must necessarily be erroneous" [cited by Hausman 1984, 1].) Of course it is not that no economist has ever said anything true about any future economic phenomenon, but just that economic theory seems to add little to the ability of a well-informed person in possession of average common sense to be right about such matters.[1] More broadly, this worry reflects a serious doubt as to whether economics is in any sense genuinely an empirical study. It is, at any rate, widely supposed that if economics were to be judged solely on the basis of the extent to which it has succeeded in making predictions about the future, then it would be seen to be a dismal failure. This empirical failure is all the more remarkable given the extent to which economics continues to be regarded as a model for the social sciences. To give just one example, Alexander Rosenberg, in a paper otherwise highly critical of the development of microeconomic theory, nevertheless describes this theory as "the most impressive edifice in social science yet erected" (1979, 47). Apparently some other grounds of scientific excellence than predictive success are being supposed. My first aim in this paper is to consider the plausibility of the claim that any such grounds could compensate for the empirical failure just noted.

One very influential economist, on the other hand, apparently sees no serious weakness in the predictive achievements of economics. Milton Friedman, in an extremely influential article (1953/1984), argues that economics should be judged, not as is often supposed, by the truth of its premises, but rather by the empirical success of its predictions. And it is clear enough that Friedman thinks that this is a test that economics will readily pass. Nevertheless, it is striking how little he says in this article to justify complacence about the predictive powers of economics. Friedman compares, for example, the hypothesis that businessmen act as if they aimed to maximize profits with the hypothesis that expert billiard players act as if they were able to make all the appropriate mathematical calculations of the trajectory of a billiard ball. The value of this latter hypothesis is sufficiently demonstrated by the successful shots of the expert billiard player, and the psychology of neither the billiard-player nor the businessman is of any relevance to the evaluation of either hypothesis. But remarkably, rather than offer any empirical evidence that, just as expert billiard players generally make their shots, businessmen, whatever they may be intending, do in fact maximize profits, Friedman offers nothing but a broken-backed a priori argument for this conclusion. "[U]nless the behavior of businessmen in some way or other approximated behavior consistent with the maximization of returns," he writes, "it seems unlikely that they would remain in business for long" (223). But the implicit argument here has no force unless it is assumed that the competitors of the non-profit-maximizing businessman are profit maximizers, which is blatantly question-begging. If no businesses, or very few businesses, are profit-maximizers, there may be very little tendency for non-profit-maximizing firms to go out of business. Perhaps survival in business is largely a matter of luck, or of access to corrupt government officials. Distinguishing between these and many other possible hypotheses would require empirical evidence of a kind that Friedman neither offers nor, apparently, sees the need for.

Another example Friedman offers is the hypothesis that a sudden increase in the money supply will produce a substantial increase in prices (216). Although the "evidence is dramatic" for this hypothesis Friedman admits regretfully that the issue remains highly contentious, an observation that might reasonably lead the external observer of economics to wonder whether this paradigm of empirical support could be quite that dramatic. It appears that in the end Friedman wants to appeal not to "any textbook list of instances in which the hypothesis has failed to be contradicted" (224), but rather to "experience from countless applications of the hypothesis to specific problems and the repeated failure of its implications to be contradicted. This evidence is extremely hard to document; it is scattered in numerous memorandums, articles, and monographs . . ." (223–24). In contemporary philosophy of science, where the constant apparent confirmation of the basic assumptions of a research program, or a paradigm during normal science, etc., has become a commonplace, this appeal is unlikely to impress. Friedman's paradigmatic though pre-Kuhnian description of a Kuhnian paradigm becomes even more striking when we read that the ability to apply economic models in the right way requires considerable professional judgment. This "is something that cannot be taught; it can be learned, but only by experience and exposure in the 'right' scientific atmosphere, not by rote. It is at this point that the 'amateur' is separated from the 'professional' in all sciences and the thin line is drawn which distinguishes the 'crackpot' from the scientist" (226). Very likely, this is the way that all sciences work. But the fact that economists know how to make their models work is even less impressive in the face of such a frank statement of the judgment that awaits those who, for whatever reason, fail to do so. In short, Friedman's manner of endorsement of the empirical success of economics tends to argue rather for the opposite conclusion.

An obvious reply to the complaint that economics has had little predictive success would be to suggest that prediction was unnecessary for the scientific status of a discipline, provided only that it be sufficiently successful at explanation. One example that might naturally be offered in support of this suggestion is that of evolutionary biology. While it is sometimes argued that evolutionary theory does have *some* predictive successes, for example with regard to (previously unexplored) aspects of the distribution and geographical relations of similar species, it is for its explanatory achievements that the theory of evolution is so widely admired. Two additional features of evolutionary theory contribute to the lack of concern about predictive deficiencies. First, it is widely perceived that there are no serious rivals to evolutionary theory as explanations of the origin and diversity of life. Thus there is some plausibility to the claim that evolution is not merely *an* explanation of these phenomena, but the only possible such explanation. In view of the complexity of the patterns of similarity between both living and fossil organisms, and the consistency of this pattern with the hypothesis of relation by descent, this claim is not implausible. I shall return to the relevance of plausible alternatives later.

The second point is of much closer relevance to traditional defenses of economics. It is that the constituent processes from which full and interesting evolutionary explanations are constructed are empirically confirmed beyond any reasonable doubt. To the extent that this is so, it is surely legitimate to accept logical elaborations of the consequences of the iteration of such processes as empirically supported explanations of observable phenomena. It is presumably in the interests of such an explanatory strategy that Darwin devotes so much space at the beginning of *The Origin of Species* to a discussion of the thoroughly empirical and moderately predictive art of animal breeding. More recently, it explains the importance of the occasional studies, such as that of the evolution of industrial melanism in the Peppered Moth (Kettlewell 1973), that really do appear to document in detail processes of natural selection in the wild. Just as such cases establish the reality of the basic mechanisms of evolution, so it is assumed that the basic postulates of economics—self-interest,[2] profit-seeking, etc.—are established by introspection or common experience. Another way of stating the point is that some, at least, of the scientific credentials attributed to evolutionary theory derive from the perception that it constitutes a successful application of what Mill described, for the case of economics, as the method a priori.

The method a priori, or the deductive method, sees economics as concerned with working out the consequences of basic postulates for which some strong independent warrant can be provided.[3] This is in sharp contrast with the hypothetico-deductive method, according to which the truth of the consequences thus deduced would provide the warrant for believing the basic postulates. Thus Lionel Robbins, a prominent advocate of the former, a priori, perspective, wrote, concerning such postulates: "We do not need controlled experiments to establish their validity: they are so much the stuff of our everyday experience that they have only to be stated to be recognized as obvious" (1935/1984, 119). But although these postulates are taken to be true, they are taken to be true only *ceteris paribus*. And it is this proviso that enables the deductive method to accommodate the falsity of most actual economic conclusions. In the real world, the factors distinguished by the basic postulates must always interact with a variety of interfering factors, a circumstance that makes it unlikely that any conclusion drawn from them will coincide with observed reality (as noted in the satirical remark of Peacock, cited above). For Mill this is precisely the reason why we must insist on the appropriateness of the a priori method rather than the a posteriori (inductive) method:

> It is vain to hope that truth can be arrived at, either in Political Economy or in any other department of the social science, while we look at the facts in the concrete, clothed in all the complexity with which nature has surrounded them, and endeavour to elicit a general law by a process of induction from a comparison of details: there remains no other method than the a priori one, or that of "abstract speculation." (Mill 1836/1984, 59)

More recently Hausman (1992), while in many ways sympathetic to the deductive method, diverges from it in one crucial respect. For Mill, predictive failures will be important in helping to identify the forces interfering with the action of those presupposed by economic theory. But if this is taken as a methodological prescription, predictive failures can never serve to cast doubt on the basic assumptions from which they are derived. Hausman considers this to be an intolerable departure from empiricist standards. Although he agrees that it is very unlikely that predictive failure will call basic assumptions into question, this is not because alternative explanations of their failure will almost always be more plausible. Thus though Hausman considers it important to remove this suggestion of dogmatism from the theoretical account of economic methodology, in practice he describes something very close to the methodology adumbrated by Mill and Robbins.

A key question to which we are led is what kind of support should be offered for these effectively unassailable basic postulates. The most basic of such postulates are those that define the behavior of (economically) rational actors and profit maximizing firms.[4] Here I shall focus only on the former. Economic rationality consists in ranking all the options available to the agent and always choosing the most highly ranked available option. These rankings must at least be complete (they must assign a position in the ranking to all options) and transitive (if A is preferred to B and B to C, A must be preferred to C). In addition the economic theory of consumer behavior requires a set of more restrictive assumptions about the kinds of options that are the subjects of people's rankings, a set of assumptions Hausman refers to as "consumerism" (1992, 30). Roughly, what this amounts to is that people should have an insatiable desire for commodities and an absolute lack of concern for the material well-being of anyone else.

It is somewhat surprising in view of the preceding discussion of methodology that these basic postulates, far from being self-evident or plainly true in the light of common experience, are quite plainly false. Certainly no one but the wholly destitute has in fact considered all the bundles of commodities available to them, let alone ranked them. There is no reason why anyone should have a definite preference between, say, a box of shredded wheat and a pound of apples and a box of Wheaties and a pound of pears. Still less need it be possible to rank a dining table against a two-week vacation in Crete, or an elephant against three camels.[5] Consumerism is, if possible, more obviously false, even in the present post-Reaganite world. And even the somewhat more plausible assumption that the most favored bundle of commodities available will be selected is massively constrained by information deficits.[6]

What then should we make of these basic postulates given that they are neither true nor, a fortiori, self-evidently true? The standard answer to this question is that they are true ceteris paribus, or that they describe what would happen in the absence of the various acknowledged interfering factors.[7] In support of this maneuver, we may begin by noting that *all* laws must be qualified to some extent by ceteris paribus conditions. Even such nomological paradigms as Newton's laws of motion are true only in the absence of interfering forces (for instance, electro-magnetic) (see Cartwright 1983). At the other extreme, not everything that could be said to be true, ceteris paribus, should count as a law. It is true, ceteris paribus, that strong earthquakes cause houses to fall down, provided that the ceteris paribus conditions include such things as that the frame was not bolted to the foundation, the walls were not reinforced, etc. Without careful specification of what it is for other things to be equal, we could claim a ceteris paribus law whenever anything has even a rarely exercised capacity to bring about some effect. Presumably the answer to this must be that ceteris paribus clauses involve what is usually, or typically, true, or at least what is typically approximately true. But this immediately seems too strong for the putative economic laws that we have been considering. For if the postulates of economic theory were typically true, or even typically approximately true, then we should surely expect that the consequences deduced from them would be typically (approximately) true. Thus if it is indeed the case that economic predictions are generally of little value, and this is because there are generally interfering factors, then the ceteris paribus clause is certainly not typically true.

No doubt the preceding argument has moved a bit too fast. One avenue that remains open is to stress the abstraction rather than the a prioricity in Mill's method. One line of thought that can be found in Mill is that good methodology consists in investigating significant forces in isolation from all interfering factors.[8] When we have a good understanding of the behavior of all the main forces we can try to see how to put them all together. Thus we need only assume, in the present case, that the economic postulates identify significant factors in the determination of social outcomes. Given the acknowledged presence of other significant forces, strict falsity both of the postulates and of deductions from them is only to be expected. This seems to me to be the strongest interpretation of economic methodology with a fighting chance of defensibility. Some important consequences of adopting this view should be noted, however. First, it would render very questionable any attempt to base policy decisions on economic theory alone. Second, it would even more clearly preclude any attempt to derive normative conclusions from economic foundations. And third, as is a major theme of Hausman (1992), giving equal status to social forces distinct from those studied by economists would be strongly opposed to the general ideology of the discipline. But rather than pursue any of these issues, I want to raise some skeptical questions as to whether even this view of economics as the production of (false) abstract models holds out much hope of being a genuine source of insight into social reality.

2. THE POSSIBILITY OF THE IMPOSSIBILITY OF ECONOMICS

Let us suppose that economic theory does consist of abstract models in the sense indicated at the end of the previous section. In what way might such models be true or useful? It will be helpful here to consider a distinction discussed by Nancy Cartwright (1990, chap. 5) between abstraction and idealization. In an idealized model, causal variables are not ignored, but rather set to convenient values. Since important variables are all included, it is supposed that idealized models are approximately true. Abstract models, on the other hand, describe only the effects of selected causal influences. The point emphasized in the preceding section, that economic models are not expected to be even approximately true, and Hausman's idea that even the ceteris paribus laws in economics are vague as to what exactly has to be equal make it clear that economic laws must be abstract rather than idealized. This leads us to the question: What are the criteria of, if not truth, then rightness or legitimacy for such abstract models?

A simple answer would be that abstract models say merely that if there were objects subject to only the forces defined in the model, then from the definition of those forces it follows that the objects would behave in the ways specified by the model. This has momentary appeal in view of the idea that forces must ultimately be defined in terms of the kind of response they produce in appropriate objects to which they are applied. But this reply is plainly too easy to help with our question. On this criterion, any model is legitimate if it is coherent. It cannot, therefore, address the issue of which models have some relevance to the real world. It is tempting, then, to resort to counterfactual claims about real objects. Thus, for example, we might interpret models of consumer behavior as asserting that if people were motivated solely by self-interest, had perfect information, complete transitive preferences, etc., they would.... To which a natural response would be, Why should we care about the hypothetical behavior of people so different from ourselves?

This entire discussion raises important philosophical difficulties that I shall, however, note only to pass over. It is questionable whether self-interest, for example, can coherently be thought of as one force competing with others in the manner of the combining or opposing forces of classical mechanics. (Though perhaps with self-interest in the guise of greed such a vision is implicit in certain, generally religious, models of moral deliberation.) However, assuming for present purposes that some adequate psychological theory can be provided to make sense of the economic model, let me consider further the proposed counterfactual. The response to the objection at the end of the last paragraph must be, presumably, that these people are not different from ourselves with regard to their dispositional structure. The claim is about how *we* would behave subject to only the forces specified. And this is not merely to serve as a definition of those forces (unless it is to be subject to the objection of triviality), but to say something contingent about our causal or dispositional structure. But even if we pass over all the notorious general difficulties with the interpretation of counterfactuals, it is still extremely doubtful whether such counterfactual claims have any relevance to real people in the real world.

We should begin with the fairly obvious point that the forces central to economic models, notably self-interest, must at least be seen as important determinants of human behavior. It is unimaginable, for example, that anyone should be much interested in a theory of some social phenomenon that treated human behavior solely insofar as it derived from embarrassment, or pique. With this assumption in mind, the question also arises, under what circumstances is self-interest a crucial or even decisive determinant of human behavior? Is the centrality of self-interest a universal feature of human nature, or something true only in a fairly specific set of social contexts?[9] Apparently training in economics has some tendency to elicit the kind of behavior assumed in economic models,[10] which suggests that such behavior might be very sensitive to details of the social environment. McCloskey (1990, 140) makes some scathing remarks about the prevalence of unrestrained self-interest among his fellow-economists.[11] Regardless of the answer to this question, even assuming that self-interest is a sufficiently crucial determining factor of human behavior to justify the development of models of its consequences, what does the counterfactual tell us about human behavior in contexts in which other important factors are also acting such as altruism, force of habit, deference to local custom, systematic "irrationality" of the kind increasingly familiar to psychologists,[12] etc.

I suggest that if there is any sense to be made of the question what is the causal force of self-interest abstracted from such other factors relevant to the determination of economic behavior, it can only be because it is possible, at least in principle, to combine the causal influences of all such forces so as to produce a resultant that fully determines, if not the relevant behavior, at least the probability of its occurrence. It is not merely that without such a possibility the abstract model can have no practical significance, but rather that this is a condition of the intelligibility of the abstract model. One way to make this point is to note that abstraction is often thought of as involving the *sub*traction of causal factors other than those under consideration (see Cartwright 1990, 197). But this metaphor only makes sense if there is some totality (the resultant of the various factors) from which this subtraction can begin, and if this really is a *sum* of the various causal factors involved.

I see no reason why these conditions should be met. The assumption that there must be some such resultant of the causal influences on social, or specifically economic, phenomena is an expression of the commitment to determinism, or at least the probabilistic successors to determinism that still insist on the existence of the complete causal truth about any sequence of events. I cannot address the difficulties with these views in any detail here.[13] However, in the absence of a metaphysical argument for the necessity of determinism, the general failure to provide empirically adequate models of social phenomena is, ipso facto, a failure to provide evidence for the truth of determinism. Evidence for determinism, I suggest, is hard to come by.

It may be easier to see what is at stake here in the context of a concrete example. I do not at all mean to deny that there are economic causes. So, for instance, a reduction in the price of a commodity will generally cause an increase in the quantity of that commodity demanded unless some other causal factor counteracts that causal tendency. Does this mean that there is some objectively real relation between price and quantity demanded of the sort expressed in a demand curve? Not necessarily. What an objectively real demand curve appears to add to the mere recognition of the causal capacity of the reduction in price is the expectation that the relation between price and quantity demanded should be quantitatively stable.[14] Thus it implies that if, on two different occasions, the price falls from P_1 to P_2, the quantity demanded should, on both occasions, rise from Q_1 to Q_2. Of course, this is not really what anyone expects. Tastes may have changed, substitutes or complements become more or less available, etc. So, as usual, this implication will be qualified as true only ceteris paribus. This, in turn, implies that if the change in quantity demanded on the second occasion is not the same as on the first, there must be some explanation of this fact—some other factor must not have been equal. But only a commitment to determinism makes it seem necessary that there be such an explanation. I see no reason why demand curves should not evolve in erratic and unpredictable ways over history so that quantitative relations between price and quantity would have very little inductive value for predicting the future. In short, the failure of economics to predict future phenomena might reflect not merely the inability of economists to take account of the numerous factors impinging on economic reality, but rather the inherent unpredictability of the phenomena concerned. And if this were the case, finally, it would be impossible even to make sense of the counterfactuals that were supposed to underwrite the abstract economic models.

There are two kinds of response to the recognition of this possibility short of abandoning entirely the traditional quantitative methodology of economics:

empirical optimism and metaphysics. By the latter, I mean that we might offer a metaphysical argument that there must be stable economic relations regardless of our inability to discover them. I have argued at length against this kind of metaphysical prejudice elsewhere (1993), and shall not discuss it further here. (Somewhat relatedly, we might suppose in the way suggested by the quote from Friedman at the beginning of this paper that the assumption of the existence of underlying order is a precondition of scientific research. I shall explain below why I think that, even in a merely methodological sense, such a presupposition is unacceptable.) In accordance with the former strategy, we may certainly continue to hope that empirical evidence will eventually sustain the stability of quantitative economic relations, though I suggest that this amounts to little more than blind faith. No doubt it may in the end turn out that, perhaps with the incorporation of specific further factors and barring gross and identifiable interventions, economic self-interest really does come close to providing an empirically adequate explanation or prediction of economic phenomena. But given the ease with which significant interfering factors can be imagined, this seems something of a forlorn hope.

It will be recalled that in offering evolutionary biology as an area of science the legitimacy of which does not seem seriously threatened by lack of predictive potential, I mentioned that the absence of serious alternatives to the theory of evolution as an explanation of the origin and diversity of life was important to its credibility. Could the same be said of economics? Surely not. As I have indicated, it seems hard to deny that the forces assumed by economic theory act, for example, on the prices of and demand for commodities. So any plausible account of the determination of these parameters will have to leave room for these causal influences. But how large a part of the story these forces will provide is another matter. In the short run, demand and supply may seldom be in equilibrium even if eventually producers will react to increasing inventories or shortages. These responses may seldom keep pace with endogenously determined changes in demand. Perhaps prices are largely determined by rule of thumb pricing techniques (cost + $X\%$, for example), imitation, or guesswork, and demand fluctuates continuously with changes in fashion, advertising strategies, and so on. And most importantly unless determinism is assumed, there need be no stable and consistent way in which these various forces combine.

An even more significant employment of the parallel between economics and evolutionary theory is the following. The TINA[15] defense of evolutionary theory is persuasive when it is deployed on behalf of broad qualitative accounts of aspects of the history of life. It is much less plausible if it is taken to legitimate detailed mathematical models of particular evolutionary phenomena. Indeed I have argued elsewhere (1993, chap. 6) that population genetics, supposedly a source of detailed mathematical models of evolutionary changes in the genetic constitution of populations of organisms, is an extremely dubious enterprise. Similarly, insofar as economics restricts itself to identifying causal influences on economic phenomena, and deploying these in narrative accounts of economic phenomena, it should be seen as constituting a generally salutary contribution to general historiography. But the mismatch between the apparent precision and scope of theoretical economic models and any reasonable assessment of the importance of the processes they describe to real economic history makes most of economic theory an unpromising candidate for epistemologically respectable science.

3. SCIENCE AND VALUES

In a suggestive article, Mary Hesse (1978) raises the question whether the criteria for scientificity generally associated with the natural sciences, especially the ability of science to facilitate prediction and control, might not be explicitly rejected by (parts of) the social sciences in favor of some more obviously value-driven goal. My suggestion that an economic theory capable of more than a quite limited degree of prediction and control might be precluded by the nature of economic, or more broadly social phenomena, provides a compelling ground for considering carefully the possibility Hesse explores.

The starting point for developing this idea is to note that given the view of a largely qualitative economics applied to a partially indeterministic domain of phenomena that I have advocated, there is no reason to think that there is any uniquely empirically best economic theory. This contrasts sharply with one of the best known metaphors for modern science, mechanism. There is one unique best theory of the workings of a machine, the theory that correctly explains how it works.[16] This is a consequence of the fact that it is (generally) entirely clear what is the behavior of a machine that calls for explanation, namely, the behavior which the machine was designed to exhibit. This is easily seen by thinking of cases where there is some departure from the normal use of the machine. Just as it is generally irrelevant to the theory of bicycles what color they are, so when the bicycle is raised off the ground and used as an exercising device it becomes irrelevant what is the condition of its tires. In more extreme cases, say the use of the bicycle as a structural element in a postmodern building, color might become the most salient feature.

The machine model often seems to be exactly what underlies pictures of the economy, and hence of the nature of economics. The economy is a machine that produces commodities. The parts of the machine are firms, consumers, and governments. Taxes or government regulations, we are often encouraged to believe, create friction and slow down the production of commodities. Ideally if we can engender enough greed to fuel the machine, and minimize all possible sources of friction, the machine will run at its best. My point here is not so much to criticize this picture—though certainly there is much to criticize— but to point to its optionality. If we reject the idea that this picture is simply descriptive—that's just what economies are for, like it or not—it is natural to see it as implicitly supporting the normative assumption that maximally efficient commodity production is what economics should be intended to promote. But if that is a correct interpretation, then surely we should object at once that it is far from obvious that this is a goal—still less *the* goal—that we should aim to promote. For some historical political economists, perhaps even economists such as Jevons at the beginning of neoclassical economics, this normative connection was explicit, with the economy as a generator of utility yoked to a theory of the good as utility-maximization. Nowadays, however, with economics concerned with utility solely in the sense of moving individuals on to ordinally higher indifference curves; with comparisons between social states generally limited to the identification of Pareto-improvements; and with moral philosophers, if not economists, widely skeptical even about genuinely substantive utilitarianism, there is no convincing possibility of providing a normative justification of the traditional economic machine. It is of course possible to undercut the need for such a justification by insisting that the machine is inevitable: that is just how things are. (This is the metaphysical version of the TINA defense.) And though such natural necessity is precisely what is often implied by the hard-headed advocates of positive economics, this is a metaphysical bullet, I have argued, that we have no reason to bite.

The myth of positive economics, an economics forced on us simply by the way things are, serves to deflect debate from the really important question, which is what goals we would like the economy to serve. Consider, for instance, the possibility of an economics premised on the assumption that the function of an economy was to maximize, and distribute as equally as possible, standard of living. Needless to say, a very substantial part of this discipline would be involved with the philosophical question of what constitutes standard of living, though as Sen (1987) has made very clear, we can be sure that this is only tangentially related either to any standard account of utility or to the accumulation of wealth. It is of course possible that the best way to do such an economics would be to graft the normative conclusions on to very much the traditional economic account of the market economy, in the manner of much contemporary welfare economics. But it seems unlikely that things would be that simple. Once we recognize the possibility that an economy can be highly productive and growing fast in current economic terms, and yet thoroughly sick in the normatively significant sense of its failure to generate high standards of living for many of its citizens, we are likely to direct our attention to somewhat different measures of economic performance and individual motivations. Recalling the observation that training in economics has some tendency to generate economically "rational" behavior, it seems likely that an economy substantially under the control of economists will promote such behavior in many different ways among all or most segments of the community. I suggest that the only remedy for this situation is to recognize that the normative part of economics, far from being hitched on to the back of a putatively positive economics, must be seen as fundamental to, and epistemologically prior to, any acceptable science of economics.

The preceding discussion should, finally, answer the objection to the indeterministic metaphysical grounding of my claims suggested by Friedman's remark about the deceptiveness of appearance through which science aims to see, the insistence that methodologically it is sheer defeatism not to assume that there is some complete, empirically adequate, theory of economics. My reply is that there are great costs to error on either side of this question. Pursuing the current, empirically generally failing, theory of economics merely on the grounds that we must hold out the hope for a physics-like, mechanistic, and empirically sufficient theory of economics may very well stand in the way of seeing the need for an admittedly partial, indeterministic, qualitative, etc. account of economics, but one driven by truly worthwhile social or political goals.

4. CONCLUSION

With a reminder that this paper is to be understood as provisional and in some respects sketchy, let me nonetheless venture some conclusions. I can best do so by returning to the question that provided the title of this paper. Though my tone has sometimes verged on the polemical, it is clear that my answer to that question is a qualified, Very probably yes. First, my skepticism has been directed to theoretical, especially mathematical constructs in economics. I do not deny that there may well be much useful, if generally loose, causal knowledge to be had in economics, though perhaps a good deal of this will be at a level accessible to no more than sophisticated common sense, and suspicions about the inductive projectibility of such causal knowledge will often be appropriate. Ultimately, my affirmative answer here may depend on some major revisions in what we take to be legitimate goals of science. Second, my skepticism about the one true economic story opens up the possibility that there might be numerous approaches to the investigation of economic phenomena, each capable of delivering partial, but nonetheless useful, insights into economic reality. And third, and most importantly, the preceding section suggests the possibility of a science of economics of real value to humanity, something which has only intermittently been true of the history of that science and seems decreasingly true of it in recent years.[17] But the precondition of this, I have argued, is the recognition that fundamental value judgments must be made at the very beginning of the project. They cannot just be tacked on at the end.[18]

NOTES

1. One economist (McCloskey 1990) seems to argue on a priori grounds that economists could not make predictions: the windfall profits that would accrue to one able to make economic predictions would motivate further economic actions that would soon falsify such predictions. Certainly economic predictions would often be an important causal influence on the phenomena they attempted to predict, though this seems only to show a further reason why prediction would be difficult, rather than that it would be impossible.

2. Strictly speaking I should not say "self-interest" but merely economic "rationality." Formally, the latter concept allows that I might prefer the well-being of others in certain respects to my own. However, I prefer to use the term "self-interest" for two reasons. First, the well-being of others can only be of concern to me through some psychological state of mine such as satisfaction, pleasure, etc. The economically "rational" man cannot aim for the good of another simply for its own sake or for moral reasons. Second, notwithstanding the

possibility of giving an extremely abstract interpretation of the formalism, a great deal of the rhetoric of economics clearly embraces a quite explicit commitment to a quite narrow sense of self-interest.

3. It is presumably in opposition to such a conception of economics that Friedman's insistence on the primacy of predictive success, discussed above, is largely motivated. No doubt this goes some way to explain some of the apparent anti-realist excesses of Friedman's position with regard to economic theory, excesses that could have been greatly vitiated by a sharper distinction between the fundamental postulates of a theory and the merely auxiliary assumptions at which his skepticism is primarily addressed. This issue is, however, largely irrelevant to my present concerns with Friedman's ideas.

4. A useful and fairly detailed discussion of the basic premises of microeconomics is provided by Hausman (1992, chaps. 1–3). My very cursory discussion leaves out some of the more technical postulates, notably the continuity of ranking and the diminishing marginal rates of substitution, for which see Hausman, op. cit.

5. I say nothing about so-called revealed preferences, since these are of no possible philosophical interest unless they do indeed reveal preferences, which is the point at issue. If they may nevertheless be of some relevance to economics, so much the worse for economics.

6. One can, of course, include the cost of information about the availability of a bundle of goods in the cost of that bundle, though this would undermine some of the main attractions of the whole theory by making the cost of a bundle of goods largely inscrutable and widely variable from person to person.

7. Hausman (1992) goes a step further and notes that these must be *vague* ceteris paribus conditions. We do not typically know exactly what are the various possible sources of interference. The presence of such vague ceteris paribus conditions defines what Hausman calls inexact laws.

8. Consideration of Mill's account of the method of difference in *A System of Logic* (1875) suggests a more empiricist interpretation of his methodology than I have here allowed. Combining his remarks on economics with his general theory of inductive reasoning in *A System of Logic* suggests a more complex and subtle methodology. Though my treatment of Mill here is therefore somewhat unfair, further exploration of this idea would be a project for another paper.

9. Economists often note that their stories apply only to conditions of scarcity. Here I have in mind no such conditions internal to the structure of economic theory, but rather suggestions such as that the behavior of 'economic man' might be uniquely characteristic of people living in market, capitalist societies, as proposed by Marx in his well-known criticisms of Smith and Ricardo in the *1844 Manuscripts* (1844/1978).

10. See Marwell and Ames 1981, cited in Hausman 1992, p. 218.

11. In accordance with his salutary emphasis on the rhetorical dimension of economics, McCloskey notes that "[t]he economist who relishes the telling of a story of greed is advocating it, whatever he may say about 'is' and 'ought'" (1990, 141).

12. See, for example, Tversky and Kahneman 1981.

13. Determinism specifically, and causal completeness generally, are criticized at length in Dupré 1993, pt. 3.

14. One economist who was apparently highly skeptical of the legitimacy of assuming inductive stability of economic parameters was John Maynard Keynes. (See the letters to Roy Harrod reprinted in Hausman 1984, 300–302.)

15. An acronym of Margaret Thatcher's famous attempt to justify a particularly unattractive economic policy, "There Is No Alternative."

16. Though for the most complicated machines it may be appropriate to approach them in various different ways, as suggested by Dennett's well-known distinctions between the physical, design, and intentional stances that can be taken towards some machines, most notably computers (Dennett 1987). Because even computers are highly deterministic we are inclined to see the physical stance as basic. Perhaps this will be less compelling for the

machines of the future, in which case the traditional associations of mechanism will have become misleading even for the understanding of machines.

17. A few proposals have been put forward for economic projects radically different from traditional neoclassical economics, especially by feminist critics of economists. Work in progress by Julie Nelson (1993), for instance, suggests that we should replace a model of economics as primarily concerned with choice with one centered on the idea of provisioning. A very important and sustained critique of a variety of fundamental economic concepts, especially those connected with the measurement and comparison of national incomes, together with some proposals for doing things differently, is provided by Waring (1988).

18. Helpful comments on an earlier draft by Regenia Gagnier and Debra Satz have led to numerous improvements in this paper.

REFERENCES

Bacon, Francis. 1620/1960. *The New Organon*. Edited by F. H. Anderson. New York.
Cartwright, Nancy. 1983. *How the Laws of Physics Lie*. Oxford.
Cartwright, Nancy. 1990. *Nature's Capacities and Their Measurement*. Oxford.
Dennett, Daniel. 1987. *The Intentional Stance*. Cambridge, Mass.
Dupré, John. 1993. *The Disorder of Things: Metaphysical Foundations of the Disunity of Science*. Cambridge, Mass.
Friedman, Milton. 1953. "The Methodology of Positive Economics." Reprinted in Hausman (1984).
Hausman, Daniel M. Ed. 1984. *The Philosophy of Economics*. Cambridge.
Hausman, Daniel M. 1992. *The Inexact and Separate Science of Economics*. Cambridge.
Hesse, Mary. 1978. "Theory and Value in the Social Sciences." In *Action and Interpretation*, edited by C. Hookway and P. Pettit. Cambridge.
Kettlewell, H.B. 1973. *The Evolution of Melanism*. Oxford.
McCloskey, Donald N. 1990. *If You're So Smart*. Chicago.
Marwell, G., and R. Ames. 1981. "Economists Free Ride. Does Anyone Else? Experiments on the Provision of Public Goods. IV." *Journal of Public Economics* 1: 237–46.
Marx, Karl. 1844. "The Economic and Philosophic Manuscripts of 1844." In *The Marx-Engels Reader*, edited by R.C. Tucker. 1978. New York.
Mill, J.S. 1836. "On the Definition of Political Economy and the Method of Investigation Proper to It." Excerpts reprinted in Hausman 1984.
Mill, J.S. 1875. *A System of Logic*. 8th ed. London.
Nelson, Julie A. 1993. *Beyond Economic Man: Feminist Theory and Economics*. Chicago.
Robbins, Lionel. 1935. *An Essay on the Nature and Significance of Economic Science*, 2d.ed. Excerpts reprinted in Hausman 1984.
Rosenberg, A. 1979. "A Skeptical History of Microeconomic Theory." In *Philosophy in Economics*, edited by J.C. Pitt, 47–61. Dordrecht.
Sen, Amartya. 1987. *The Standard of Living*. Edited by G. Hawthorn. Cambridge.
Tversky, A., and D. Kahneman. 1981. "The Framing of Decisions and the Psychology of Choice." *Science* 211: 453–58.
Waring, Marilyn, 1988. *If Women Counted: A New Feminist Economics*. San Francisco.

MIDWEST STUDIES IN PHILOSOPHY, XVIII (1993)

Function and Design

PHILIP KITCHER

I

The organic world is full of functions, and biologists' descriptions of that world abound in functional talk. Organs, traits, and behavioral strategies all have functions.[1] Thus the function of the *bicoid* protein is to establish anterior-posterior polarity in the *Drosophila* embryo; the function of the length of jackrabbits' ears is to assist in thermoregulation in desert environments; and the function of a male baboon's picking up a juvenile in the presence of a strange male may be to appease the stranger, or to protect the juvenile, or to impress surrounding females. Ascriptions of function have worried many philosophers. Do they presuppose some kind of supernatural purposiveness that ought to be rejected? Do they fulfil any explanatory role? Despite a long, and increasingly sophisticated, literature addressing these questions, I believe that we still lack a clear and complete account of function-ascriptions. My aim in what follows is to take some further steps towards dissolving the mysteries that surround functional discourse.

I shall start with the idea that there is some unity of conception that spans attributions of functions across the history of biology and across contemporary ascriptions in biological and non-biological contexts. This unity is founded on the notion that the function of an entity S is *what S is designed to do*. The fundamental connection between function and design is readily seen in our everyday references to the functions of parts of artifacts: the function of the little lever in the mousetrap is to release the metal bar when the end of the lever is depressed (when the mouse takes the cheese) for that is what the lever is designed to do (it was put there to do just that). I believe that we can also recognize it in pre-Darwinian perspectives on the organic world, specifically in the ways in which the organization of living things is taken to reflect the intentions of the Creator: Harvey's claim that the function of the heart is to pump the blood can be understood as proposing that the wise and beneficent designer foresaw the need for a circulation of blood and assigned to the heart the job of pumping.

Now examples like these are precisely those that either provoke suspicion of functional talk or else prompt us to think that the concept of function has been altered in the course of the history of science. Even though we may retain the idea of the "job" that an entity is supposed to perform in contexts where we can sensibly speak of systems fashioned and/or used with definite intentions—paradigmatically machines and other artifacts—it appears that the link between function and design must be broken in ascribing functions to parts, traits, and behaviors of organisms. But this conclusion is, I think, mistaken. On the view I shall propose, the central common feature of usages of function— across the history of inquiry, and across contexts involving both organic and inorganic entities—is that the function of S is what S is designed to do; design is not always to be understood in terms of background intentions, however; one of Darwin's important discoveries is that we can think of design without a designer.[2]

Contemporary attributions of function recognize two sources of design, one in the intentions of agents and one in the action of natural selection. The latter is the source of functions throughout *most* of the organic realm—there are occasional exceptions as in cases in which the function of a recombinant DNA plasmid is to produce the substance that the designing molecular biologist intended. But, as I shall now suggest, the links to intentions and to selection can be more or less direct.

II

Imagine that you are making a machine. You intend that the machine should do something, and that is the machine's function. Recognizing that the machine will only be able to perform as intended if some small part does a particular job you design a part that is able to do the job. Doing the job is the function of the part. Here, as with the function of the whole machine there is a direct link between function and intention: the function of X is what X is designed to do, and the design stems from an explicit intention that X do just that.

It is possible that you do not know everything about the conditions of operation of your machine. Unbeknownst to you, there is a connection that has to be made between two parts if the whole machine is to do its intended job. Luckily, as you were working, you dropped a small screw into the incomplete machine and it lodged between the two pieces, setting up the required connection. I claim that the screw has a function, the function of making the connection. But its having that function cannot be grounded in your explicit intention that it do that, for you have no intentions with respect to the screw. Rather, the link between function and intention is much less direct. The machine has a function grounded in your explicit intention, and its fulfilling that function poses various demands on the parts of which it is composed. You recognize some of these demands and explicitly design parts that can satisfy them. But in other cases, as with the luckily placed screw, you do not see that a demand of a particular type has to be met. Nevertheless, whatever satisfies that demand has the function of so doing. The function here is grounded in the contribution that is made towards the performance of the whole machine and in the link between the performance and the explicit intentions of the designer.

Pre-Darwinians may have tacitly relied on a similar distinction in ascribing functions to traits and organs. Perhaps the Creator foresaw all the details of the grand design and explicitly intended that all the minutest parts should do particular things. Or perhaps the design was achieved through secondary causes: organisms were equipped with abilities to respond to their needs, and the particular lines along which their responses would develop were not explicitly identified in advance. So the Creator intended that jackrabbits should have the ability to thrive in desert environments, and explicitly intended that they should have certain kinds of structures. However, it may be that there was no explicit intention about the length of jackrabbits' ears. Yet, because the length of the ears contributes to the maintenance of roughly constant body temperature, and because this is a necessary condition of the organism's flourishing (which is an explicitly intended effect), the length of the ears has the function of helping in thermoregulation.

Understanding this distinction enables us to see how earlier physiologists could identify functions without engaging in theological speculation.[3] Operating on the presupposition that organisms were designed to thrive in the environments in which they are found, physiologists could ask after the necessary conditions for organisms of the pertinent types to survive and multiply. When they found such necessary conditions, they could recognize the structures, traits, and behaviors of the organisms that contributed to satisfaction of such conditions as having precisely such functions—without assuming that the Creator explicitly intended that those structures, traits, and behaviors perform just those tasks.

I have introduced this distinction in the context of machine design and of pre-Darwinian biology because it is more easily grasped in such contexts. I shall now try to show how a similar distinction can be drawn when natural selection is conceived as the source of design, and how this distinction enables us to resolve important questions about functional ascriptions.

III

We can consider natural selection from either of two perspectives. The first, the organism-centered perspective, is familiar. Holding the principal traits of members of a group of organisms fixed, we investigate the ways in which, in a particular environment or class of environments, variation with respect to a focal trait, or cluster of focal traits, would affect reproductive success. Equally, we can adopt an environment-centered perspective on selection. Holding the principal features of the environment fixed, we can ask what selective pressures are imposed on members of a group of organisms. In posing such questions we suppose that some of the general properties of the organisms do not vary, and consider the obstacles that must be overcome if organisms with those general properties are to survive and reproduce in environments of the type that interests us.

So, for example, we might consider the selection pressures on mammals whose digestive systems are capable of processing vegetation but not meat (or carrion) in an environment in which the accessible plants have tough cellulose outer layers. Holding fixed the very general properties of the animals that determine their need to take in food and the more particular features of their digestive systems, we recognize that they will not be able to survive to maturity (and hence not able to reproduce) unless they have some means of breaking down the cellulose layers of the plants in their environments. Thus the environments impose selection pressure to develop some means of breaking down cellulose. Organisms might respond to that pressure in various ways: by harboring bacteria that can break down cellulose or by having molars that are capable of grinding tough plant material. If our mammals do not have an appropriate colony of intestinal bacteria, but do have broad molars that break down cellulose, we may recognize the molars as their particular response to the selection pressure and ascribe them the function of processing the available plants in a way that suits the operation of their digestive systems. At a more fine-grained level, we may hold fixed features of the dentition, and identify properties of particular teeth as having functions in terms of their contributions to the breakdown of cellulose.

This illustration can serve as the prototype of a style of functional analysis that is prominent in physiology and in general zoological and botanical studies. One starts from the most general evolutionary pressures, stemming from the competition to reproduce and concomitant needs to survive to sexual maturity, to produce gametes, to identify and attract mates, and so forth. In the context of general features of the organisms in question and of the environments they inhabit, we can specify selection pressures more narrowly, recognizing needs to process certain types of food, to evade certain kinds of predators, to produce particular types of signals, and so forth. We now appreciate that certain types of complex structures, traits, and behaviors enable the organisms to satisfy these more specific needs. *Their* functions are specified by noting the selection pressures to which they respond. The functions of their constituents are understood in terms of the contributions made to the functioning of the whole. Here, I suggest, we have a mixture of evolutionary and mechanistic analysis. There is a link to selection through the environment-centered perspective from which we generate the selection pressures that determine the functions of complex entities, and there is a mechanistic analysis of these complex entities that displays the ways in which the constituent parts contribute to total performance.

I claim that understanding the environment-centered perspective on selection enables us to draw an analogous distinction to that introduced in section II, and thus to map the diversity of ways in which biologists understand functions.

However, before offering an extended defense of this claim, two important points deserve to be made.

First, the environment-centered perspective has obvious affinities with the idea that organisms face selective "problems," posed by the environment, an idea that Richard Lewontin has recently criticized.[4] According to Lewontin, there is a "dialectical relationship" between organism and environment that renders senseless the notion of an environment prior to and independent of the organism to which "problems" are posed. Lewontin's critique rests on the correct idea that there is no specifying which parts of the universe are constituents of an organism's environment, without taking into account properties of the organism. In identifying the environment-centered perspective, I have explicitly responded to this point, by proposing that the selection pressures on organisms arise only when we have held fixed important features of those organisms, features that specify limits on those parts of nature with which they causally interact. Quite evidently, if we were to hold fixed properties that could easily be modified through mutation (or in development), we would obtain an inadequate picture of the organism's environment and, consequently, of the selection pressures to which it is subject. If, however, we start from those characteristics of an organism that would require large genetic changes to modify—as when we hold fixed the inability of rabbits to fight foxes—then our picture of the environment takes into account the evolutionary possibilities for the organism and offers a realistic view of the selection pressures imposed.

Second, as we shall see in more detail below, recognizing a trait, structure, or behavior of an organism as responding to a selection pressure imposed by the environment (in the context of other features of the organism that are viewed as inaccessible to modification without severe loss of fitness) we do not necessarily commit ourselves to claiming that the entity in question originated by selection or that it is maintained by selection. For it may be that genetic variation in the population allows for alternatives that would be selectively advantageous but are fortuitously absent. Thus the entity is a response to a genuine demand imposed on the organism by the environment even though selection cannot be invoked to explain why it, rather than the alternative, is present. In effect, it is the analogue of the luckily placed screw, answering to a real need, but not itself the product of design. I shall be exploring the consequences of this point below.

IV

The simplest way of developing a post-Darwinian account of functions is to insist on a direct link between the design of biological entities and the operation of natural selection. The function of X is what X is designed to do, and what X is designed to do is that for which X was selected. Since the publication of a seminal article by Larry Wright, etiological accounts of function have become extremely popular.[5] Wright claimed that the function of an entity is what explains why that entity is there. This simple account proved vulnerable to counterexamples: if a scientist conducting an experiment becomes unconscious because gas escapes from a leaky valve, then the presence of the gas in the room is explained by the fact that the scientist is unconscious (for otherwise she would have turned off the supply), but the function of the gas is not to asphyxiate scientists.[6] Such objections can be avoided by restricting the form of explanations to explanations in terms of selection, so that identifying the function of X as that for which X was selected enables us to preserve Wright's idea that functions play a role in explaining the presence of their bearers without admitting those forms of nonselective explanation that generate counterexamples.[7] However, this move forfeits one of the virtues of Wright's analysis, to wit, its recognition of a common feature in attributions of functions to artifacts and to organic entities.

There are other issues that etiological analyses of functional ascriptions must confront, issues that arise from the character of evolutionary explanations. First is the question of the *time* at which the envisaged selection regime is supposed to act. Second we must consider the *alternatives* to the entity whose presence is to be explained and the extent of the role that selection played in the singling out of that entity.[8] If these issues are neglected—as they frequently are—the consequence will be either to engage in highly ambiguous attributions of function or else to fail to recognize the demands placed on functional ascription.

Selection for a particular property may be responsible for the original presence of an entity in an organism or for the maintenance of that entity.[9] In many instances, selection for P explains the initial presence of a trait *and* the subsequent maintenance of that trait: the initial benefit that led to the trait's increase with respect to its rivals also accounts for its superiority over alternatives that arose after the original process of fixation. But as a host of well-known examples reveals, this is by no means always the case. To cite one of the most celebrated instances, feathers were apparently originally selected in early birds (or their dinosaur ancestors) for their role in thermoregulation; after the development of appropriate musculature (and other adaptations for flight) the primary selective significance of feathers became one of making a causal contribution to efficient flying.

Faced with examples in which the properties for which selection initially occurs are different from those for which there is selection in maintaining a trait, behavior, or structure, the etiological analysis must decide which of the following conditions is to govern functional attributions:

(1) The function of X is Y only if the initial presence of X is to be explained through selection for Y,

(2) The function of X is Y only if the maintenance of X is to be explained through selection for Y,

(3) The function of X is Y only if both the initial presence of X and the maintenance of X are to be explained through selection for Y.

But deciding among these three conditions is only the beginning of the enterprise of disambiguating the etiological analysis of function. Just as the properties important in initiating selection may not be those that figure in maintaining selection, it is possible that an entity may be *maintained* by selection for different properties at different times. Hence, both (2) and (3) require us to specify the appropriate period at which the maintenance of X is to be considered. I believe that there are two plausible candidates with respect to (2), namely the present and the recent past, and that the most well-motivated version of (3) requires that the character of the selective regime is constant across all times. Thus we obtain:

(2a) The function of X is Y only if selection of Y has been responsible for maintaining X in the recent past,

(2b) The function of X is Y only if selection for Y is currently responsible for maintaining X,

(3) The function of X is Y only if selection for Y was responsible for the initial presence of X and for maintaining X at all subsequent times up to and including the present.

A consequence of adopting (1)—which effectively takes functions to be *original* functions—is that two of Tinbergen's famous four why-questions are conflated: there is now no distinction between the "why" of evolutionary origins and the "why" of functional attribution.[10] In those biological discussions in which an etiological conception of function is most apparent (ecology, and especially behavioral ecology), Tinbergen's distinction seems to play an important role. Thus I doubt that an etiological analysis based on (1) reflects much that is significant in biological practice.

Etiological analyses clearly based on (3) can sometimes be found in the writings of those who are critical of unrigorous employment of the notion of function. So, for example, Stephen Jay Gould's and Elisabeth Vrba's contrast between functions and "exaptations" seems to me to thrive on the idea that specification of functions must rest on the presupposition that selection has been operating in the same way in originating and maintaining traits (and, indeed, that traits maintained by selection were originally fashioned by selection).[11] Because there is frequently no available evidence for this presupposition, adoption of etiological conception based on (3) can easily fuel skepticism about ascriptions of function.

I suspect that some biologists do tacitly adopt an etiological conception of function founded on (3), and that their practice of ascribing functions is subject to Gould's strictures. Others plainly do not. Thus, Ernst Mayr explicitly recognizes the possibility of change of function over evolutionary time, suggesting that he acknowledges *two* notions of function, one ("original function") founded upon (1) and another ("present function") based on some version of (2).[12] For biologists who draw such distinctions, Gould's criticisms will seem to claim novelty for a point that is already widely appreciated. (Of course, one of the most prominent features of the debates about adaptationism is the opposition between those who believe that the criticisms tiresomely remind the evolutionary community of what is already well known and those who contend that what is professed under attack is ignored in biological practice.)[13]

The most prevalent concept of function among contemporary ecologists is, I believe, an etiological concept founded on some version of (2). Claims about functions are founded on measurements or calculations of fitness, and the measurements and calculations are made on *present* populations. Faced with the question, "Do you believe that the properties for which selection is now occurring are those that originally figured in the fixation of the trait (structure, behavior)?" sophisticated ecologists would often plead agnosticism. Their concern is with what is currently occurring, and they are happy to confess that things may have been different in a remote past that is beyond their ability to observe and analyze in the requisite detail. Hence the concept of function they employ is founded on the link between functions and contemporary processes of selection that maintain the entities in question, a link recorded in (2).

But which version of (2) should they endorse? Here, I believe, philosophical analyses reveal unresolved ambiguities in biological practice. An account of functions that effectively endorses (2b) has been proposed by John Bigelow and Robert Pargetter (who, idiosyncratically it seems to me, attempt to distance themselves from Wright and other etiological theorists).[14] My own prior discussions of functional ascriptions presuppose a concept based on (2a), and this notion of function has been thoroughly articulated by Peter Godfrey-Smith.[15] On what basis can we decide among these accounts?

As Godfrey-Smith rightly notes, a "recent history" notion of function, committed to (2a), gives functional ascriptions an explanatory role. Identifying the function of an entity outlines an explanation of why the entity is now present by indicating the selection pressures that have maintained it in the recent past. Arguing that philosophers ought to identify a concept that does some explanatory work, he concludes that (2a) represents the right choice. But this seems to me to be too quick. The conception of function defended by Bigelow and Pargetter, founded on (2b), is perhaps most evident in those biological discussions in which the recognition that a trait is functional supports a prediction about its future presence in the population. Yet the "forward-looking" conception also allows ascriptions of function to serve as explanations of why the trait will continue to be present. There is still an explanatory project, but the *explanandum* has been shifted from current presence to future presence.

Biological practice seems to me to be too various for definitive resolution of these differences. Sometimes attributions of function outline explanations of current presence, sometimes offer predictions about the course of selection in the immediate future, sometimes sketch explanations of the presence of traits in succeeding generations. Moreover, since it is often reasonable to think that the environmental and genetic conditions are sufficiently constant to ensure that the operation of selection in the recent past was the same as the selection seen in the present, it will be justifiable to combine the main features of the "recent past" and "forward-looking" accounts to found a notion of function on a combination of (2a) and (2b)

> (2c) The function of X is Y only if selection of Y is responsible for maintaining X both in the recent past and in the present.

In situations in which there is reason to think that the action of selection has been constant across the relatively short time periods under consideration, use of a notion of function founded on (2c) will allow functional attributions to play a role in all the explanatory and predictive projects I have considered.

If biological practice overlooks potential ambiguities with respect to the timing of the selection processes that underlie attributions of function, it is even more silent on issues about the competition involved in such processes. What are the alternatives to the biological entity whose presence is due to selection? And to what extent is selection the *complete* explanation of the presence of that entity?

Ecologists working on pheromones in insects or on territory size in birds can sometimes specify rather exactly the set of alternatives they consider. Holding fixed certain features of the organisms they study, features that would, they suppose, only be modifiable by enormous genetic changes that render rivals effectively inaccessible, they can impose necessary conditions that define a set of rival possibilities: pheromones must have such-and-such diffusion properties, territories must be able to supply such-and-such an amount of food, and so forth. In light of these constraints, they may be able to construct a mathematical model showing that the entity actually found in the population is optimal (or, more realistically, "sufficiently close" to the optimum).[16] A different strategy is to consider alternatives that arise by mutation in populations that can be observed and to measure the pertinent fitness values. Either of these approaches will support claims about selection processes that have occurred/are occurring in the recent past or the present. In both instances there may be legitimate concern that unconsidered alternatives might have figured in historically more remote selection processes, either because the organisms were not always subject to the constraints built into the mathematical model or because the genetic context in which mutations are now considered is quite different from the genetic contexts experienced by organisms earlier in their evolutionary histories. So far this simply underscores our previous conclusions about the greater plausibility of analyses based on some version of (2).

But now let us ask how exactly selection is supposed to winnow the alternatives. Suppose we ascribe a function to an entity X, basing that function on a selection process with alternatives $X_1, \ldots, X_n$. Must it be the case that organisms with X have higher fitness than organisms with any of the X_i? On a strict etiological analysis of functional discourse, this question should be answered affirmatively: where selection is the *complete* foundation of the design that underlies X's function, X is favored by selection over *all* its rivals. Thus on the strongest version of an etiological conception, functional ascriptions should be based either on recognition that X has greater fitness than all the alternatives arising by mutation in current populations, or on an analysis that shows X to be strictly optimal. I believe that some biologists—particularly in ecology and behavioral ecology—make functional claims in this strong sense and attempt to back them up with careful and ingenious observations and calculations.[17] Nonetheless, there is surely room for a less demanding account of biological function.

Consider two possibilities. First, our optimality analysis shows that, while X is reasonably close to the optimum, it is theoretically suboptimal. We do not know enough about the genetics and developmental biology of the organisms under study to know whether mutations providing a genetic basis for superior rivals could arise in the population. Under these circumstances, one cannot claim that the presence of X is entirely due to the operation of selection. It may be that X is present because theoretically possible mutants have not (recently) arisen, and selection, acting on a limited set of alternatives, has fixed X. Second, we may be able to identify actual rivals to X that are indeed superior in fitness but that have fortuitously been eliminated from the population. During the period that concerns us (present or recent past) organisms bearing some entity X_i have arisen, and these have had greater fitness than organisms bearing X. By chance, however, such organisms have perished. Here, we can go further than simply recognizing an inability to support the strong claim about optimality—we recognize that X is definitely suboptimal, and that its presence is not the result of selection alone.

Nevertheless, many biologists would surely be uninterested in these possibilities or actualities, regarding X as having the function associated with the selective process, even if it were possibly, even definitely, suboptimal. There are various ways of weakening the requirement that X's fitness be greater than those of alternatives. We might demand that X be fitter than *most* alternatives, that it be fitter than the *most frequently occurring* alternatives, and so forth. It requires only a little imagination to devise scenarios in which an entity is inferior in fitness to most of its rivals and/or to its most frequently occurring rivals, even though it may still be ascribed the function associated with the selection process.

Imagine that there is a species of moth that is protected from predatory birds through a camouflaging wing pattern that renders it hard to perceive when it rests on a common environmental background. We observe the population and discover a number of rival wing colorations, none of which ever occurs in substantial numbers. Less than half of these alternatives are absolutely disastrous, and organisms with them are vulnerable to predation, and quickly eliminated. Investigating the other, we find, to our surprise, that they prove slightly superior to the prevalent form, in affording improved camouflage, without any deleterious side effects. However, as the result of various events that we can identify—disruptions of habitat, increased concentrations of predators in areas in which there is a high frequency of the mutants—these alternatives are eliminated as the result of chance. Nonetheless, although it is somewhat inferior to most of its rivals, the common wing pattern still has the function of protecting the moth from predation.

I think that it is obvious what we should say about this and kindred scenarios. The impulse to recognize X as having a function can stem from recognition that X is a response to an identifiable selection pressure, *whether or not the presence of X is completely explicable in terms of selection*. Thus, instead of trying to weaken the conditions on etiological conceptions of function, I suggest that we can accommodate cases that prove troublesome by drawing on the distinctions of sections II and III. I shall now try to show how this leads to a rich account of functional ascriptions that will cover practice in physiology as well as in those areas in which the etiological conception finds its most natural home.

V

Entities have functions when they are designed to do something, and their function is what they are designed to do. Design can stem from the intentions of a cognitive agent or from the operation of selection (and, perhaps, recognizing how unintuitive the notion of design without a designer would have seemed before 1859, from other sources that we cannot yet specify). The link between function and the source of design may be direct, as in instances of agents explicitly intending that an entity perform a particular task, or when the entity is present because of selection for a particular property (that is, its presence is completely explained in terms of selection for that property). Or the link may be indirect, as when an agent intends that a complex system perform some task and a component entity makes a necessary causal contribution to the performance, or when organisms experience selection pressure that demands some complex response of them and one of their parts, traits, or behaviors makes a needed causal contribution to that response. As noted in the previous section, there are also ambiguities about the time period throughout which the selection process is operative. It would be easy to tell a parallel story about agents and their intentions.

I have noted that the strong etiological conception—that based on a direct link between function and the underlying source of design (in this case, selection)—is very demanding. While some ecologists undoubtedly aim to find functions in the strong sense, much functional discourse within ecology, as well as in other parts of biology is more relaxed. Imagine practicing biologists accompanied by a philosophical Jiminy Cricket, constantly chirping doubts about whether selection is *entirely* responsible for the presence of entities to which functions are ascribed. Many biologists would ignore the irritating cavils, contending that the attribution of function is unaffected by the possibilities suggested by philosophical conscience. It is enough, they would insist, that genuine demands on the organism have been identified and that the entities to which they attribute functions make causal contributions to the satisfaction of those demands. What is wrong with the relaxed attitude?

Functional attributions in the strong sense have clear explanatory work to do. They indicate the lines along which we should account for the presence of the entities to which functions are ascribed. To say that the function of X is F is to propose that a complete explanation of the presence of X (at the appropriate time) should be sought in terms of selection for F. Once we relax the demands on functional ascriptions, the role of selection is no longer clear; indeed, a biologist may explicitly allow that selection has not been responsible for maintaining X (or, at least, not completely responsible). But there is a different type of explanatory project to which the more lenient attributions contribute. They help us to understand the causal role that entities play in contributing to complex effects.

Here we encounter a central theme of the main philosophical rival to the etiological conception, lucidly articulated in an influential article by Robert Cummins.[18] For Cummins, functional analysis is about the identification of constituent causal contributions in complex processes. This style of activity is prominent in physiological studies, where the apparent aim is to decompose a complex "organic function" and to recognize how it is discharged. I claim that Cummins has captured an important part of the notion of biological function, but that his ideas need to be integrated with those of the etiological approach, not set up in opposition to it.

When we attribute functions to entities that make a causal contribution to complex processes, there is, I suggest, always a source of design in the background. The constituents of a machine have functions because the machine, as a whole, is explicitly intended to do something. Similarly with organisms. Here selection lurks in the background as the ultimate source of design, generating a hierarchy of ever more specific selection pressures, and the structures, traits, and behaviors of organisms have functions in virtue of their making a causal contribution to responses to those pressures.

Without recognizing the background role of the sources of design, an account of the Cummins variety becomes too liberal. Any complex system can be subjected to functional analysis. Thus we can identify the "function" that a particular arrangement of rocks makes in contributing to the widening of a river delta some miles downstream, or the "functions" of mutant DNA sequences in the formation of tumors—but there are no genuine functions here, and no functional analysis. The causal analysis of delta formation does not link up in any way with a source of design; the account of the causes of tumors reveals *dysfunctions*, not functions.

Recognizing the liberality of Cummins-style analyses, proponents of the etiological conception drag evolutionary considerations into the foreground. In doing so they make *all* projects of attributing functions focus on the explanation of the presence of the bearers of those functions. However, important though the theory of evolution by natural selection undoubtedly is to biology, there are other biological enterprises, some even continuous with those that occupied pre-Darwinians, which can be carried out in ignorance of the details of selective regimes. Thus the conscience-ridden biologists who offer more relaxed attributions of function can quite legitimately protest that the niceties of selection processes are not their primary concerns: without knowing what alternatives there were to the particular valves that help the heart to pump blood, they can recognize both that there is a general selection pressure on vertebrates to pump blood and that particular valves make identifiable contributions to the pumping. Selection, they might say, is the background source of design here, but it need not be dragged into the foreground to raise questions that are irrelevant to the project they set for themselves (understanding the mechanism through which successful pumping is achieved).

I believe that the account I have offered thus restores some unity to the concept of function through the recognition that each functional attribution rests on some presupposition about design and a pertinent source of design. But it allows for a number of distinct conceptions of function to be developed, based on sources of design (intention versus selection), time relation between source of design and the present, and directness of connection between source of design and the entity to which functions are ascribed. This pluralism enables us to capture the insights of the two main rival philosophical conceptions of function, and to do justice to the diversity of biological projects.

Does it go too far? In their original form, etiological accounts were vulnerable to counterexample, and the resolution invoked selection *ad hoc*. Am I committed to supposing that the leaky valve that asphyxiates the scientist has the function of so doing? No. For there is no explaining the presence of the valve in terms of selection for ability to asphyxiate scientists, nor is there any selection pressure on a larger system to whose response the action of the valve makes a causal contribution. Even though the account I have offered is more inclusive than traditional etiological conceptions, it does not seem to fall victim to the traditional counterexamples.

VI

I have tried to motivate my account of function and design by alluding to some quickly sketched examples. This strategy helps to elaborate the approach, but invites concerns to the effect that a more thorough investigation of biological practice would disclose less ambiguity than I have claimed. To alleviate such concerns I now want to look at some cases of functional attribution in a little more detail.

I shall start with two examples that are explicitly concerned with evolutionary issues. The first concerns a "functional analysis of the egg sac" in golden silk spiders.[19] The orb-weaving spider *Nephila clavipes* lays its eggs under the leaf canopy, covers them with silk, and weaves a loop of silk around twig and branch which holds the sac in place. The authors of the study (T. Christenson and P. Wenzl) investigate the functions of components of the egg-laying behavior. I shall concentrate on the spinning of the loop.

Christenson and Wenzl write:

The functions of the silk loop around the attachment branch were assessed by examining clutches that fell to the ground. We found 19 of the 59 egg sacs that fell due to naturally occurring twig breakage; 84.2% (16) failed to produce spiderlings, 13 because of ground moisture and subsequent rotting, and 3 because of predation. . . . The remaining three sacs had fallen a few weeks prior to the normal time of spring emergence; the spiderlings appeared to disperse and inhabit individual orbs.[20] In contrast to those that fell, sacs that remained in the tree were dry and appeared relatively safe from predation. Only 4.5% (15 of 353) showed unambiguous signs of predation, that is, some damage to the silk such as a tear or a bore hole.[21]

I interpret this passage as demonstrating a marked fitness difference between spiders who perform the looping operation that attaches the egg sac to twig and branch and those who fail to do so. Christenson and Wenzl are tacitly comparing the normal behavior of *N. clavipes* with mutants whose ability to weave an attachment loop was somehow impaired. Their emphasis on evolutionary considerations is evident not only in their detailed measurements of survivorships, but also in the framing of their analysis and in their final discussion. The authors begin by noting that "[f]unctional analyses of behaviours are often speculative due to the difficulty of demonstrating that the behavior contributes to the individual's reproductive success, and what the relevant selective agents might be."[22] They conclude by contending that "Female *Nephila* maximize their reproductive efforts, in part, through the construction of an elaborate egg sac."[23] This study is thus naturally interpreted as deploying the strong etiological conception of function, linking function directly with selection and proposing that the entities bearing functions are optimal.

Similarly, a study of the function of roaring in red deer by T. Clutton-Brock and S. Albon explicitly connects the attribution of function to claims about selection.[24] The authors begin by examining a traditional proposal:

A common functional explanation is that displays serve to intimidate the opponent. . . . This argument has the weakness that selection should favour individuals which are not intimidated unnecessarily and which adjust their behaviour only to the probability of winning and the costs and benefits of fighting. . . .[25]

Here it seems that a necessary condition on the truth of an ascription of function is that there should not be possible mutants that would be favored by selection. The same strong conception of function is apparent later in the discussion, when Clutton-Brock and Albon consider the hypothesis that roaring serves as an advertisement enabling stags to assess others' fighting ability. Although their careful observations indicate that stags rarely defeat those by whom they have been out-roared, they recognize that their data leave open other possibilities for the relation between roaring and fighting ability. They suggest that fighting and roaring may both draw on the same groups of muscles, so that roaring serves as an "honest advertisement" to other stags. But they note that this depends on assuming that "selection could not produce a mutant which was able to roar more frequently without increasing its strength or stamina in fights."[26] I interpret the caution expressed in their discussion to be grounded in recognition of the stringent conditions that must be met in showing that a form of behavior maximizes reproductive success, and thus their reliance on the strong version of the etiological conception.

I now turn to two physiological studies in which the connection to evolution is far less evident. Here, there are neither detailed measurements of the fitnesses (or proxies such as survivorships) of rival types of organism (as in the study of golden silk spiders) or connections with mathematical models of a selection process (as in the investigation of the roaring of stags). Instead, the authors undertake a mechanistic analysis of the workings of a biological system. Consider the following discussion of digestion in insects.

> Food in the midgut is enclosed in the peritrophic membrane, which is secreted by cells at the anterior end of the midgut in some insects or formed by the midgut epithelium in most. It is secreted continuously or in response to a distended midgut, as in biting flies. It is likely that the peritrophic membrane has several functions, although the evidence is not conclusive. It may protect the midgut epithelium from abrasion by food or from attack by microorganism or it may be involved in ionic interactions within the lumen. It has a curious function in some coleopterous larvae, where, in various ways, it is used to make the cocoon.[27]

The interesting point about this passage is that it could easily be accepted by a biologist ignorant of or hostile to evolutionary theory. So long as one has a sense of the overall life of an insect and of the conditions that must be satisfied for the insect to thrive, one can view the peritrophic membrane as making a causal contribution to the organism's flourishing. Of course, Darwinians will view these conditions as grounded in selection pressures to which insects must respond, but physiology can keep this Darwinian perspective very much in the background. It is enough to recognize that insects must have a digestive system capable of processing food items, that the passage of food through the system must not abrade the cells lining the gut, and so forth. I suggest that this, like so many other physiological discussions, presupposes a background picture of the selection pressures on the organisms under study and analyzes the causal mechanisms that work to meet those pressures, without attending to the fitness of alternatives that would have to be considered to underwrite a claim about the operation of selection.

Finally, I turn to a developmental study of sexual differentiation in *Drosophila*.[28] The problem is to understand simultaneously how an embryo with two X chromosomes becomes a female, how an embryo with one X chromosome becomes a male, and how the organism compensates for the extra chromosomal material found in females. The author (M. Kaulenas) summarizes a complex causal story, as follows:

> The primary controlling agent in sex determination and dosage compensation is the ratio between the X chromosomes to sets of autosomes (the X:A ratio). This ratio is "read" by the products of a number of genes; some of which function as numerator elements, while others as denominator elements. Two of the numerator genes have been identified [*sisterless a (sis a)* and *sisterless b (sis b)*] and others probably exist. The denominator elements are less clearly defined. The end result of this "reading" is probably the production of DNA-binding proteins, which, with the cooperation of the *daughterless (da)* gene product (and possibly other components) activate the *Sex lethal (sxl)* gene. This gene is the key element in regulating female differentiation. One early function is autoregulation, which sets the gene in the functional mode. Once functional, it controls the proper expression of the *doublesex (dsx)* gene. The function of *dsx* in female somatic cell differentiation is to suppress male differentiation genes. *Dsx* needs the action of the *intersex (ix)* gene for this function. Female differentiation genes are not repressed, and female development ensues.[29]

Here is a causal story about how female flies come to express the appropriate proteins in their somatic cells. The elements of the story concern the ways in which particular bits of DNA code for proteins that either activate the right genes or block transcription of the wrong ones. In the background is a general picture of how selection acts on sexually reproducing organisms, a picture that recognizes the selectively disadvantageous effects of failing to suppress one set of genes (those associated with the distinctive reactions that occur in male somatic cells) and of failing to activate the genes in another set (those whose action is responsible for the distinctive reactions of female somatic cells). The functions of the specific genes identified by Kaulenas are understood in terms of the causal contributions they make in a complex process. There is no attempt to canvass the genetic variation in *Drosophila* populations or to argue that the specific alleles mentioned are somehow fitter than their rivals. The discussion takes for granted a particular type of selection pressure—thus adopting the environment-centered perspective on evolution—and considers only the causal interactions that result in a response to that selection pressure. The causal analysis is vividly presented in a diagram (reproduced in the figure 1), which shows the kinship between the type of mechanistic approach adopted in this study and the analysis of complex systems designed by human beings. Selection furnishes a context in which the overall design is considered, and, within that context, the physiologist tries to understand how the system works.

I offer these four examples as paradigmatic of two very types of biological practice offering ascriptions of function. I hope that it is evident how introducing the strong etiological conception within the last two would distort the character of the achievement, rendering it vulnerable to skeptical worries about

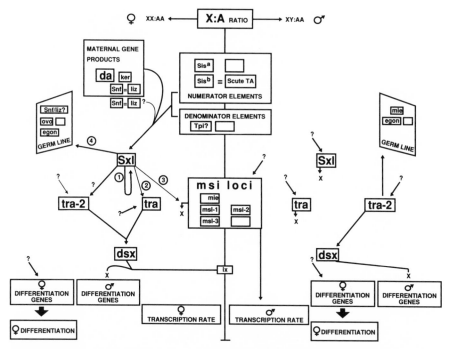

FIGURE Diagram illustrating the interrelationships of the genes involved in the control of sexual differentiation and dosage compensation in *Drosophila*, (from M. Kaulenas, *Insect Accessory Reproductive Structures: Function, Structure and Development*, New York: Springer, 1992, 18).

the operation of selection that are in fact quite irrelevant. By the same token, it is impossible to appreciate the line of argument offered in the explicitly evolutionary studies without recognizing the stringent requirements that the strong etiological conception imposes. There are undoubtedly many instances in which the notion of function intended is far less clear. I believe that keeping our attention focused on paradigms will be valuable in the work of disambiguation.

VII

Philosophical discussions of function have tended to pit different analyses and different intuitions against one another without noting the pluralism inherent in biological practice.[30] On the account I have offered here, there is indeed a unity in the concept of function, expressed in the connection between function and design, but the sources of design are at least twofold and their relation to the bearers of function may be more or less direct. This means, I believe, that the insights of the main competitors, Wright's etiological approach and Cummins's account of functional analysis, can be accommodated (and, as the discussion in section IV indicates, variants of the etiological approach can also be given their due).

The result is a general account of functions that covers both artifacts and organisms. I believe that it can also be elaborated to cover the apparently mixed case of functional ascriptions to social and cultural entities, in which both explicit intentions and processes of cultural selection may act together as sources of design. But working out the details of such impure cases must await another occasion.

NOTES

I am extremely grateful to the Office of Graduate Studies and Research at the University of California–San Diego for research support, and to Bruce Glymour for research assistance. My thinking about functional attributions in biology has been greatly aided by numerous conversations with Peter Godfrey-Smith. Despite important residual differences, I have been much influenced by Godfrey-Smith's careful elaboration and resourceful defense of an etiological view of functions.

1. I shall sometimes identify the bearers of functions simply as "entities," sometimes, for stylistic variety, talk of traits, structures, organs, behaviors as having functions. I hope it will be obvious throughout that my usage is inclusive.

2. This aspect of Darwin's accomplishment is forcefully elaborated by Richard Dawkins in *The Blind Watchmaker* (London, 1987). Although I have reservations about Dawkins's penchant for seeing adaptation almost everywhere in nature, I believe that he is quite correct to stress Darwin's idea of design without a designer.

3. The fact that the intentions of the Creator are in the remote background in much pre-Darwinian physiological work is one of the two factors that allow for continuity between pre-Darwinian physiology and the physiology of today. As I shall argue later, appeals to selection as a source of design are kept in the remote background in contemporary physiological discussions.

4. Richard Lewontin, "Organism and Environment" (manuscript), and Lewontin and Richard Levins, *The Dialectical Biologist* (Cambridge, Mass., 1987).

5. Wright, "Functions," *Philosophical Review* 82 (1973): 139–68. For further elaboration, see Ruth Millikan, *Language, Thought, and Other Biological Categories* (Cambridge, Mass., 1984); Karen Neander "Functions as Selected Effects: The Conceptual Analyst's Defense," *Philosophy of Science* 58, 168–84; and Peter Godfrey-Smith, "Functions," forthcoming.

6. This example stems from Christopher Boorse, "Wright on Functions," *Philosophical Review* 85 (1976): 70–86.

7. This way of evading the trouble is due to Millikan, *op.cit.*

8. These issues are broached by Godfrey-Smith in his forthcoming paper. He and I are in broad agreement about questions of timing and diverge in our approaches to the second cluster of questions.

9. Here, and in the ensuing discussion, I permit myself an obvious shorthand. In speaking of the origination of an entity in an organism I do not, of course, mean to refer to the mutational and developmental history that lies behind the emergence of the entity in an individual organism but in the process that culminates in the initial fixation of that entity in members of the population. I hope that this abbreviatory style will not cause confusions.

10. See N. Tinbergen, "On War and Peace in Animals and Man," *Science* 160 (1968): 1411–18.

11. Gould and Vrba, "Exaptation—A Missing Concept in the Science of Form," *Paleobiology* 8 (1982): 4–15.

12. Mayr, "The Emergence of Evolutionary Novelties" in *Evolution and the Diversity of Life* (Cambridge, Mass., 1976).

13. The point that biologists often ignore in practice the strictures on adaptationist claims that they recognize in theory is very clearly expressed in Gould and Lewontin "The Spandrels

of San Marco and the Panglossian Paradigm: A Critique of the Adaptationist Programme," *Proceedings of the Royal Society* B 205 (1979): 581–98.

14. Bigelow and Pargetter, "Functions," *Journal of Philosophy* 84 (1987); 181–96.

15. See his "Functions" (forthcoming). For my own commitments to a similar view see "Why Not The Best?" in John Dupre, *The Latest on the Best: Essays on Evolution and Optimality* (Cambridge, Mass., 1988), and "Developmental Decomposition and the Future of Human Behavioral Ecology," *Philosophy of Science* 57 (1990): 96–117.

16. See, for example, the discussion of Geoffrey Parker's ingenious and sophisticated work on copulation time in male dungflies in my *Vaulting Ambition: Sociobiology and the Quest for Human Nature* (Cambridge, Mass., 1985), chapter 5.

17. See the examples given in section VI below.

18. Robert Cummins, "Functional Analysis," *Journal of Philosophy* 72 (1975): 741–65.

19. T. Christenson and P. Wenzl, "Egg-Laying of the Golden Silk Spider, *Nephila clavipes* L. (Araneae, Araneidae): Functional Analysis of the Egg Sac," *Animal Behaviour* 28 (1980): 1110–18.

20. I should note here that spiderlings typically overwinter in the egg sac, so that the period of a few weeks represents a fall only a *short* time before the usual time of emergence. Thus the successful instances are those in which the normal course of development is only slightly perturbed.

21. Christenson and Wenzl, *op.cit.*, 1114.

22. Ibid., 1110.

23. Ibid., 1115.

24. T. Clutton-Brock and S. Albon, "The Roaring of Red Deer and the Evolution of Honest Advertisement," *Behaviour* 69 (1979): 145–68.

25. Ibid., 145.

26. Ibid., 165.

27. J. McFarlane, "Nutrition and Digestive Organs," in M. Blum, *Fundamentals of Insect Physiology* (Chichester, 1985), 59–90. The quoted passage is from p. 64.

28. M. Kaulenas, *Insect Accessory Reproductive Structures: Function, Structure, and Development* (New York, 1992), Section 2.3 "Genetic Control of Sexual Differentiation."

29. Ibid., 17.

30. As I have argued elsewhere, biological practice is pluralistic in its employment of concepts of gene and species and in its identification of units of selection. See my essays, "Genes" (*British Journal for the Philosophy of Science* 33 [1982]: 337–59) and "Species" (*Philosophy of Science* 51 [1984]: 308–33), and Kim Sterelny and Philip Kitcher, "The Return of the Gene" (*Journal of Philosophy* 85 [1988]: 335–58).

Contributors

Peter Achinstein, Department of Philosophy, Johns Hopkins University
Phillip Bricker, Department of Philosophy, University of Massachusetts, Amherst
John Dupré, Department of Philosophy, Stanford University
John Earman, Department of History and Philosophy of Science, University of Pittsburgh
Arthur Fine, Department of Philosophy, Northwestern University
Paul Horwich, Department of Linguistics and Philosophy, Massachusetts Institute of Technology
R. I. G. Hughes, Department of Philosophy, University of South Carolina
Martin R. Jones, Department of Philosophy, University of California, Berkeley
Philip Kitcher, Department of Philosophy, University of California, San Diego
Elisabeth Lloyd, Department of Philosophy, University of California, Berkeley
Peter Railton, Department of Philosophy, University of Michigan
Alexander Rosenberg, Department of Philosophy, University of California, Riverside
Steven Savitt, Department of Philosophy, University of British Columbia
Lawrence Sklar, Department of Philosophy, University of Michigan
Brian Skyrms, Department of Philosophy, University of California, Irvine
Elliott Sober, Department of Philosophy, University of Wisconsin, Madison
Howard Stein, Department of Philosophy, University of Chicago
Patrick Suppes, Department of Philosophy, Stanford University
Paul Teller, Department of Philosophy, University of California, Davis
Linda Wessels, Department of History and Philosophy of Science, Indiana University
Mark Wilson, Department of Philosophy, Ohio State University

Peter A. French is Lennox Distinguished Professor of Philosophy at Trinity University in San Antonio, Texas. He has taught at the University of Minnesota, Morris, and has served as Distinguished Research Professor in the Center for the Study of Values at the University of Delaware. His books include *The Scope of Morality* (1980), *Collective and Corporate Responsibility* (1980), and *Responsibility Matters* (1992). He has published numerous articles in the philosophical journals. **Theodore E. Uehling, Jr.**, is professor of philosophy at the University of Minnesota, Morris. He is the author of *The Notion of Form in Kant's Critique of Aesthetic Judgment* and articles on the philosophy of Kant. He is a founder and past vice-president of the North American Kant Society. **Howard K. Wettstein** is chair and professor of philosophy at the University of California, Riverside. He has taught at the University of Notre Dame and the University of Minnesota, Morris, and has served as a visiting associate professor of philosophy at the University of Iowa and Stanford University. He is the author of *Has Semantics Rested on a Mistake? and Other Essays* (1992).